Plant Growth and Climate Change

Biological Sciences Series

A series which provides an accessible source of information at research and professional level in chosen sectors of the biological sciences.

Series Editor:

Professor Jeremy A. Roberts, Plant Sciences Division, School of Biosciences, University of Nottingham. UK.

Titles in the series:

Biology of Farmed Fish Edited by K.D. Black and A.D. Pickering

Stress Physiology in Animals Edited by P.H.M. Balm

Seed Technology and its Biological Basis Edited by M. Black and J.D. Bewley

Leaf Development and Canopy Growth Edited by B. Marshall and J.A. Roberts

Environmental Impacts of Aquaculture Edited by K.D. Black

Herbicides and their Mechanisms of Action Edited by A.H. Cobb and R.C. Kirkwood

The Plant Cell Cycle and its Interfaces Edited by D. Francis

Meristematic Tissues in Plant Growth and Development Edited by M.T. McManus and B.E. Veit

Fruit Quality and its Biological Basis Edited by M. Knee

Pectins and their Manipulation Edited by Graham B. Seymour and J. Paul Knox

Wood Quality and its Biological Basis Edited by J.R. Barnett and G. Jeronimidis

Plant Molecular Breeding Edited by H.J. Newbury

Biogeochemistry of Marine Systems Edited by K.D. Black and G. Shimmield

Programmed Cell Death in Plants Edited by J. Gray

Water Use Efficiency in Plant Biology Edited by M.A. Bacon

Plant Lipids – Biology, Utilisation and Manipulation Edited by D.J. Murphy

Plant Nutritional Genomics Edited by M.R. Broadley and P.J. White

Plant Abiotic Stress Edited by M.A. Jenks and P.M. Hasegawa

Gene Flow from GM Plants Edited by G.M. Poppy and M.J. Wilkinson

Antioxidants and Reactive Oxygen Species in Plants Edited by N. Smirnoff

Plant Growth and Climate Change Edited by J.I.L. Morison and M.D. Morecroft

Plant Growth and Climate Change

Edited by

JAMES I.L. MORISON
Department of Biological Sciences
University of Essex
Colchester, UK

and

MICHAEL D. MORECROFT
Centre for Ecology & Hydrology
Maclean Building
Wallingford, UK

Blackwell
Publishing

Editorial Offices
Blackwell Publishing Ltd, 9600 Garsington Road, Oxford OX4 2DQ, UK
Tel: +44 (0)1865 776868
Blackwell Publishing Professional, 2121 State Avenue, Ames, Iowa 50014-8300, USA
Tel: +1 515 292 0140
Blackwell Publishing Asia Pty Ltd, 550 Swanston Street, Carlton, Victoria 3053, Australia
Tel: +61 (0)3 8359 1011

First published 2006 by Blackwell Publishing Ltd

ISBN: 978-14051-3192-6

Library of Congress Cataloging-in-Publication Data

Plant growth and climate change / edited by James I.L. Morison and Michael D. Morecroft.
p. cm.
Includes bibliographical references and index.
ISBN-13: 978-1-4051-3192-6 (hardback : alk. paper)
ISBN-10: 1-4051-3192-6 (hardback : alk. paper) 1. Climate changes. 2. Crops and climate. 3. Growth (Plants) I. Morison, James I.L. II. Morecroft, Michael D.
S600.7.C54P52 2006
632′.1—dc22

2006009717

A catalogue record for this title is available from the British Library

The publisher's policy is to use permanent paper from mills that operate a sustainable forestry policy, and which has been manufactured from pulp processed using acid-free and elementary chlorine-free practices. Furthermore, the publisher ensures that the text paper and cover board used have met acceptable environmental accreditation standards.

For further information on Blackwell Publishing, visit our website:
www.blackwellpublishing.com

Contents

The colour plate section appears after page 50

List of Contributors

Dr James A. Bunce	Crops Systems and Global Change Laboratory, ARS, USDA, 10300 Baltimore Avenue, Bldg 046A BARC-West, Beltsville, MD 20705-2350, USA
Dr Maria-Conceição Caldeira	Departamento de Engenharia Florestal, Instituto Superior de Agronomia, Tapada da Ajuda, 1399 Lisboa Codex, Portugal
Dr Maria-Manuela Chaves	Departamento de Botenica e Engenharia Biologica, Instituto Superior de Agronomia, Tapada da Ajuda, 1399 Lisboa Codex, Portugal
Dr Alexandre V. Correia	Departamento de Engenharia Florestal, Instituto Superior de Agronomia, Tapada da Ajuda, 1399 Lisboa Codex, Portugal
Professor William J. Davies	Department of Biological Sciences, I.E.N.S., Lancaster University, Lancaster LA1 4YQ, UK
Professor John Grace	School of Geosciences, University of Edinburgh, Crew Building, Mayfiel Road, Edinburgh EH9 3JN, UK
Professor Christian Körner	Institute of Botany, University of Basel, Schönbeinstrasse 6, CH-4056 Basel, Switzerland
Dr Ross McMurtrie	School of Biological, Earth and Environmental Sciences, University of New South Wales, Sydney 2052, New South Wales, Australia
Dr Belinda Medlyn	School of Biological, Earth and Environmental Sciences, University of New South Wales, Sydney 2052, New South Wales, Australia
Professor Annette Menzel	Lehrstuhl für Ökoklimatologie, Technical University of Munich, Am Hochanger 13, D-85354 Freising, Germany
Dr Michael D. Morecroft	Centre for Ecology & Hydrology, Maclean Building, Crowmarsh Gifford, Wallingford, OX10 8BB, UK

Dr James I.L. Morison	Department of Biological Sciences, University of Essex, Wivenhoe Park, Colchester CO4 3SQ, UK
Mr James S. Paterson	Environmental Change Institute, Oxford University Centre for the Environment, Dyson Perrins Building, South Parks Road, Oxford, OX1 3QY, UK
Dr David Pepper	School of Biological, Earth and Environmental Sciences, University of New South Wales, Sydney 2052, New South Wales, Australia
Professor Joao Pereira	Departamento de Engenharia Florestal, Instituto Superior de Agronomia, Tapada da Ajuda, 1399 Lisboa Codex, Portugal
Dr Tim Sparks	Centre for Ecology and Hydrology, Monks Wood, Abbots Ripton, Huntingdon, Cambridgeshire PE28 2LS, UK
Dr David Viner	Climatic Research Unit, University of East Anglia, Norwich NR4 7TJ, UK
Dr Craig Wallace	Climatic Research Unit, University of East Anglia, Norwich NR4 7TJ, UK
Dr Ying-Ping Wang	CSIRO Marine and Atmospheric Research, PMB #1, Aspendale, Victoria 3195, Australia
Dr Rui Zhang	School of Geosciences, University of Edinburgh, Crew Building, Mayfiel Road, Edinburgh EH9 3JN, UK
Dr Lewis H. Ziska	Crops Systems and Global Change Laboratory, ARS, USDA, 10300 Baltimore Avenue, Bldg 046A BARC-West, Beltsville, MD 20705-2350, USA

Preface

Evidence grows daily of the rapid changes in climate due to human activities and their impact on plants and animals. Plant function is inextricably linked to climate and atmospheric carbon dioxide concentration. On the shortest and smallest scales the climate affects the plant's immediate environment and thus directly influence physiological processes. On longer and larger time and space scales climate influ ences species distribution and community composition and determines what crops can be viably produced in managed agricultural, horticultural and forestry ecosystems. Plant growth also influence the local, regional and global climate through the exchanges of energy and gases between the plants and the air around them. This book examines the major aspects of how anthropogenic climate change is affecting plants, covering the wide range of scales molecular and cellular through organ and plant, up to biome and global.

Anthropogenic climate change poses major scientifi challenges for plant scientists. Firstly, we need to expand and apply our understanding of plant responses to the environment so that we can predict the impacts of climate change on plant growth for crops and natural ecosystems. This understanding in turn needs to be built into assessments of the global climate system, in order to correctly quantify the numerous feedbacks between plants, the atmosphere and the climate. Understanding plant growth responses to climate change is also important to allow society to respond. Plant production has to be maximised, to overcome the new or altered climatic constraints on food and fibr production, in the face of the continuing population growth. The sustainability of agricultural and forestry production needs to be improved by reducing greenhouse gas emissions from land use and fossil fuel use and by reducing water and nutrient consumption. Conservation policies and the management of natural and seminatural areas have to be adjusted to conserve biodiversity in the changing environmental conditions. The contributions in this volume exemplify work that addresses many of these challenges.

In planning this book, we felt that the literature has often been rather divided between books on the effects of climate change on plants in agricultural and other managed systems and those examining effects on natural ecosystems. In addition, more fundamental aspects of plant physiology were often missing. It is clear that climate impacts research is informed by all aspects of plant physiology and ecology and we sought to organise a book that looked across the range of plant growth (although restricting the scope to terrestrial and vascular plants). We also wanted to show the range of scientifi questions that exercise the wide variety of plant scientists involved in climate change research.

This book therefore tackles the main aspects of climate change and focuses on several key determinants of plant growth: atmospheric CO_2, temperature, water availability and their interaction. Although atmospheric CO_2 might not strictly be considered an aspect of climate, we felt it was essential that it was included as it is the main driver of climate change. The book demonstrates the plethora of techniques used across plant science: detailed physiology in controlled environments; observational studies based on long-term data sets; fiel manipulation experiments and modelling. Chapter 1 provides an overview of the processes in climate change, summarising the evidence for recent changes to temperature, precipitation and solar radiation and outlining the likely scenarios for change produced in the IPCC reports. In Chapter 2, Ziska and Bunce review what is known about plant responses to the increased atmospheric CO_2, looking across the spectrum of scales from gene expression to whole ecosystems. They draw attention to difficultie in understanding at the two extremes of this spectrum and emphasise the point that CO_2 change is not a single factor, but must be considered with other environmental variables.

The themes of timescales and the need for combining fiel and controlled environment work in order to understand the effects of temperature on plant growth is taken up by Körner in Chapter 3. He explores the paradoxes in plant short-term response and medium term acclimation to temperature and the very different issues of continuous effects of temperature compared to threshold responses. He shows us the difficultie in bridging from the single species physiological scale to ecosystems and the interactions with other variables such as soil nutrient and water supply and day length. In Chapter 4 Menzel and Sparks demonstrate the sensitivity of plant development to temperature and show many examples ranging from grapes and cereals to trees of how recent temperature changes have altered phenological development. Their examples emphasise the importance of long records, both from traditional observations and from newer technologies such as satellite NDVI remote sensing and they discuss some of the methodological problems in assessing phenological environmental relationships.

Warmer conditions and changed precipitation patterns will alter water availability for plants. In Chapter 5 Davies reviews cell and plant water relations and the signalling processes that coordinate the response of plants to water availability. He points out that exploitation of understanding of these physiological processes is already leading to improvements in crop production. This review is complemented by the larger scale examination of the relationship of plant productivity to water availability in Chapter 6 by Pereira and colleagues. They demonstrate the importance of considering the longer timescales for drought resistance and resilience in the response of perennial vegetation to drought. They discuss the interrelationship between productivity, drought and fires especially in the Mediterranean environment. Temperature and water interactions are considered at the community scale in Chapter 7. Morecroft and Paterson assess the fiel experiments that have highlighted the sensitivity of plant community composition to climate changes, particularly temperature and precipitation. These changes often hinge on the interaction with soil and nutrients, although our information is dominated by studies in temperate and arctic or alpine environments.

The fina two chapters illustrate the essential use of models to synthesise our understanding from the physiological and ecological experimental work, to test hypotheses and to make predictions on large spatial and temporal scales. Wang and colleagues (Chapter 8) demonstrate that increased CO_2 concentration cannot be considered alone in modelling plant productivity, because of the interaction with nutrients, especially nitrogen. Thus plant models have to be intimately linked to soil decomposition and mineral cycling models on longer timescales, which are also dependent on temperature and water availability. Chapter 9 examines the measurement and modelling of global plant productivity and the carbon cycle (Grace & Zhang). They demonstrate how the modelling of production depends on temperature responses of respiration and photosynthesis and thus highlight the importance of a full assessment of physiological responses of plants, on the correct timescales, to fiel conditions, as identifie in the earlier chapters.

Much plant physiology has been founded on an experimental paradigm of investigating responses to one factor at a time, over short time periods, whereas much ecological work has been based in experimental manipulations in the field over longer periods. Climate change impacts research has brought these two disciplines very closely together and the contributors to this volume admirably demonstrate the resulting synergies. We thank them for all their time and efforts in responding to our challenge.

James Morison and Mike Morecroft

1 Recent and future climate change and their implications for plant growth

David Viner, James I.L. Morison and Craig Wallace

1.1 Introduction

The geographic distribution of plant species, vegetation types and agricultural cropping patterns demonstrate the very strong control that climate has on plant growth. Solar radiation, temperature and precipitation values and seasonal patterns are key determinants of plant growth through a variety of direct and indirect mechanisms. Other climatic characteristics are also major influences such as wind speed and storm frequency. There is a rapidly growing number of well-documented instances of change in ecosystems due to recent (and probably anthropogenic) climate change (Walther *et al.*, 2002). For example, there are several lines of evidence in the Arctic, ranging from indigenous people's knowledge to satellite images, that show that species distributions have changed, with growing shrub cover and increasing primary productivity (Callaghan *et al.*, 2004). Another example is that plant species composition in the mountains of central Norway has changed over a 70-year period, with lowland species coming in and snow-bed and high-altitude species disappearing (Klanderud & Birks, 2003). Meta-analyses of data for well-studied alpine herbs, birds and butterflie by Parmesan and Yohe (2003) found a mean range shift of approximately 6 km per decade towards the poles or 6 m per decade in elevation, and that the date of the start of spring has advanced by 2 days per decade. In agriculture, there are clear examples of recent climate change affecting plant growth and cropping potential or performance. For example, in Alberta (Canada) the potential maize-growing zone, define by temperature limits, has shifted north by 200–300 km over the last century (Shen *et al.*, 2005). However, climate change is not just affecting temperate zones. For example, in some arid zones there have been increases in precipitation, leading to increased shrub density, and changes in the rest of the ecosystem (e.g. Brown *et al.*, 1997). Overall, the Intergovernmental Panel on Climate Change (IPCC, 2001b) concluded that 'from collective evidence, there is high confidenc that recent regional changes in temperature have had discernible impacts on many physical and biological systems'. These recent climate changes are likely to accelerate as human activities continue to perturb the climate system, and many reviews have made predictions of serious consequences for ecosystems (e.g. Izaurralde *et al.*, 2005) and for food supplies and food security (e.g. Reilly *et al.*, 2003; Easterling & Apps, 2005). This chapter outlines recent past and future anthropogenic climate change. Much of the relevant research has already been drawn together, reviewed and summarised by the many contributors to the IPCC reports (IPCC, 2001a–c), and we have therefore relied heavily on that authoritative source of information.

1.2 The climate system

The recent and future anthropogenic changes to the climate have to be considered in the context of natural climate changes. The Earth's climate results from the complex interaction of many components: the ocean, atmosphere, geosphere, cryosphere and biosphere. Although the climate system is ultimately driven by the external solar energy, changes to any of the internal components, and how they interact with each other, as well as variability in the solar radiation received can lead to changes in climatic conditions. These influence are often considered as 'forcings', changes to the energy inputs and outputs that result in modification in the climate. Therefore there are many causes of climate change that operate on a variety of timescales. On the longest timescales are mechanisms such as geological processes and the changes in the Earth's orbit around the sun (Milankovitch-Croll effect). The latter is believed to be the mechanism underlying the cycle of ice ages and interglacials. Geological processes resulting from the movement of tectonic plates and consequent major changes in physical relief, continental distributions and ocean basin shape and connectivity clearly have influence global climate patterns. Geological processes can also work on a much shorter timescale through volcanism. Large, explosive volcanic eruptions can inject millions of tons of soot and ash into the middle atmosphere where they reflec solar radiation, creating a 'global soot veil'. The Tambora eruption in Southeast Asia in 1815 caused extensive global cooling and 'the year without a summer' in Europe (e.g. Engvild, 2003; Oppenheimer, 2003). The climate impacts of such volcanic events usually decay after 1 or 2 years (as in the Mt. Pinatubo eruption of June 1991, which caused 0.25–5°C drop in mean temperature for 1–2 years in several parts of the world; Hansen *et al.*, 1996). However, some research has suggested that very infrequent, regional so-called supereruptions can alter the climate for enough time to cause radical species loss (Rampino, 2002), although this is much debated.

In addition to geological and orbital changes, the climate system is sensitive to inherent and periodic internal variability in any one of its components, such as ocean currents. These can be on decadal timescales, such as the Interdecadal Pacifi Oscillation. Or the variations can be on near-interannual timescales, such as the well-documented El-Niño/Southern Oscillation (ENSO) and North Atlantic Oscillation (NAO). During ENSO events when the ocean upwelling in the eastern equatorial Pacifi is weaker than normal, the resulting changes to sea surface temperatures and to the wind patterns dramatically affect climate and consequently impact the biosphere across the region. For example, in El Niño years, maize yields in China are decreased by 5% (Tao *et al.*, 2004) and in Australia wheat crop yields are closely related to the severity of rainfall reductions (e.g. Nicholls, 1985). The NAO has wide ecological effects (e.g. reviewed by Ottersen *et al.*, 2001), such as determining the length of the growing season in Europe, as evident in extensive phenological observations (see Chapter 4, Menzel, 2003). Also correlations of the NAO index have been found with crop yields in Europe and North America (e.g. Gimeno *et al.*, 2002). The overall effects of such internal changes on climate are difficul to predict, because of the feedbacks between the climate system components. For example, an ocean current change might warm a high-latitude region, leading to reduced snow cover, which in turn leads to more land surface exposure and more solar energy absorption which results in a positive feedback.

1.3 Mechanisms of anthropogenic climate change

Although most public discussion on climate change currently focuses on fossil fuel combustion, CO_2 emissions and the enhanced 'greenhouse effect', it must be noted that there are other components of human-induced climate change. Human activity has modified and continues to modify, the Earth's surface on a very large scale, through deforestation, afforestation, cultivation, mineral extraction, irrigation, drainage and flooding These large alterations in land cover change the surface short-wave reflect vity and hydrological and thermal properties of the land surface. Thus, replacing forest with pasture changes the surface energy balance and increases the proportion of radiant energy going into heating the air and reducing evaporation, as many studies have shown (e.g. von Randow *et al.*, 2004). Conversely, the very large expansion of irrigation in previously dry areas changes land cover and solar radiation absorption and increases energy partitioning into evaporation, as well as changing the seasonal pattern of surface–atmosphere exchanges (e.g. Adegoke *et al.*, 2003).

The crux of the enhanced greenhouse effect is that human modificatio of the atmospheric concentration of the key radiation-absorbing gases – CO_2, CH_4, N_2O and various halocarbons – has resulted in a radiative forcing of the climate system. These gases have been released primarily as a result of industrial, transport and domestic activities and to a lesser extent from agricultural activities and land use changes (IPCC, 2001a). Direct and indirect determination of CO_2, CH_4 and N_2O in the atmosphere over the past 1000 years show marked and unprecedented increases in concentrations in recent times (Figure 1.1). The start of these increases coincides with the rapid industrialisation of the Northern Hemisphere during the late eighteenth and nineteenth centuries, and so since 1750, the global mean atmospheric concentration of CO_2 has increased by 31%; approximately 75% of this increase has come from fossil fuel combustion and 25% from land use change (IPCC, 2001a). Analysis of extended data sets from ice cores indicates that the current atmospheric concentration of CO_2 is the highest for the past 420 000 years, and is likely to be the highest within the last 20 million years (IPCC, 2001a). The percentage increase in methane concentrations is greater, having risen by 151% since 1750, whilst the concentration of nitrous oxide has increased by 17% over the same period (IPCC, 2001a). The estimated radiative forcing associated with the increased concentrations

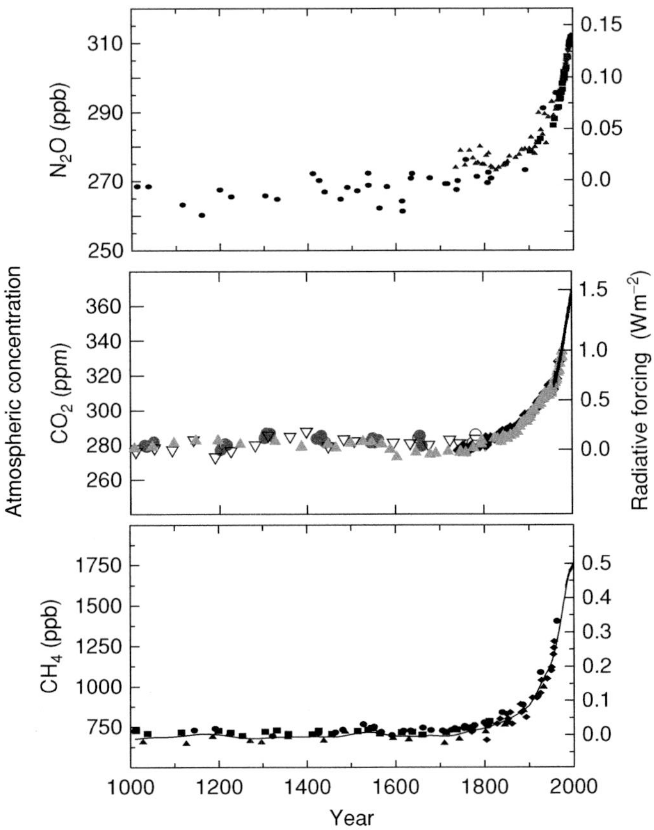

Figure 1.1 Changes in the atmospheric concentrations of CO_2, CH_4 and N_2O over the last 100 years. Data from Antarctic and Greenland ice cores and recent direct air samples. The estimated positive radiative forcing of the climate system is indicated on the right-hand scale. (From IPCC, 2001a.)

of the three main greenhouse gases is shown on the right-hand axis of Figure 1.1. In total, increased atmospheric concentrations of CO_2, CH_4, N_2O and halocarbons are estimated to have placed an additional 2.4 W m^{-2} of radiative forcing onto the climate system since 1750 (IPCC, 2001a). At the same time there have been other changes in radiative forcing, particularly from changes in carbon and sulphate aerosols, also produced by fossil fuel combustion and biomass burning. While there is still some uncertainty over their direct and indirect effects through cloud modification it is widely agreed that sulphate aerosol pollution has had a net negative forcing, resulting in a cooling effect, particularly in source regions (IPCC, 2001a). Therefore, as aerosol pollution is now declining in some areas, e.g. Europe and North America, the effect of the positive forcing due to the increased greenhouse gases may become more marked in those regions (IPCC, 2001a). Clearly, future net forcing will be affected by the amount of sulphate emissions and what intensity and technology of fossil fuel combustion is adopted.

The change in temperature resulting from the various forcings is termed the climate sensitivity and clearly depends upon many components of the climate system, not all of which are well understood. Nonetheless, computer simulations of the Earth's climate indicate that the level of observed global warming evident in the instrumental record is consistent with the estimated response to the anthropogenic radiative forcing. It is this, and the geographical pattern of the observed warming, that led the IPCC to conclude in 2001 that 'in the light of new evidence and taking into account the remaining uncertainties, most of the observed warming over the past 50 years is likely to have been due to the increase in greenhouse gas concentrations' (IPCC, 2001a). The continuing huge international scientifi efforts since that Third Assessment Report (TAR) have largely confirme this work, and the forthcoming Fourth Assessment Report of the IPCC due in 2007 is likely to agree and strengthen this conclusion while providing further advances in our understanding of human influence on the climate system.

1.4 Recent climate changes

Clearly, the changes in the Earth's climate in the past have been well documented by palaeoclimatologists. Analysis of oceanic and lake sediment cores has established that during the course of the past 800 000 years Earth has experienced a number of warm interglacial and cold glacial periods, each of which lasted several (and maybe tens of) thousands of years. We are currently experiencing a warm interglacial period which began approximately 10 000–12 000 years ago and which marks the start of the current epoch, the Holocene (e.g. Lamb, 1977). The changes in temperature that accompanied the switch from the last glacial to the present interglacial period were not smooth and varied greatly over the planet. For example, work focusing on the British Isles has estimated that between 13 300 and 12 500 years before the present time, the mean temperature rose by 8°C in summer and approximately 20°C in winter (Atkinson *et al.*, 1987).

Historical records suggest some substantial changes over the past one or two millennia, with century-length colder and warmer periods (e.g. Lamb, 1977). Climate reconstructions based upon proxy records (particularly tree-ring widths) permit a quantitative examination of the last 1000 years (Colour Plate 1). The last millennium is generally accepted to have experienced three main climatic epochs. The 'Medieval Warm Period' (MWP) characterised the climate of the twelfth and thirteenth centuries, and was followed in the sixteenth and seventeenth centuries by the 'Little Ice Age'. The third, recent climatic event has been 'Post-industrial Warming'.

The dates of the firs two events are debated because much of the evidence varies in timing for different parts of the planet. Indeed, whether or not the terms are actually applicable in describing the average climatic conditions of the time is also increasingly questioned. For example, Jones (2002) has questioned the validity of

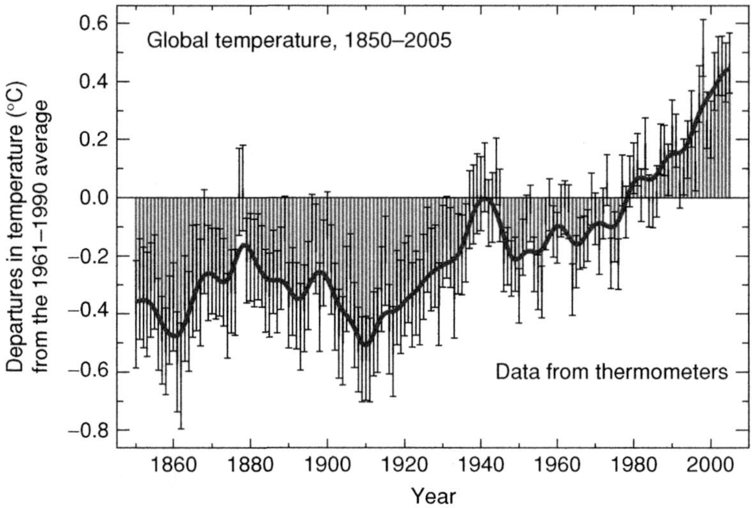

Figure 1.2 The global surface temperature record from 1850 to 2005, expressed as departures from the 1961–1990 mean. The solid line is a filtere curve to show interdecadal variations. (Source: The HadCRUT3 data set, the UK Meteorological Office Brohan *et al.*, in press.)

the MWP, pointing to a lack of a distinct rise in the proxy temperature record for the Northern Hemisphere average at this time. What is evident from many of the curves in Colour Plate 1 is the existence of a cooler period during the sixteenth and seventeenth centuries. Glacial advances within Europe have been shown to be widespread, and many reconstructed climate records indicate that the coldest annual temperature for the Northern Hemisphere in the last 1000 years occurred in 1601 (Jones, 2002). Nonetheless, the validity of the Little Ice Age label has, like the MWP, come under question itself. Some researchers point to the fact that many individual years during the Little Ice Age period saw temperatures as warm as present levels (Jones, 2002) and glacial advances occurred at different times during the supposed 'cold' centuries (Matthews & Briffa, 2005).

The third climatic event of the last 1000 years, Post-industrial Warming, can clearly be seen in the observed instrumental record (the black curve in Colour Plate 1 and a more detailed curve in Figure 1.2) and is key evidence of human-induced climate change. Two warming events are apparent and these constitute the *only* statistically significan events of the instrumental record (Jones, 2002). The firs warming period occurred between 1920 and 1945; the second since 1975. It is clear that globally the 1990s have been the warmest decade of the last 1000 years, and that 1998 was the warmest individual year. The global curve in Figure 1.2 shows that compared to temperatures representative of the late nineteenth century, 1998 was approximately 0.8°C warmer.

However, to understand effects of temperature on plant growth, we need more information than just data on changes to the mean global annual temperature highlighted by Colour Plate 1 and Figure 1.2. The instrumental record also shows (a) that the Post-industrial Warming has affected the mid to high latitudes of the Northern Hemisphere the most, (b) that winter months have warmed more rapidly than summer months and (c) that night-time temperatures are more affected than the day time temperatures (IPCC, 2001a). In addition, there has been a reduction in the frequency of extreme low monthly and seasonal average temperatures across much of the globe and a small increase in the frequency of extreme high temperatures (IPCC, 2001a). Other temperature changes that are probably of major importance to plant growth are 10–15% reductions in the number of days with air frosts (minimum air temperature $< 0°C$) found across the Northern Hemisphere (e.g. Frich *et al.*, 2002), and reductions in the spring snow cover extent since the 1960s (IPCC, 2001a).

Although the dramatic recent changes in the mean global temperature are easy to depict (e.g. Colour Plate 1 and Figure 1.2), it is harder to generalise the overall changes in precipitation, as there is substantial temporal and spatial variation (IPCC, 2001a). In the mid to high latitudes of the Northern Hemisphere, precipitation increased by approximately 10% (30–85°N) over the twentieth century, and these increases correlate with various reports of increased stream fl w and increased soil moisture in some areas within these latitudes (IPCC, 2001a). There is also compelling evidence that intense winter precipitation events in some mid-latitude areas are becoming more common already (Osborn and Hulme, 2002), which has serious consequences for erosion and flooding In the tropics and subtropics, patterns of precipitation change have been much more regional and variable over decadal timescales (IPCC, 2001a). For example, in West Africa the rainfall during the last 30 years of the century was on average 15–40% lower than during the previous 30 years (Nicholson, 2001).

In addition to these changes in temperature and precipitation, there have been substantial changes in solar irradiance. The pioneering work of Stanhill drew attention to these changes when he carefully analysed the rather few high-quality solar measurement records and found a gradual decline in solar irradiance of approximately 3% per decade over the period 1950–2000 (0.5 W m^{-2} $year^{-1}$; Stanhill & Cohen, 2001). Support for this also comes from several regional analyses of evaporation pan records in both the Northern and Southern hemispheres, which show annual reductions of 2–4 mm $year^{-1}$ (e.g. Roderick & Farquhar, 2002; Liu & Zeng, 2004). These solar radiation changes are probably because of increases in anthropogenic aerosols affecting atmospheric and cloud optical properties, and they could have substantial direct effects on plant growth (Stanhill & Cohen, 2001). However, recent work has questioned the persistence and magnitude of the 'global dimming' effect. One suggestion is that it may be due to the bias of measurement sites for densely populated locations (declines of 0.41 W m^{-2} $year^{-1}$), while sites in sparsely populated areas showed only 0.16 W m^{-2} $year^{-1}$ (Alpert *et al.*, 2005). More evidence comes from an analysis of global satellite data, which showed that there was a decrease from 1983 to 1990 followed by an increase up to 2001, amounting to an overall increase of 0.16 W m^{-2} $year^{-1}$ (Pinker *et al.*, 2005). Newly analysed surface observations also suggest an increase since the late 1980s (Wild *et al.*, 2005). Therefore, it is clear that solar radiation receipt at the surface has varied substantially over decadal timescales, and will change in the future with changes in cloud and aerosol load. The effect of this on plant growth is rarely directly considered.

1.5 Future changes in anthropogenic forcing of climate

Projections of future climate change can be developed by computer simulation of the Earth's climate system, given different scenarios of future changes to both natural and anthropogenic radiative forcing. In the Special Report on Emissions Scenarios (SRES; Nakicenovic & Swart, 2000), the IPCC devised six possible future scenarios of greenhouse gas emissions through to the year 2100 based upon changes that may occur in global population growth, degree of globalisation, technological change and use of sustainable energy sources. The six SRES scenarios ranged from those likely to produce high anthropogenic climate forcing because of heavy use of fossil fuels (e.g. scenario A1FI) to those with low forcing because of reduced consumption and introduction of resource-efficien technologies (e.g. B1; IPCC, 2001a).

1.5.1 Future global climate scenarios

The aforementioned SRES scenarios have been used in global circulation models (GCMs) to make projections of future climate change during the present century. GCMs are mathematical approximations of the real physical climate system and model the atmospheric circulation and the exchange of energy between the main climate system components. All GCMs used by the IPCC to develop climate change scenarios for the TAR had interactive atmospheric and oceanic components (atmosphere–ocean general circulation models, AOGCMs), including representation of seasonal sea ice, and most of the GCMs also had an interactive land surface scheme which simulated the moisture and energy flu es between the ground and the atmosphere. However, the uncertainties associated with GCM results should be acknowledged. In particular, some real-world climate system components are poorly understood, and so their approximation by mathematical equations is difficult A good example, and a major continuing debate in climate change, is the effect of changing cloud characteristics (altitude, water content, droplet or crystal size) as well as the scale at which they are considered in the models (IPCC, 2001a). Uncertainties in climate projections also arise because of the constraints of the current level of computing power, which can limit how realistically some physical processes can be incorporated at the large geographic scale required for model practicability. While specifi regional climate models have been developed that simulate processes on a fine geographical scale, they are very costly to run and have more uncertainty in long-term predictions.

Because of the rapid and comparatively recent rise in atmospheric concentrations of greenhouse gases (some of which have very long lifetimes), the climate system is not in equilibrium, and thus temperature increases must be anticipated, even if further anthropogenic emissions could be immediately stopped. Of course this is impossible, and so the SRES scenarios provide outlines for more likely changes in anthropogenic forcing to drive the GCMs. The global mean temperature response to each scenario (Colour Plate 2) is different, reflectin the extent to which greenhouse gas emissions either stabilise, decrease or rise during the twenty-firs century. For example, the temperature response in a fossil fuel intensive scenario, (A1FI; red small-dotted line in Colour Plate 2) by the year 2100, could be anywhere between 3.0 and 5.8°C above the mean 1961–1990 conditions. However, if a B1-type scenario is followed (green line in Colour Plate 2) then the temperature response, although positive, may be somewhat lower, in the range of 1.4–2.6°C above the 1961–1990 'normal' conditions. Acknowledging this range, the IPCC concluded that 'the globally averaged surface temperature is projected to increase by 1.4 to 5.8°C over the period 1990–2100' (IPCC, 2001a). Furthermore, the warming over land will be larger than the global mean, particularly in higher latitudes in the cold season. With this increase in mean surface air temperature, there are expected to be more frequent extreme high temperature events, and a lower frequency of extreme low temperature events, because of the upward shift in mean temperatures (IPCC, 2001a). There is still much uncertainty whether there will be more variability in climate, which might also contribute to changes in extremes. Clearly, increased variability could have major implications for plant growth and for agriculture and forestry (e.g. Salinger, 2005).

The projected temperature increases in the IPCC TAR were larger than those previously estimated (e.g. IPCC, 1995). This is due to the lower projected sulphur emissions in the SRES scenarios than in their predecessors, and the sulphate aerosols are also responsible for the small differences in the projected temperature increases between the SRES scenarios for the next 50 years or so, as depicted in Colour Plate 2. In fossil fuel intensive scenarios (e.g. A1FI) the rise in greenhouse gases is also accompanied by an increase in sulphate emissions (the greenhouse gas warming is therefore partly offset). Conversely, in scenarios where emissions of atmospheric pollutants decrease, lower levels of greenhouse gases are matched by lower levels of sulphur emissions (and the offsetting is lower). The net temperature changes in the firs few decades are therefore broadly similar. It is not until the second half of the twenty-firs century that the longer lived greenhouse gases such as CO_2 dominate over the sulfur emissions and the temperature responses diverge (IPCC, 2001a).

Globally averaged precipitation is projected to rise during the twenty-firs century, because of the increases in temperature, which will increase evaporation rates and increase the amount of moisture the atmosphere can hold. The average global precipitation response under the A2 scenario for the fina 30 years of the twenty-firs century is 3.9% higher compared to mean 1961–1990 conditions, with a range of 1.3–6.8%. The B2 scenario, having a lower anthropogenic forcing, responds with a lower increase of precipitation, 3.3%, with a range of 1.2–6.1% for the same time period (IPCC, 2001a). The IPCC (2001a) concluded that by the second half of the century, precipitation would have increased over northern mid latitudes in winter, and in northern high latitudes in winter and summer. At low latitudes there will be regional variation with some decreases and some increases. The IPCC also concluded that increased levels of precipitation will be accompanied by an increase in year-to-year variation in precipitation (IPCC, 2001a).

1.5.2 Future regional climate scenarios

When viewed globally, the predicted climate response to the SRES forcing scenarios can be summarised fairly simply: a substantially warmer and slightly wetter world seems likely within this century. However, for predicting biotic impacts much more spatial and temporal detail is obviously required. When the predictions of different AOGCMs were analysed by broad region (IPCC, 2001a), they were not always consistent in the relative magnitudes of warming or precipitation change (Figures 1.3a and 1.3b), although they did agree in key features. In particular, for scenarios A2 and B2, they agreed that warming over land will be larger than the global mean, especially in higher latitudes in the cold season there will be large (>20%) increases in high-latitude rainfall and the precipitation will decrease over Australia (between 5% and 20%) and the Mediterranean (>20%) in their respective summers.

Clearly, such general statements are of limited use in analysing the impact on plant growth, which will be affected by the local and microclimatic changes. Substantial effort has gone into developing methods to 'downscale' information from global and regional models, in order to assess the local changes to climate that will affect plant growth, and particularly the regional changes that would affect agriculture (e.g. Harrison *et al.*, 1995; Downing *et al.*, 2000; Smith *et al.*, 2005). One example of such a downscaling approach was in the Europe ACACIA Project (Hulme & Carter, 2000; Parry, 2000). Europe forms a good case study, as it illustrates the climate change over an oceanic–continental cline and over a substantial latitudinal gradient with very different seasonality of plant growth. Colour Plate 3 shows examples of the ACACIA output that summarise projected changes in summertime (June, July and August) temperature and precipitation for Europe, under the B2 SRES scenarios for three periods: 2020, 2050 and 2080, relative to the mean 1961–1990 period. The rate of annual warming is projected to be between 0.1 and 0.4°C per decade. The largest predicted warming occurs over southern Europe, where summers 4.5°C warmer than the climatological norm are expected by the end of the century. Summer warming over northern Europe, although smaller in magnitude, still amounts to approximately 2.0°C in places. In winter, eastern Europe and western Russia warm the quickest (0.15–0.6°C per decade; IPCC, 2001b), although by the 2080s over the whole of Europe 'cold winters' (those calculated to occur 1 in every 10 years during 1961–1990) virtually cease to occur (IPCC, 2001b). There is, however, some level of uncertainty associated with these projections and the values in the right-hand panels are the absolute range of all eight GCM simulations used to assemble the projections.

GCM projections of European rainfall agree that wetter winters are probable over northern Europe in both the A2 and B2 scenarios (Figure 1.3b). The rate of this change is estimated to be between 1 and 4% per decade. The change is smaller over southern Europe, where the main response appears to be a drier summer climate.

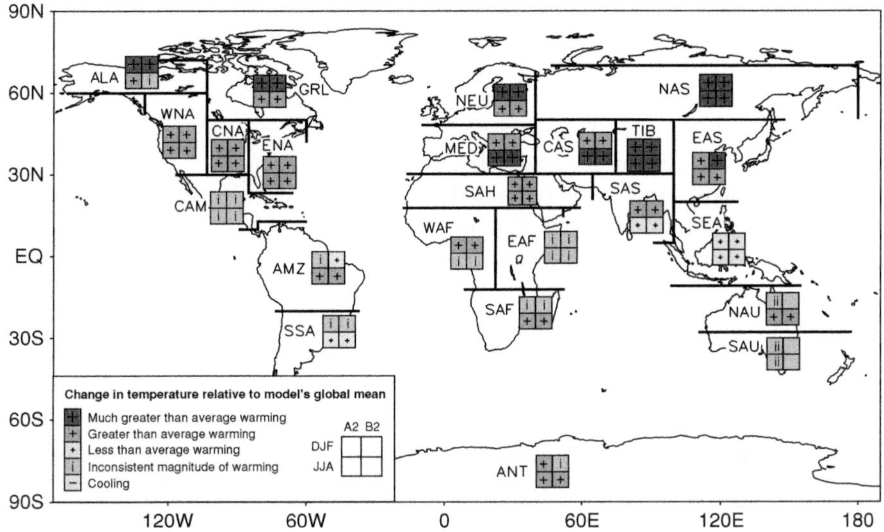

(a)

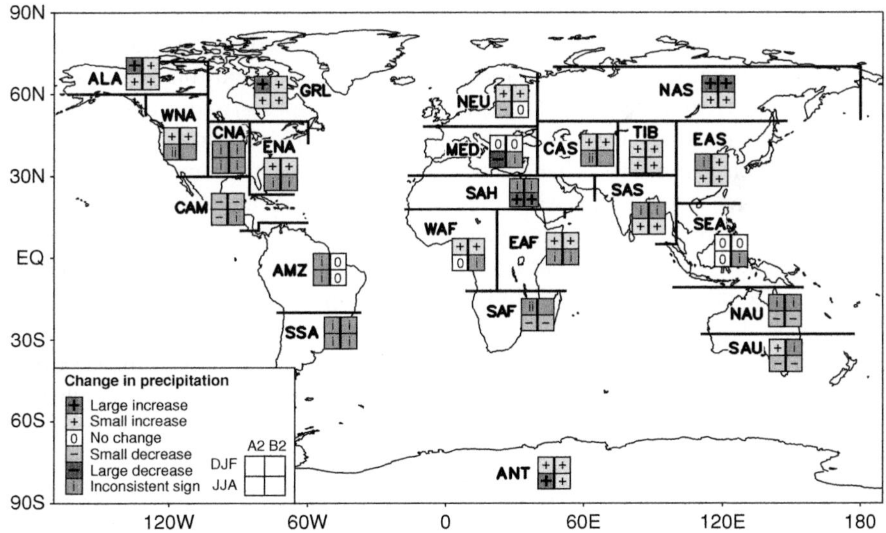

(b)

Figure 1.3 Predictions of regional (a) temperature and (b) precipitation changes from AOGCMs compared to the global mean change with scenarios A2 and B2 from SRES. Changes are identifie using nine climate circulation models, forced with the appropriate changes in atmospheric greenhouse gas composition for each scenario. DJF is for December, January and February, and JJA is for June, July and August. (From IPCC, 2001a, Chapter 10.)

However, accurate simulation of rainfall by GCMs is difficult and this point is well illustrated in Colour Plate 3b, showing projected changes in summer under the B2 scenario. Firstly, the lack of values in some of the grid boxes indicates that the projected changes are not statistically significan from the variability in rainfall which is experienced within the climate model when it is run under 'normal' conditions with no change to future climate forcing. It is not until the 2080s that significan changes are visible (Colour Plate 3), and even then the range of the projected changes is often greater than the median change, indicating that even the sign of the median change may be incorrect.

It is also probable that the incidence of extreme weather events (as judged by today's norms) over Europe will increase with global warming. This is especially so for intense winter precipitation events and for hot summer events; the frequency of extreme cold events will fall.

There has been considerable discussion over the possibility of abrupt climate change in Europe triggered by changes to the Atlantic thermohaline circulation (THC) responsible for maintaining the Atlantic Gulf Stream. The THC is a key determinant of climate conditions in Europe and North America, and possibly through various teleconnections to substantial regional climate changes elsewhere (Vellinga & Wood, 2002). A recent model suggests that a decrease in THC strength of 50% would increase the maritime influenc on climate in Europe but decrease the overall temperatures and precipitation (Jacob *et al.*, 2005). A complete collapse of the THC, although unlikely, may be possible if the anthropogenic forcing undergoes marked increases in the coming centuries (e.g. a quadrupling) and is applied to the climate system for long enough (Manabe and Stouffer, 1994; Wood *et al.*, 2003). The impacts of such major THC changes on ecosystems have been explored in recent modelling exercises with timescales of a few centuries and are usually severe (e.g. Higgins & Schneider, 2005). More plausible, however, is a weakening of the THC of around 20–50% during the next 100 years due to the influ of freshwater into the North Atlantic from increased precipitation and ice melt (Dixon *et al.*, 1999; Wood *et al.*, 2003). The IPCC TAR (IPCC, 2001a) concluded that the amount of cooling that might be associated with a THC weakening would not be sufficien to negate the direct greenhouse warming, and so the net effect would be warming in Europe.

1.6 Concluding comments

Clearly, the recent and future changes in greenhouse gas concentrations and consequent changes in climate have numerous direct implications for plant growth. The mechanisms and possible consequences of these effects will be considered in detail in later chapters. There will also be indirect changes such as those of rising sea level, which may be severe in some places, although we have not considered them here. Management of plant growth in the face of such climate change will be a major challenge, and there is therefore renewed urgency in the search by plant scientists for ways to improve crop growth for food and fibr production and to minimise natural resource use. For those studying the function of natural ecosystems, the impact of recent and future climate change is also a key, all-pervasive aspect.

In addition, the concern over greenhouse gas accumulation in the atmosphere has another implication for primary production in managed ecosystems: attempts to mitigate gas emissions by modificatio to agriculture and other forms of land use. Commitments by governments to reduce greenhouse gas emissions involve net reductions in fossil fuel consumption and other emissions. Agriculture is a substantial fossil fuel user and, in particular, is responsible for some 20% of all anthropogenic greenhouse gas emissions (mainly in the form of methane and nitrous oxide; IPCC, 2001b). Changes in agriculture and other managed land use could help to mitigate greenhouse gas emissions through change in a number of agricultural practices outlined in the 2001 report of the Third Working Group (IPCC, 2001c). For instance, a reduction in land use intensity and employing conservation tillage techniques would both act to increase soil carbon. Rice paddy field are a major source of methane, and so a shift towards rice varieties that can be grown under drier conditions would reduce methane emissions. Significan reductions in agricultural emissions of nitrous oxide could be achieved by altering fertilising methods, through replacing inorganic nitrogen sources with organic manures, or by increasing legume use.

At the process level, plants actually respond to the 'weather', the short-term aerial conditions around them, through physical exchanges of energy and gases with the surroundings. The longer term averaged conditions is what is meant by climate, and for climatologists this is often taken as the mean values over a standard 30-year period, as used in the data sets discussed above. However, it is important to recognise that the climate of a location includes not just period mean conditions (annual, monthly, decad) and normal seasonality, but also the typical variability in conditions, such as extremes of temperature and interannual variation in precipitation regime. Recent assessments of the impact of climate change on agriculture have started to examine this (e.g. Downing *et al.*, 2000; IPCC, 2001b) and it is critically important (Salinger, 2005). Variation in climatic conditions is as critical to plant growth as the normal conditions. For example, dry conditions for just one spring and summer season can have a large effect on species composition in temperate grasslands (Dunnett *et al.*, 1998; Morecroft *et al.*, 2004). Thus assessments of the impact of climate change on plant growth need to examine changes in the mean values, changes in seasonality and changes in variability. It is these sorts of changes and their interactions with other environmental factors (such as day length and nutrient supply) that affect plant growth and development that will determine the form of vegetation, agriculture and forestry in the rest of this century.

References

Adegoke, J.O., Pielke, R.A., Eastman, J., Mahmood, R. & Hubbard, K.G. (2003) Impact of irrigation on midsummer surface flu es and temperature under dry synoptic conditions: a regional atmospheric model study of the U.S. High Plains. *Mon. Weather Rev.*, **131** (3), 556–564.

Alpert, P., Kishcha, P., Kaufman, Y.J. & Schwarzbard, R. (2005) Global dimming or local dimming? Effect of urbanization on sunlight availability. *Geophys. Res. Lett.*, **32** (17), Art. No. L17802.

Atkinson, T.C., Briffa, K.R. & Coope, G.R. (1987) Seasonal temperatures in Britain during the past 22,000 years, reconstructed using beetle remains. *Nature*, **325**, 587–592.

Brohan, P., Kennedy, J.J., Haris, I., Tett, S.F.B & Jones, P.D. (in press) Uncertainty estimates in regional and global observed temperature changes: a new dataset from 1850. *J. Geophys. Res. (Atmospheres)*.

Brown, J.H., Valone, T.J. & Curtin, C.G. (1997) Reorganization of an arid ecosystem in response to recent climate change. *Proc. Natl. Acad. Sci. USA*, **94** (18), 9729–9733.

Callaghan, T.V., Bjorn, L.O., Chernov, Y., Chapin, T., Christensen, T.R., Huntley, B., Ims, R.A., Johansson, M., Jolly, D., Jonasson, S., Matveyeva, N., Panikov, N., Oechel, W., Shaver, G., Elster, J., Jonsdottir, I.S., Laine, K., Taulavuori, K., Taulavuori, E. & Zockler, C. (2004) Responses to projected changes in climate and UV-B at the species level. *Ambio*, **33** (7), 418–435.

Dixon, K.W., Delworth, T.L., Spelman, M.J. & Stouffer, R.J. (1999) The influenc of transient surface flu es on North Atlantic overturning in a coupled GCM climate change experiment. *Geophys. Res. Lett.*, **26**, 2749–2752.

Downing, T.E., Harrison, P.A., Butterfield R.E. & Lonsdale, K.G. (eds) (2000) *Climate Change, Climatic Variability and Agriculture in Europe. An Integrated Assessment*, 445 pp. Environmental Change Institute, Oxford.

Dunnett, N.P., Willis, A.J., Hunt, R. & Grime, J.P. (1998) 38-year study of relations between weather and vegetation dynamics in road verges near Bibury, Gloucestershire. *J. Ecol.*, **86** (4), 610–623.

Easterling, W. & Apps, M. (2005) Assessing the consequences of climate change for food and forest resources: a view from the IPCC. *Clim. Change*, **70**, 165–189.

Engvild, K.C. (2003) A review of the risks of sudden global cooling and its effects on agriculture. *Agric. For. Meteorol.*, **115**, 129–139.

Frich, P., Alexander, L.V., Della-Marta, P., Gleason, B., Haylock, M., Tank, A.M.G.K. & Peterson, T. (2002) Observed coherent changes in climatic extremes during the second half of the twentieth century. *Clim. Res.*, **19** (3), 193–212.

Gimeno, L., Ribera, P., Iglesias, R., de la Torre, L., Garcia, R. & Hernandez, E. (2002) Identificatio of empirical relationships between indices of ENSO and NAO and agricultural yields in Spain. *Clim. Res.*, **21** (2), 165–172.

Hansen, J., Ruedy, R., Sato, M. & Reynolds, R. (1996) Global surface air temperature in 1995: return to pre-Pinatubo level. *Geophys. Res. Lett.*, **23** (13), 1665–1668.

Harrison, P.A., Butterfield R.E. & Downing, T.E. (eds) (1995) *Climate Change and Agriculture in Europe. Assessment of Impacts and Adaptations*, 413 pp. Environmental Change Unit, Oxford.

Higgins, P.A.T. & Schneider, S.H. (2005) Long-term potential ecosystem responses to greenhouse gas-induced thermohaline circulation collapse. *Glob. Change Biol.*, **11**, 699–709.

Hulme, M. & Carter, T.R. (2000) The changing climate of Europe. In: *Assessment of Potential Effects and Adaptations for Climate Change in Europe* (ed. M.L. Parry). Jackson Environment Institute, Norwich, UK.

IPCC (1995) *Climate Change 1995: The Science of Climate Change* (eds J.T. Houghton, L.G. Meiro Filho, B.A. Callander, N. Harris, A. Kattenberg & K. Maskell), 572 pp. Cambridge University Press, Cambridge, UK.

IPCC (2001a) *Climate Change 2001: The Scientific Basis* (eds J.T. Houghton, Y. Ding, D.J. Griggs, M. Noguer, P.J. van der Linden, X. Dai, K. Maskell & C.A. Johnson), 881 pp. Cambridge University Press, Cambridge, UK.

IPCC (2001b) *Climate Change 2001: Impacts, Adaptation, and Vulnerability* (eds J.J. McCarthy, O.F. Canziani, N.A. Leary, D.J. Dokken & K.S. White), 1032 pp. Cambridge University Press, Cambridge, UK.

IPCC (2001c) *Climate Change 2001: Mitigation* (eds B. Metz, O. Davidson, R. Swart & J. Pan), 752 pp. Cambridge University Press, Cambridge, UK.

Izaurralde, R.C., Thomson, A.M., Rosenberg, N.J. & Brown, R.A. (2005) Climate change impacts for the conterminous USA: an integrated assessment. Part 6: Distribution and productivity of unmanaged ecosystems. *Clim. Change*, **69**, 107–126.

Jacob, D., Goettel, H., Jungclaus, J., Muskulus, M., Podzun, R. & Marotzke, J. (2005) Slowdown of the thermohaline circulation causes enhanced maritime climate influenc and snow cover over Europe. *Geophys. Res. Lett.*, **32** (21), Art. No. L21711.

Jones, P.D. (2002) Changes in climate and variability over the last 1000 years. In: *Meteorology at the Millennium* (ed. R.P. Pearce), pp. 133–142. Academic Press, New York.

Klanderud, K. & Birks, H.J.B. (2003) Recent increases in species richness and shifts in altitudinal distributions of Norwegian mountain plants. *Holocene*, **13** (1), 1–6.

Lamb, H.H. (1977) *Climate: Past, Present and Future*, Vol. 2. Methuen, London.

Liu, C.M. & Zeng, Y. (2004) Changes of pan evaporation in the recent 40 years in the Yellow River Basin. *Water Int.*, **29** (4), 510–516.

Manabe, S. & Stouffer, R.J. (1994) Multiple-century response of a coupled ocean–atmosphere model to an increase of atmospheric carbon dioxide. *J. Clim.*, **7**, 5–23.

Matthews, J.A. & Briffa, K.R. (2005) The 'Little Ice Age': re-evaluation of an evolving concept. *Geog. Ann. Ser. A Phys. Geogr.*, **87A** (1), 17–36.

Menzel, A. (2003) Plant phenological anomalies in Germany and their relation to air temperature and NAO. *Clim. Change*, **57** (3), 243–263.

Morecroft, M.D., Masters, G.J., Brown, V.K., Clarke, I.P., Taylor, M.E. & Whitehouse, A.T. (2004) Changing precipitation patterns alter plant community dynamics and succession in an ex-arable grassland. *Funct. Ecol.*, **18** (5): 648–655.

Nakicenovic, N. & Swart, R. (eds) (2000) *Special Report on Emissions Scenarios*, 612 pp. Cambridge University Press, Cambridge, UK.

Nicholls, N. (1985) Impact of the Southern Oscillation on Australian crops. *J. Climatol.*, **5** (5), 553–560.

Nicholson, S.E. (2001) Climatic and environmental change in Africa during the last two centuries. *Clim. Res.*, **17** (2), 123–144.

Oppenheimer, C. (2003) Climatic, environmental and human consequences of the largest known historic eruption: Tambora volcano (Indonesia) 1815. *Prog. Phys. Geogr.*, **27**, 230–259.

Osborn, T.J. & Hulme, M. (2002) Evidence for trends in heavy rainfall events over the UK. *Philos. Trans. R. Soc.*, **360** (A), 1313–1325.

Ottersen, G., Planque, B., Belgrano, A., Post, E., Reid, P.C. & Stenseth, N.C. (2001) Ecological effects of the North Atlantic Oscillation. *Oecologia*, **128** (1), 1–14.

Parmesan, C. & Yohe, G. (2003) A globally coherent fingerprin of climate change impacts across natural systems. *Nature*, **421** (6918), 37–42.

Parry, M.L. (ed.) (2000) *Assessment of Potential Effects and Adaptations for Climate Change in Europe*. The Europe ACACIA Project, Jackson Environment Institute, Norwich, UK.

Pinker, R.T, Zhang, B. & Dutton, E.G. (2005) Do satellites detect trends in solar radiation? *Science*, **308**, 850–854.

Rampino, M. (2002) Threats to civilisation from impacts and super-eruptions. In: *Conference Proceedings of Environmental Catastrophes and Recoveries*. Brunel University, Uxbridge, UK.

Reilly, J., Tubiello, F., McCarl, B., Abler, D., Darwin, R., Fuglie, K., Hollinger, S., Izaurralde, C., Jagtap, S., Jones, J., Mearns, L., Ojima, D., Paul, E., Paustian, K., Riha, S., Rosenberg, N. & Rosenzweig, C. (2003) US agriculture and climate change: new results. *Clim. Change*, **57** (1–2), 43–69.

Roderick, M.L. & Farquhar, G.D. (2002) The cause of decreased pan evaporation over the past 50 years. *Science*, **298** (5597), 1410–1411.

Salinger, M.J. (2005) Climate variability and change: past, present and future – an overview. *Clim. Change*, **70** (1–2), 9–29.

Shen, S.S.P., Yin, H., Cannon, K., Howard, A., Chetner, S. & Karl, T.R. (2005) Temporal and spatial changes of the agroclimate in Alberta, Canada from 1901 to 2002. *J. Appl. Meteorol.*, **44** (7), 1090–1105.

Smith, S.J., Thomson, A.M., Rosenberg, N.J., Izaurralde, R.C., Brown, R.A. & Wigley, T.M.L. (2005) Climate change impacts for the conterminous USA: an integrated assessment. *Clim. Change*, **69**, 7–25.

Stanhill, G. & Cohen, S. (2001) Global dimming: a review of the evidence for a widespread and signifi cant reduction in global radiation with discussion of its probable causes and possible agricultural consequences. *Agric. For. Meteorol.*, **107** (4), 255–278.

Tao, F.L., Yokozawa, M., Zhang, Z., Hayashi, Y., Grassl, H. & Fu, C.B. (2004) Variability in climatology and agricultural production in China in association with the East Asian summer monsoon and El Nino Southern Oscillation. *Clim. Res.*, **28** (1), 23–30.

Vellinga, M. & Wood, R.A. (2002) Global climatic impacts of a collapse of the Atlantic thermohaline circulation. *Clim. Change*, **54,** 251–267.

von Randow, C., Manzi, A.O., Kruijt, B., de Oliveira, P.J., Zanchi, F.B., Silva, R.L., Hodnett, M.G., Gash, J.H.C., Elbers, J.A., Waterloo, M.J., Cardoso, F.L. & Kabat, P. (2004) Comparative measurements and seasonal variations in energy and carbon exchange over forest and pasture in South West Amazonia. *Theor. Appl. Climatol.*, **78** (1–3), 5–26.

Walther, G.R., Post, E., Convey, P., Menzel, A., Parmesan, C., Beebee, T.J.C., Fromentin, J.M., Hoegh-Guldberg, O. & Bairlein, F. (2002) Ecological responses to recent climate change. *Nature*, **416** (6879), 389–395.

Wild, M., Gilgen, H., Roesch, A., Ohmura, A., Long, C.N., Dutton, E.G., Forgan, B., Kallis, A., Russak, V. & Tsvetkov, A. (2005) From dimming to brightening: decadal changes in solar radiation at Earth's surface. *Science*, **308** (5723), 847–850.

Wood, R.A., Vellinga, M. & Thorpe, R. (2003) Global warming and thermohaline circulation stability. *Philos. Trans. R. Soc. Lond. A Math. Phys. Eng. Sci.*, **361** (1810), 1961–1974.

2 Plant responses to rising atmospheric carbon dioxide

Lewis H. Ziska and James A. Bunce

2.1 Introduction

2.1.1 Overview of plant biology

Currently, there is unprecedented scientifi and societal emphasis on assessing future anthropogenic changes in global temperature and the subsequent impacts on managed and unmanaged systems (Houghton *et al.*, 2001). Yet, the principle anthropogenic gas associated with this potential warming, carbon dioxide, is also one of the four abiotic requirements necessary for plant growth (i.e. light, nutrients, water, and CO_2). Any change in the availability of these abiotic parameters, particularly on a global scale, will impact not only plant biology but also all living systems.

Records of carbon dioxide concentration ($[CO_2]$) obtained from the Mauna Loa observatory in Hawaii have shown an increase in $[CO_2]$ of about 22% from 311 to approximately 380 parts per million (ppm) since the late 1950s (cdiac.esd.ornal.gov/home.html). The current annual rate of $[CO_2]$ increase ($\sim$0.5%) is expected to continue with concentrations exceeding 600 ppm by the end of the twenty-firs century (Houghton *et al.*, 2001). Interestingly, because the observatory at Mauna Loa and other global monitoring sites sample air at high elevations, away from anthropogenic sources, actual ground-level $[CO_2]$ can be significantl higher. This suggests that while the Mauna Loa data may reflec $[CO_2]$ for the globe as a whole, regional increases in $[CO_2]$ may already be occurring as a result of urbanization (Idso *et al.*, 1998) (Figure 2.1). Recent data indicate that plants may already be responding to both diurnal and urban-induced differences in atmospheric CO_2 (Ziska *et al.*, 2001, 2003, respectively). Such studies emphasize that carbon dioxide may be increased nonuniformly and illustrate the critical need for research that increases our fundamental understanding of how plant biology will respond to changing CO_2 environments.

However, given the complexity of the subject, we cannot address all aspects of plant function affected by carbon dioxide in a single synopsis, and the reader is referred to a number of excellent books on the subject (e.g. Koch & Mooney, 1996; Murray, 1997; Luo & Mooney, 1999; Reddy & Hodges, 2000; Karnosky *et al.*, 2001). In the current appraisal we (a) review areas of recent progress, (b) examine what is known at, and between, given levels of functional organization, and (c) illustrate critical areas for future research. Our intention is to provide a timely synthesis of what is currently known about the role of CO_2 in the biology of vascular plants, using temporal and spatial biological scales as an initial framework (Figure 2.2).

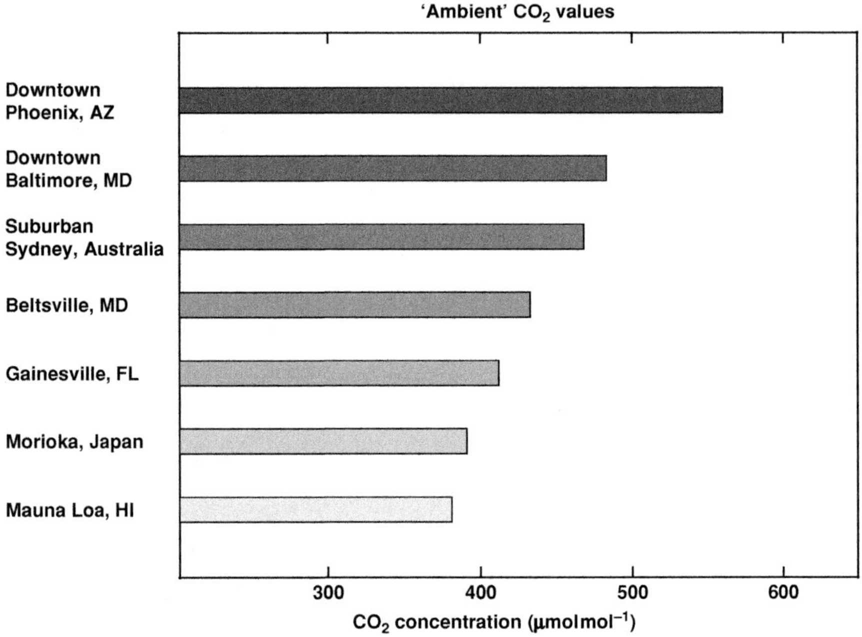

Figure 2.1 Values of 24-h ambient CO_2 concentration (μmol mol^{-1}) as a function of urbanization relative to the Mauna Loa, Hawaii, standard. Data are from Ziska *et al.* (2001) except the Mauna Loa data which are taken from the cdiac.esd.ornal.gov Web site and the Phoenix data which are derived from Idso *et al.* (1998).

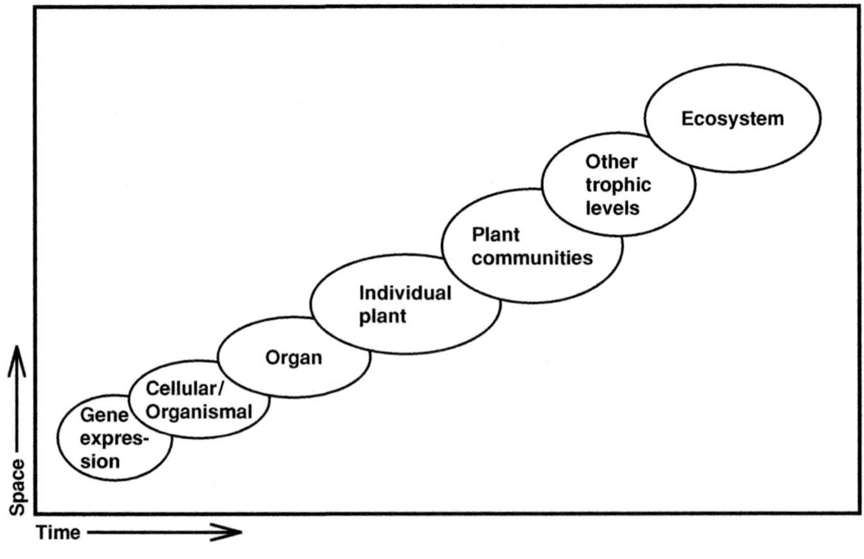

Figure 2.2 Theoretical construct of spatial and temporal scales definin different levels of organization within plant biology.

2.1.2 A word about methodology

As new methodologies become available for doing CO_2 fumigation, it is tempting to focus only on results obtained from newer methods and to ignore or reinterpret previous findings We would caution that all methodologies used to ascertain plant responses to CO_2 have both positive and negative attributes, and data obtained from a given experiment should not be judged 'superior', based solely on methodology. For example, environmental growth chambers (EGCs) are useful in evaluating the impact of preambient CO_2 concentrations on whole plant development (e.g. Ziska *et al.*, 2004), but an EGC environment will differ significantl from in situ conditions. Conversely, free air CO_2 enrichment (FACE) allows assessment of plant communities, but rapidly fluctuatin [CO_2] within elevated FACE rings may underestimate the fertilization effect of enriched CO_2 on plant growth (Holtum & Winter, 2003).

In general, the cost and complexity of methodologies increases with spatial and temporal scales. As a consequence, most of what is known concerning rising CO_2 and plant function is at the level of single leaves or whole plants (e.g. Curtis & Wang, 1998). These levels of organization represent the most experimentally accessible data, while less is known for either very large (e.g. ecosystem) or very small (e.g. genetic regulation, proteomics) bioprocesses. Ultimately, appropriate technologies should be determined by the specifi level(s) of organization that the researcher wishes to investigate.

2.2 Gene expression and carbon dioxide

The influenc of projected, future increases in [CO_2] on gene expression, particularly for photosynthetic regulation of the small subunit of Ribulose-1,5-bisphosphate carboxylase (rubisco), have been examined in a number of studies (Cheng *et al.*, 1998; Moore *et al.*, 1999; Makino *et al.*, 2000). In these instances genetic regulation is thought to be mediated by increased sugar levels resulting from exposure to future CO_2 concentrations (e.g. Cheng *et al.*, 1998); others, however, have argued that any high CO_2-induced decline in photosynthetic gene transcripts is due to a temporal shift in leaf ontogeny (Ludewig & Sonnewald, 2000).

Although a number of reviews have examined how carbohydrate accumulation may modify genetic regulation of both photosynthetic and non-photosynthetic genes (Sheen, 1994; Koch, 1996), the specifi function of [CO_2] in the subsequent change in carbohydrate signaling has not been fully elucidated. Unpublished data for maize grown in SPAR (soil-plant-atmosphere research) units indicated that approximately 5% of the genome responded to elevated CO_2 (750 ppm) (Soo-Hyung Kim, 2005; personal communication). However, it was difficul to link this result to physiological responses, partly because many of these genes encoded unknown or putative functional proteins. For *Arabidopsis thaliana* exposed to higher [CO_2] in a growth chamber experiment, transcription levels of some non-photosynthetic genes for growth, development, and stress were increased (Bae & Sicher, 2004). In contrast, a separate experiment on *A. thaliana* indicated that different environmental conditions between growth chambers and FACE systems resulted in greater changes in gene expression than were observed with increased [CO_2] alone (Miyazaki *et al.*, 2004). However, the latter experiment examined *A. thaliana* plants that had encountered severe stress symptoms following transplantation to a FACE ring, and so it was unclear if the observed changes in gene expression, particularly the large increase in stress proteins, is an experimental artifact (Miyazaki *et al.*, 2004).

Changes in gene expression may provide crucial insights into specifi mechanisms or cellular systems that may be regulated by changes in atmospheric CO_2, but the mechanistic basis for such changes, whether they involve carbohydrate accumulation or accelerated ontogeny are unclear. Nevertheless, analyses of transcript profile from microarray experiments, particularly from plants grown from seed in the fiel over a range of carbon dioxide values, may be of particular benefi for breeding programs. While *A. thaliana* remains, at present, the only vascular plant species where the entire transcriptome is available for analysis, it is hoped that similar approaches will be possible in evaluating the CO_2 response of agronomic staples such as barley, corn, rice, soybean, and wheat.

2.3 Cellular processes: photosynthetic carbon reduction (PCR) and carbon dioxide

2.3.1 *C_3 photosynthesis*

Plants evolved at a time when the atmospheric [CO_2] appears to have been four or fi e times the present values (Bowes, 1996). Because CO_2 remains the sole source of carbon for plant photosynthesis, and because at present [CO_2] is less than optimal, as atmospheric [CO_2] increases, photosynthesis at the biochemical level will be stimulated accordingly. Elevating [CO_2] stimulates net photosynthesis in plants with the C_3 photosynthetic pathway by raising the CO_2 concentration gradient from air to leaf and by reducing the loss of CO_2 through photorespiration (photosynthetic carbon oxidation, PCO; see Section 2.4). The increase in carbon uptake resulting from increasing CO_2 concentration and suppression of the PCO requires no additional light, water, or nitrogen. The stimulation of C_3 photosynthesis is one of the most established aspects of rising CO_2 concentration, and it has been described in numerous studies and reviews (e.g. Bowes, 1996, inter alia).

2.3.2 *C_4 photosynthesis*

The development of the C_4 pathway ($\sim$4% of all known plant species) may well be a photosynthetic modificatio that evolved in response to declining CO_2 and warmer climates that exacerbated photorespiratory losses through the PCO cycle in C_3 plants (e.g. Ehleringer & Monson, 1993). Because C_4 plants have a mechanism for concentrating CO_2 around rubisco, increases in external [CO_2] should have little effect on net photosynthesis in C_4 plants (for reviews see Bowes, 1996; Ghannoum *et al.*, 2000).

Yet, a number of researchers have observed enhanced photosynthesis in C_4 plants in response to elevated CO_2, even under optimal abiotic conditions (e.g. Sionit & Patterson, 1984; Morgan *et al.*, 1994; Ziska & Bunce, 1997a). A limited number of studies suggest that leakiness of the bundle sheath is not associated with elevated CO_2 responsiveness (Ziska *et al.*, 1999), although rising CO_2 can decrease the thickness of the bundle sheath cell walls in sorghum (Watling *et al.*, 2000). Alternatively, the development pattern of C_4 expression may be sensitive to increasing atmospheric CO_2. Recent data for sorghum have indicated that rubisco accumulated before phosphoenolpyruvate carboxylase (PEPc) during cellular development, suggesting potentially greater CO_2 sensitivity in younger leaves (Cousins *et al.*, 2003). Overall, however, many of the details regarding how the C_4 biochemical and cellular mechanism responds to elevated CO_2 remain unclear.

2.3.3 Crassulacean acid metabolism photosynthesis

Photosynthetic rate is also stimulated in a number of Crassulacean acid metabolism (CAM) species (~1% of all plant species) by high [CO_2]. This stimulation may be related to the ability of some CAM species to switch to C_3 photosynthesis when water is available; however, nonfacultative CAM plants may also show increased CO_2 uptake early or late in the day (Poorter & Navas, 2003). In addition, elevated CO_2 may also stimulate CO_2 uptake by PEPc during the night, with a subsequent increase in nocturnal malate accumulation (Drennan & Nobel, 2000). Drennan and Nobel (2000) also reported that elevated CO_2 decreased chlorophyll content and rubisco/PEPc activities, but that the activated percentage of rubisco increased and the Michaelis–Menten constant (K_m) decreased for PEPc. However, our present understanding of the biochemical/cellular responses to high [CO_2] and CAM photosynthesis are based on few experiments (see Poorter & Navas, 2003).

2.3.4 Photosynthetic acclimation to rising CO_2

Although photosynthesis is stimulated in the short-term by elevated [CO_2], over time photosynthetic rates often decline relative to plants grown at current [CO_2] when measured at a common [CO_2]. This phenomenon, termed photosynthetic acclimation or down regulation, was initially thought to occur in response to restricted root volumes associated with plants in small pots (e.g. Arp, 1991; Thomas & Strain, 1991). However, acclimation has been confirme in a variety of plant species even under fiel conditions.

At the cellular/biochemical level, there are at least four potential mechanisms associated with photosynthetic acclimation at elevated CO_2: (a) sugar accumulation and gene repression (gene repression of the *D*1 and *D*2 genes of photosystem II, cyt f, the small and large rubisco subunits, and carbonic anhydrase (e.g. Krapp *et al.*, 1993; Sheen, 1994; van Oosten & Besford, 1995); (b) insufficien N uptake (e.g. Geiger *et al.*, 1999; Stitt & Krapp, 1999); (c) a tie-up of inorganic phosphate with carbohydrate accumulation and a subsequent limitation in RuBP (ribulose bisphosphate) regeneration capacity (e.g. Sharkey, 1985; Socias *et al.*, 1993); and (d) a potential direct effect on photosynthesis through increased saccharide content (e.g. Lewis *et al.*, 2002).

Whether different mechanisms of acclimation are associated with a given photosynthetic type is unclear, and has not, to our knowledge, been studied. Overall, at the cellular/biochemical level, there does not appear to be one ubiquitous mechanism associated with acclimation and, in fact, carbohydrate accumulation can occur independently of photosynthetic acclimation (Chu *et al.*, 1992; Bunce & Sicher, 2001). Furthermore, acclimation is not a ubiquitous response, even in the long-term (Ainsworth *et al.*, 2003) and may vary with weather conditions (Bunce & Sicher, 2003).

2.4 Cellular processes: photosynthetic carbon oxidation (PCO) and carbon dioxide

If rubisco fi es oxygen rather than carbon dioxide, the PCO cycle is initiated. This cycle results in the release of CO_2, which is called photorespiration. Because CO_2 is released, the net rate of CO_2 fixatio (i.e. photosynthesis) is reduced. CO_2 and O_2 are competitive inhibitors, and increasing the [CO_2] at the site of rubisco either metabolically (as in C_4 metabolism) or abiotically (as with increased atmospheric CO_2) reduces the rate of oxygenation and photorespiration with a subsequent increase in net photosynthetic rates.

The reaction of O_2 with RuBP results in 2-phoshoglycolate and 3-phosphoglycerate. The 3-phosphoglycerate enters into the normal photosynthetic carbon reduction (PCR) cycle, but the 2-phosphoglycolate is metabolized to glycolate and enters the peroxisome, where it is metabolized to glycine, an amino acid. In the mitochondrion, two glycine molecules can be combined to form serine, with the release of CO_2 and ammonia. The ammonia is reassimilated into amino acids in the chloroplast. Therefore, by reducing photorespiration, increasing [CO_2] may result in large decreases in leaf concentrations of glycine, serine, and ammonium (Ferrario-Mery *et al.*, 1997; Geiger *et al.*, 1998). Although a connection between decreased pool sizes of glycine and serine and lower nitrogen and protein in leaves developed at elevated CO_2 seems logical, any connection may be indirect, since both amino acids can be synthesized by other pathways besides the PCO cycle (Stitt & Krapp, 1999). Reduced photorespiration also decreases the rate of nitrate photoreduction (Rachmilevitch *et al.*, 2004), and this may contribute to lower protein content in leaves that develop at elevated CO_2.

2.5 Single leaf response to CO_2

2.5.1 Leaf carbon dynamics

Overall, our understanding of the impact of rising CO_2 on the PCO and PCR cycles provides a good predictive indicator of the photosynthetic response of single leaves to increasing atmospheric CO_2, at least in the short-term (days, hours) (e.g. Acock & Allen, 1985). Longer term leaf responses to CO_2, however, may be tempered by other abiotic variables such as nitrogen availability (Weerakoon *et al.*, 1999) or changes in PAR (photosynthetically active radiation) (Sims *et al.*, 1999). For example, for C_3 leaves, increasing temperature favours the PCO cycle, and suppression of this cycle by additional CO_2 results in a greater stimulation of photosynthesis with increasing CO_2 as temperature increases (e.g. Long, 1991). However, over longer timescales (weeks), photosynthetic acclimation to temperature can obscure or eliminate this synergy (Bunce, 2000; Ziska, 2001a).

2.5.2 *Inhibition of dark respiration*

The release of CO_2 during the oxidation of organic compounds in the mitochondria is termed *dark respiration*. It is thought that dark respiration may be slower in the light than in darkness in photosynthetic tissue, but methods to quantify dark respiration occurring simultaneously with photosynthesis remain equivocal (Pinelli & Loreto, 2003). Uncertainties arise because the amount of CO_2 efflu during dark respiration varies with the substrates oxidized, and because the degree of involvement of the alternative (uncoupled) respiratory pathway varies, which is poorly understood. A further complication is that fixatio of CO_2 by PEPc may occur even at night, and so CO_2 exchange rates in the dark may not solely reflec dark respiration.

Although it is generally acknowledged that very high concentrations of CO_2 (i.e. thousands of ppm) often drastically reduce rates of dark respiration (Palta & Nobel, 1989), there has been a considerable debate about whether the changes in [CO_2] concentration anticipated with anthropogenic change can directly inhibit specifi leaf respiration rates. Methodological problems with gaskets in small clamp-on leaf cuvettes in photosynthesis systems (Pons & Welschen, 2002) may compromise respiration measurements and may account for some reports of direct effects of CO_2 on dark respiration. However, inhibition of cytochrome c-oxidase and succinate dehydrogenase activities can occur with [CO_2] projected to occur in the near future (Gonzalez-Meler *et al.*, 1996). No effects of carbon dioxide on oxygen exchange have been detected using gas phase oxygen sensors (Davey *et al.*, 2004), although such sensors are still an order of magnitude less sensitive than CO_2 sensors, while liquid phase oxygen measurements have found CO_2 effects (Kaplan *et al.*, 1977; Reuveni *et al.*, 1993b). Further complicating the issue, Gonzalez-Meler *et al.* (2004) found that under certain metabolic conditions, coupled respiration may be decreased by elevated CO_2 without any effect on the rate of oxygen or carbon dioxide exchange. Effects of CO_2 concentration during the night on the rates of translocation and nitrate reduction (Bunce, 2004b), processes dependent on dark respiration, are indirect evidence that elevated CO_2 may sometimes reduce respiration at the leaf level.

2.5.3 *Leaf chemistry*

As a result of the cellular/biochemical impacts of [CO_2] on the PCR and PCO cycles, it has often been observed that the ratio of carbon to nitrogen within the leaf

(C/N) increases with increasing CO_2 (e.g Bazzaz, 1996; Drake *et al.*, 1997). This change in C/N ratio may be accompanied by concomitant increases in carbohydrate, lignin, and/or cellulose content (Bazzaz, 1996). Consequently, there are a number of anticipated changes that include changed decomposition rates, with reductions in nitrogen recycling (but see Billings *et al.*, 2003), as well as potential changes in leaf-freezing resistance (e.g. Obrist *et al.*, 2001).

Perhaps one of the best studied phenomenon related to CO_2-induced changes in C/N ratio is the association between the production of secondary chemicals and plant–herbivore interactions. For example, it has been widely observed that herbivore feeding is strongly influence by leaf allelochemicals as well as by leaf nutritional quality (e.g. Lincoln & Couvet, 1989). A number of studies have shown that the level of secondary (carbon-based) products tend to increase with enhanced $[CO_2]$ (Lindroth *et al.*, 1993; Lavola & Julkunen-Titto, 1994; Lindroth & Kinney, 1998), although this response is not ubiquitous (e.g. Kerslake *et al.*, 1998).

2.5.4 *Stomatal response and CO_2*

Observed reductions in stomatal conductance with increases in $[CO_2]$ are widespread, but not universal. While the mechanism by which carbon dioxide alters stomatal opening can be considered at the cellular or biochemical level (e.g. Assman, 1999), the overall impact is especially relevant to whole leaf processes, particularly stomatal limitation of photosynthesis and changes in water use. A number of studies have addressed the former question, arguing that reductions in stomatal aperture and conductance might reduce CO_2 availability with a subsequent negative impact on photosynthesis, independently of any direct change in carbon availability. However, a review of these studies suggests that while stomatal conductance is generally reduced by increasing CO_2, stomatal limitation of photosynthesis decreases (e.g. Drake *et al.*, 1997). Similarly, a number of studies have examined the impact of rising CO_2 on stomatal conductance and transpiration (e.g. Jones, 1998), concluding that rising CO_2 increases the water-use efficien y (WUE) of the leaf, usually define as the ratio of leaf carbon uptake to water loss. Carbon dioxide-induced improvements in leaf WUE have been suggested to either increase or maintain photosynthesis and carbon uptake indirectly for C_3 plants in water-stressed environments (in addition to any direct effect of CO_2 availability). Improved WUE and leaf water content with elevated CO_2 is also thought to be a significan factor in increased leaf photosynthesis in C_4 species with either increased salinity or decreased water availability (e.g. Drake & Leadley, 1991).

In spite of what is known regarding the effect of CO_2 on stomatal conductance, not all aspects have been fully elucidated. For example, it is not clear whether stomata on different leaf surfaces respond similarly, whether other abiotic parameters (e.g. vapor pressure deficit alter stomatal CO_2 sensitivity, or whether long-term CO_2 exposure results in significan changes in leaf stomatal number (see Morison, 1998 for a review).

2.6 Whole plant responses to rising CO_2

2.6.1 Plant development

One of the most documented effects of increasing [CO_2] is stimulation in plant growth relative to current [CO_2] (e.g. Kimball, 1993; Ghannoum *et al.*, 2000). However, this simple observation may reflec a number of complex developmental changes in addition to any leaf-level effect of [CO_2]. For example, a number of herbaceous-plant studies have shown that an approximate doubling of current [CO_2] could enhance seed germination (Esashi *et al.*, 1989; Ziska & Bunce, 1993), as could CO_2 concentrations above low Pleistocene levels (i.e. 180, 270, 360, and 600: gol mol^{-1}) (Mohan *et al.*, 2004). While stimulation of germination is not a ubiquitous response (e.g. Garbutt *et al.*, 1990), increasing CO_2 may interact with or increase the production of ethylene, a plant growth regulator that stimulates seed germination (e.g. Esashi *et al.*, 1987). Following germination and emergence, vegetative development may be particularly sensitive to increased CO_2. For example, in both C_3 and C_4 grasses, there is a strong response of tiller formation to rising CO_2 (e.g. sorghum, Ottman *et al.*, 2001; wheat, Ziska *et al.*, 2004), as well as a stimulation in leaf formation, growth and size in herbaceous and woody C_3 species (Bazzaz, 1996) and some C_4 species (e.g. Ziska & Bunce, 1997a; Seneweera *et al.*, 2001). Root growth may also be stimulated by increasing CO_2 during early development with observed increases in root length (Rogers *et al.*, 1992; Ziska *et al.*, 1996) as well as root diameter and cortex width (Rogers *et al.*, 1992). Change in root production may also be associated with increased fine-roo colonization of arbuscular mycorrhizal fungi (Olesniewicz & Thomas, 1999) as well as with increased nodule formation (Temperton *et al.*, 2003). Increased [CO_2] affects reproduction as flora number and pollen production may increase (e.g. Reekie *et al.*, 1997; Ziska & Caulfield 2000), as well as seed and fruit size, number, and quality (Garbutt & Bazzaz, 1984; Curtis *et al.*, 1994; Ward & Strain, 1997). Reductions in seed nitrogen for non-leguminous plants have also been observed (Jablonski *et al.*, 2002). Asexual production may also increase in response to CO_2 (Ziska, 2003a).

However, these documented CO_2 effects involve differential responses for a specifi plant organ, and do not necessarily consider changes in either growth form (morphology) or phenology (development rate). Allocation of additional carbon acquired in increased CO_2 may reflec shifts in biomass allocations during development to structures that are associated with a limiting resource. For example, if growth at elevated CO_2 increases the demand for nutrients, then additional carbon may go to root growth. A review of root/shoot (R/S) ratio in crop species grown in elevated CO_2 did demonstrate a significan increase in approximately 60% of the species studied (Rogers *et al.*, 1994). However, plants grown at different CO_2 concentrations are likely to differ in size, and it is unclear if reported changes in R/S ratio, many of which are one-time measurements, reflec seasonal changes or allometric shifts during development (e.g. Tissue *et al.*, 1997; but see Ziska, 2003a for Canada thistle). Changes in carbon allocation for reproduction have obvious implications with respect to long-term species success. Reproduction is often increased in response to rising CO_2 as additional carbon is allocated both to fl wers and to increased nodes and branches (see Ward & Strain, 1999 for a review). Increasing CO_2 can also alter developmental rate at the whole plant level with slower (Carter & Peterson, 1983), faster (St. Omer & Horvath, 1983), or similar (Garbutt & Bazzaz, 1984) rates being observed. In common ragweed (*Ambrosia artemisiifolia*), time to reproduction was altered, in part, by faster growth rates (Ziska *et al.*, 2003); however, for other species, elevated CO_2 altered the size at which plants initiated reproduction (e.g. Reekie & Bazzaz, 1991). Elevated CO_2 may also alter plant senescence, increasing it in some cases (St. Omer & Horvath, 1983; Sicher, 1998; Jach & Ceulemans, 1999), delaying it in others (e.g. Hardy & Havelka, 1975).

What are the links between physiological processes at the genetic/cellular/leaf level and whole plant phenology and/or morphology, as CO_2 increases? What determines carbon allocation between plant organs or rate of development? It seems unlikely that the observations reported here are strictly related to leaf-level impacts; e.g. increases in relative growth rate at elevated CO_2 can occur before leaf maturation (e.g. Ziska & Bunce, 1995), and early exposure to elevated CO_2 is associated with tiller production in agronomic grasses independent of changes in leaf area (Christ & Körner, 1995). Unfortunately, with few exceptions (e.g. Masle, 2000), almost nothing is known about the link between anatomical or physiological processes and developmental response at the whole plant level as a function of $[CO_2]$. Yet, understanding such links has both pragmatic implications for managed systems (e.g. selecting the most CO_2 responsive cultivars) and natural plant communities (e.g. understanding plant-to-plant interactions and competitive outcomes).

2.6.2 Carbon dynamics

Given the large number of studies regarding the response of single leaves to rising CO_2, and our understanding of carboxylation kinetics (e.g. Bowes, 1996), how well does the single leaf response predict the degree of photosynthetic stimulation or acclimation at the whole plant level? Surprisingly, while only a handful of studies have examined single leaf and whole plant photosynthetic responses to CO_2 simultaneously, these data indicate that single leaf responses are a poor predictor of whole plant photosynthesis or growth (e.g. Amthor, 1994). This suggests that the degree of photosynthetic stimulation/acclimation in response to CO_2 may differ as a function of scale. If, however, the degree of stimulation or acclimation is a function of sources and sinks of carbon (e.g. Stitt, 1991), then how might CO_2-induced changes in morphology and/or development at the whole plant level alter the response of single leaf photosynthesis?

Clearly, CO_2 may induce temporal changes in plant development, and these changes in turn may alter sink availability. For example, in kohlrabi and sugar beet reversibility of leaf acclimation to elevated $[CO_2]$ is observed following the development of additional carbon sinks (e.g. Ziska *et al.*, 1995; Bunce & Sicher, 2003). Conversely, if CO_2 increases the rate of senescence, then remobilization of nitrogen

Table 2.1 Net CO_2 assimilation rates (A) of leaves and whole plants of *Phaseolus vulgaris* L. cv. Red Kidney measured at 30°C and 1800 μmol m^{-2} s^{-1} PPFD (photosynthetic photon flu density) at 29 days after sowing*

CO_2 (μmol mol^{-1})		A (μmol m^{-2} s^{-1})		
Growth	Measurement	First trifoliolate leaf	Second trifoliolate leaf	Whole plant
270	270	21.3a	23.0a	14.7a
370	270	17.4b	22.8a	14.5a
720	270	11.9c	21.9a	13.6a
270	370	27.9a	31.7a	20.5a
370	370	25.3b	31.0a	20.4a
720	370	21.5c	30.0a	18.9a
270	720	46.2a	47.5a	29.2a
370	720	42.0b	46.5a	30.0a
720	720	35.1c	44.2a	28.3a

* For each common measurement CO_2 concentration, numbers followed by different letters are signifi cantly different at $P = 0.05$. (Bunce, unpublished.)

at the leaf level may exacerbate the degree of acclimation (Sicher, 1998). Sinks may also respond separately to other abiotic parameters, with subsequent effects on the ability of single leaves to respond photosynthetically to CO_2. For example, if air temperatures exceeds the optimum for pollen formation, but not leaf function, then the resulting increase in flora sterility may limit reproductive sinks with a subsequent decline in photosynthetic response to CO_2 (e.g. Lin *et al.*, 1997). Whole plant response to CO_2 may alter carbon sources as well. If acclimation occurs only late in leaf development, then acclimation may be apparent at the leaf but not at the whole plant level (Table 2.1). If enhanced leaf development increases the degree of self-shading (relative to [CO_2]), and leaf photosynthetic acclimation also occurs, then whole plant photosynthesis would decline in response to rising CO_2, a situation observed with increasing CO_2 and temperature in soybean (Ziska & Bunce, 1997b). Self-shading and nitrogen redistribution within whole plant canopies could, potentially, underlie the degree of photosynthetic acclimation to elevated CO_2 in leaves of wheat and poplar at different depths in the canopy (Adam *et al.*, 2000; Takeuchi *et al.*, 2001). Sources may also respond separately to other abiotic inputs, limiting plant photosynthetic response to increasing CO_2. For example in rice, nitrogen limits any stimulation in tillering in response to elevated CO_2, with a subsequent decline in whole plant photosynthesis (Ziska *et al.*, 1996). Overall, the temporal and species-specifi nature of plant development, subsequent changes in morphology, and potential interactions with other environmental variables such as light and nutrient availability will determine the regulation of carbon sources and sinks and feedback inhibition at the leaf level. Therefore, while we possess a thorough understanding of how rising CO_2 alters leaf carbon dynamics (e.g. Long, 1991), our ability to utilize this knowledge to predict whole plant responses is limited. Indeed, even simple associations between nutrient status and CO_2 responsiveness of whole plants remain a subject of controversy (e.g. Poorter, 1998 vs Lloyd & Farquhar, 1996).

What is the potential impact of increasing CO_2 on whole plant dark respiration? Reuveni and Gale (1985) were the firs to demonstrate that elevated CO_2 only at night (950 ppm) resulted in a significantl greater net carbon gain for *Medicago sativa* seedlings, suggesting an elevated CO_2-induced inhibition of dark respiration. Since this initial study, numerous reports have argued both for (e.g. Bunce, 1994; Wullschleger *et al.*, 1994; Drake *et al.*, 1999) and against (Amthor, 2001; Jahnke & Krewitt, 2002; Davey *et al.*, 2004) any inhibition of dark respiration. Yet, a number of studies have repeated the original Reuveni and Gale experiment, with elevated CO_2 only given during the dark, and significan increases in whole plant growth have been observed (Reuveni *et al.*, 1993a; Bunce, 1995a; Ziska *et al.*, 2001). If growth is increasing with only night-time increases in CO_2, then either CO_2 is altering respiration, is being fi ed directly, or is having some other, uncharacterized, indirect effect on carbon uptake. Interestingly, the ratio of whole plant respiration to photosynthesis declines at elevated CO_2 in developing soybean plants, suggesting a reduction in respiratory cost per unit tissue (Ziska & Bunce, 1998). This result is consistent with canopy data showing that respiration does not increase proportionally to increases in biomass in response to elevated CO_2 (Gonzalez-Meler *et al.*, 2004). Overall, despite its importance for plant growth, the issue of how rising CO_2 alters dark respiration remains unresolved.

2.6.3 Stomatal regulation and water use

Although the direct effect of elevated CO_2 on reducing stomatal conductance and water use of single leaves is generally acknowledged, there are several factors that affect our ability to predict changes in WUE at the whole plant level. One of these factors is whether the short-term effect of elevated carbon dioxide on stomatal conductance may become larger or smaller after prolonged exposure to elevated carbon dioxide. Because changes in photosynthetic capacity are often paralleled by changes in stomatal conductance, acclimation of photosynthesis to elevated carbon dioxide was anticipated to produce a parallel acclimation of stomatal conductance. This does occur in some species, such as wheat and potato (Bunce, 2001). However, acclimation of stomatal conductance to elevated carbon dioxide in barley did not parallel photosynthetic acclimation, and in sorghum acclimation of stomatal conductance occurred without acclimation of photosynthesis (Bunce, 2001). We are not aware of data indicating that plants acclimate to elevated carbon dioxide by increasing stomatal conductance, rather it is unchanged or reduced compared with plants grown at lower concentrations, when all are measured at elevated carbon dioxide. Second, if plants growing at elevated carbon dioxide have increased leaf area, this can offset reductions in water loss per unit of area. This is frequently observed in controlled environments, but increases in leaf area per plant at elevated [CO_2] seem to be much less common under fiel conditions, at least for crop plants (Bunce, 2004a). Third, even for a single leaf, a reduction in stomatal conductance produces a less than proportional reduction in water loss, because of the presence of the leaf boundary layer resistance to water loss, and because of feedback through leaf energy balance effects (Jarvis & McNaughton, 1986). For whole plants and canopies, aerodynamic resistances become increasingly important, especially for short canopies, and the effect of reduced stomatal conductance on water loss decreases with increasing scale. Furthermore, reactions of stomatal conductance to the changes in leaf temperature and the humidity of the air adjacent to the leaves caused by lower conductance at elevated carbon dioxide produce other feedback effects on the relationship between changes in stomatal conductance and whole plant transpiration (Wilson *et al.*, 1999).

2.7 Plant-to-plant interactions

It is sometimes assumed that because different plant species do not compete for carbon dioxide directly, CO_2 is less important in plant to plant interactions than other abiotic parameters (e.g. nutrients or water). However, any resource that affects the growth of an individual alters its ability to compete. Hence, competition not only occurs in response to limited resources, but also occurs when species respond differently to resource enhancement. While not all plant–plant interactions are competitive (e.g. some are facultative or neutral; see Bazzaz, 1996), it is the competitive aspect of plant–plant interactions that has received the most attention (e.g. Poorter & Navas, 2003). It is particularly important to understand the effects of elevated CO_2 on plant–plant interactions since the response of individual plants to increasing CO_2 differs considerably from plants grown in competition (e.g. Bazzaz *et al.*, 1995). Furthermore, competitive outcomes with increasing [CO_2] cannot always be predicted based on plant functional types or photosynthetic pathway (e.g. Bazzaz & McConnaughay, 1992; Owensby *et al.*, 1993).

2.7.1 Plant competition: managed systems

Because the C_4 photosynthetic pathway is overly represented in troublesome weedy species, many experiments and most reviews concerned with weed competition and rising [CO_2] in managed systems have reported on C_3 crop–C_4 weed interactions (Patterson *et al.*, 1984; Patterson, 1986). However, crop–weed competition varies significantl by region; consequently, depending on temperature, precipitation, soil, etc. C_3 and C_4 crops will interact with C_3 and C_4 weeds. In addition, a C_3 crop vs C_4 weed interpretation does not address weed–crop interactions where the photosynthetic pathway is the same. Yet, many of the worst/troublesome weeds for a given crop are genetically similar, and frequently possess the same photosynthetic pathway (e.g. sorghum and Johnson grass, both C_4; oat and wild oat, both C_3).

Overall, data regarding the competitive outcomes of crops and weeds as a function of increasing [CO_2] remain scarce (Table 2.2). The majority of studies involving different photosynthetic pathways have focused on a C_3 crop in competition with a

Table 2.2 Summary of studies examining whether weed or crops were 'favoured' as a function of elevated [CO_2]*

Crop	Weed	Increasing [CO_2] favours?	Environment	Reference
A. C_4 crops/C_4 weeds				
Sorghum	*Amaranthus retroflexus*	Weed	Field	Ziska (2003b)
B. C_4 crops/C_3 weeds				
Sorghum	*Xanthium strumarium*	Weed	Glasshouse	Ziska (2001b)
Sorghum	*Albutilon theophrasti*	Weed	Field	Ziska (2003b)
C. C_3 crops/C_3 weeds				
Soybean	*Chenopodium album*	Weed	Field	Ziska (2000)
Lucerne	*Taraxacum officinale*	Weed	Field	Bunce (1995b)
Pasture	*Taraxacum and Plantago*	Weed	Field	Potvin and Vasseur (1997)
Pasture	*Plantago lanceolatae*	Weed	Chamber	Newton *et al.* (1996)
D. C_3 crops/C_4 weeds				
Fescue	*Sorghum halapense*	Crop	Glasshouse	Carter and Peterson (1983)
Soybean	*Sorghum halapense*	Crop	Chamber	Patterson *et al.* (1984)
Rice	*Echinochloa glabrescens*	Crop	Glasshouse	Alberto *et al.* (1996)
Pasture	*Paspalum dilatatum*	Crop	Chamber	Newton *et al.* (1996)
Lucerne	Various grasses	Crop	Field	Bunce (1993)
Soybean	*Amaranthus retroflexus*	Crop	Field	Ziska (2000)

*'Favoured' indicates whether elevated [CO_2] produced significantl more crop or weed biomass. 'Pasture' refers to a mix of C_3 grass species.

C_4 weed (Table 2.2). In those comparisons, increasing CO_2 increased the crop/weed biomass ratio, consistent with the known biochemical/cellular/leaf response. However, it is interesting to point out that biomass and/or yield of grain sorghum (C_4 crop) was reduced by high [CO_2] when grown in the presence of either velvetleaf (*Albutilon theophrasti*) or cocklebur (*Xanthium strumarium*), both C_3 weeds. Most comparisons with the same photosynthetic pathway for the vegetative growth of crops and weeds resulted in significan decreases in crop/weed biomass when weed and crop emerged simultaneously (Table 2.2). Only two studies have actually quantifie changes in crop seed yield with weedy competition as a function of rising [CO_2] (Ziska, 2000, 2003b). In these studies, two crop species, one C_3 (soybean) and one C_4 (dwarf sorghum), were grown with lamb's-quarters (C_3) and redroot pigweed (C_4) and velvetleaf (C_3) and redroot pigweed, respectively, at a density of two weeds per meter of row. Although soybean yield losses were less from pigweed, all other crop–weed interactions resulted in increased yield loss in elevated [CO_2]. Interestingly, in these later studies, the presence of any weed species negated the ability of the crop to respond either vegetatively or reproductively to enhanced [CO_2]. This may be significan since CO_2 enhancement studies of crop yield rarely consider crop–weed competition. However, additional field-base studies are needed to confir and amplify the results presented here.

2.7.2 Plant competition: unmanaged systems

Less is known regarding the influenc of rising CO_2 on plant competition among unmanaged systems, in part, because competition in plant communities involves multispecies comparisons and is best considered in an ecosystem context (see Section 2.8). In addition, it may be difficul to separate the impact of elevated [CO_2] from competition for other abiotic resources such as water or nutrients. However, there are circumstances in unmanaged systems where only a handful of species are competing at a given time. For example, during early succession, competition between species can be altered as a function of CO_2 (Bazzaz, 1996). For forest systems, vines and slower growing trees exhibit differential responses to rising CO_2, with positive effects on vine biomass (e.g. Granados & Körner, 2002) and subsequent effects on vine–tree competition (Phillips *et al.*, 2002). For C_3 and C_4 comparisons, elevated CO_2 was shown to favour the biomass production of a C_3 sedge (*Scirpus olneyi*) over the production of a C_4 grass (*Spartina patens*) in a marsh system (Curtis *et al.*, 1989). In contrast, a dominant C_4 grass was favoured over a dominant C_3 species in response to elevated CO_2 (Owensby *et al.*, 1993) in a dry, tallgrass prairie system because of higher drought tolerance for the C_4 species.

2.7.3 How does CO_2 alter plant-to-plant interactions?

There is no question that ongoing increases in atmospheric CO_2 will change plant competition and composition (Bazzaz & McConnaughay, 1992). Given the economic and/or environmental importance of predicting competitive outcomes in plant systems, do we, in fact, know what specifi aspects of plant growth and development are associated with increased competitive success as CO_2 increases? At the cellular/leaf level we could argue that the differential CO_2 sensitivities of the C_3 and C_4 photosynthetic pathways could be used to predict competitive outcomes; yet, as we have seen, this is not always a reliable predictor. Furthermore, it does not address competitive outcomes where photosynthetic pathway is the same. At the whole plant level we could argue that fast-growing species are more responsive to CO_2 or that nutrient stress enhances CO_2 response; yet, these are also not a good predictor of competitive outcomes (e.g. Poorter & Perez-Soba, 2001).

At present, there does not appear to be any 'one size fit all' explanation of competitive success in response to CO_2. This is not too surprising, given the complexities of competition per se. There is a great need for detailed studies of plant-to-plant interactions that help resolve species-specifi temporal and spatial exploitation of either aboveground (e.g. light) or belowground (e.g. nitrogen, water) resources in response to increasing CO_2.

2.8 Plant communities and ecosystem responses to CO_2

2.8.1 Managed plant systems

Because of the importance of food security, much of the early focus regarding the impact of rising CO_2 was on agricultural crops (e.g. Acock & Allen, 1985). However, many of these studies were of individual plants, and field-base evaluations of crop systems are only evident from the 1990s (e.g. Kimball *et al.*, 1995). In general, CO_2 concentrations (usually 200–400 ppm above current ambient levels) have been found to stimulate the growth and yield of C_3 (rice, wheat), but not C_4 cereals (corn, sorghum); and stimulate leguminous (soybean) and tuberous crops (potatoes) as well as numerous leafy vegetables (see Reddy & Hodges, 2000 for a review). Fibre crops, such as cotton, may also show a strong growth and boll response to elevated CO_2 (Kimball & Mauney, 1993). Alternatively, pastures do not always show a strong response to CO_2 (e.g. Körner, 1997).

Although the response of annual crops in managed systems has been well studied, less is known regarding managed perennial species. Forest plantations in the world now total approximately 130 Mha with annual rates of establishment of about 10.5 Mha (Janssens *et al.*, 2000). In the United States, commercially planted loblolly pine (*Pinus taeda*) remains a major source for wood products (Jokela *et al.*, 2004). While the response of mature loblolly has been examined in response to CO_2 in unmanaged systems (e.g. DeLucia *et al.*, 1999), the response of cultivated and fertilized stands is unknown. Similarly, the CO_2-induced changes in the productivity of stone or tropical fruits have been largely unexamined (Janssens *et al.*, 2000).

2.8.2 Water use in managed systems

Stomatal and leaf areas responses to elevated CO_2 have best been characterized in annual crops (Bunce, 2004b). But does the single leaf response translate to a decrease in water use at the community level in managed systems? Developmentally, increased CO_2 can result in an increase in leaf area and plant size as well as changes in the R/S ratio, foliage anatomy, and the growth of conductive tissue in the shoot (Tyree & Alexander, 1993). Hence, it is unclear if any savings of water or increase in WUE at the leaf level is observed within crop communities.

Although crops have among the largest reductions in stomatal conductance at elevated [CO_2], the relative day-to-day reduction is inconsistent and may vary considerably with other abiotic inputs (e.g. light, temperature) with a wide range of stomatal variability (Bunce, 2004a). Overall, using both simulations and direct measurements in FACE systems, any large reductions in crop stomatal conductance would only translate into small reductions in community evapotranspiration, in part, because of the direct effect of CO_2 on increasing canopy temperature and decreasing humidity (Bunce, 2004a). These later changes may also have important impacts on crop yields (e.g. Matsui *et al.*, 1997). However, at the system level, even a reduction of a few percent in evapotranspiration could be important both to crop yield and to the economics of crop production.

2.8.3 *Unmanaged plant systems*

Methodological changes, particularly the advent of FACE technology in the 1990s, spurred interest in addressing the potential impact of rising CO_2 on community level responses (Hendrey & Kimball, 1994). However, the response of unmanaged systems is complicated, since unlike managed agriculture, abiotic inputs such as water or nutrients can be extremely variable. Although unmanaged systems can show increases in productivity and changes in plant species composition in response to elevated CO_2 (e.g. Smith *et al.*, 2000), there is a wide range of specifi predictions regarding how elevated CO_2 will alter community level processes.

Early evaluations of CO_2 responses in arctic tundra systems, for example, exhibited little change in productivity (Grulke *et al.*, 1990). Grassland communities have shown a mixed growth response to [CO_2], with communities with a greater degree of species richness showing a larger response (e.g. Reich *et al.*, 2001), possibly as a result of highly CO_2 responsive species not present in the less diverse communities (e.g. Grunzweig & Körner, 2000). For desert ecosystems, the extent of elevated [CO_2] impacts was correlated with rainfall events (i.e. increased water and nutrients) with a subsequent increase in community productivity (Smith *et al.*, 2000). Conversely, plants within a wetland system (e.g. *S. olneyi*, a C_3 marsh species) continue to show species-specifi positive growth responses to elevated CO_2 after 17 years of exposure (Rasse *et al.*, 2005).

Among unmanaged systems, the impact of rising CO_2 on forest productivity is of particular interest, given the role of forests in sequestration of terrestrial carbon. In general, a review of long-term experiments with young trees does indicate a significan increase in growth (~30%) with a doubling of [CO_2] from current levels (Medlyn *et al.*, 2001). However, it is unclear, given the differences in macroclimate between open and closed canopies, whether a similar response will be observed temporally for the growth and net primary productivity (NPP) of more mature stands. To date, much of the experimental evidence does suggest that there may be a permanent CO_2 effect even if photosynthetic acclimation does occur, but additional information, particularly on belowground carbon allocation, (Zak *et al.*, 2003) is needed.

2.8.4 *Water use in unmanaged plant systems*

For grasslands and deserts, there is a clear interaction between indirect (stomatal) effects of [CO_2] on water use and stimulation of plant growth at the community level. Morgan *et al.* (2004) recently reviewed several fiel experiments including Kansas tallgrass prairie, Colorado shortgrass steppe, and Swiss calcareous grasslands, with all systems showing a greater [CO_2] enhancement in dry years. In contrast, a Texas C_3/C_4 grassland and a New Zealand pasture were unaffected by yearly variation in soil water, while plant growth in the Mojave desert was only stimulated by elevated [CO_2] during wet years. While the interaction between [CO_2] and water availability is apparent within these ecosystems, to date, no systematic separation of CO_2 enrichment responses vs indirect water-driven responses has been done experimentally (Morgan *et al.*, 2004). Similarly, no long-term evaluations separating CO_2 fertilization from water use effects are available for forest communities. Longer term evaluations of hydrologic balance for loblolly pine for the Duke FACE facility indicate that no direct effect of elevated CO_2 on water savings was discernable (after 3.5 years); rather, the forest transpired progressively more water, possibly as a result of reduced soil evaporation due to the additional litter buildup at the high [CO_2] (Schafer *et al.*, 2002).

2.8.5 Other trophic levels

Any consideration of ecosystem responses to increasing [CO_2] should include not only plant productivity but also potential impacts on higher trophic levels. For example, it is probable that herbivore biology will be impacted by the physiological effects of elevated CO_2 on host plant metabolism. Specifi CO_2-induced changes at the leaf level would include increased C/N ratio, altered concentrations of defensive compounds, increased starch and fibr content, and increased water content (e.g. Lincoln & Couvet, 1989). What is less clear, however, is whether the response observed at the leaf or plant level is consistent with the response of plant communities. For example, there are compensatory changes in leaf production that could, potentially, overcome insect-related damage (Hughes & Bazzaz, 1997). For scrub oak and marsh ecosystems, less infestation of leaf-eaters was observed at elevated CO_2 (Thompson & Drake, 1994; Stiling *et al.*, 2002). Recent data for gypsy moth in a mature forest suggest that species-specifi changes in leaf chemical composition induced by high [CO_2] may lead to contrasting herbivore responses (Hattenschwiler & Schafellner, 2004). Preferential herbivore feeding on one species may, in turn, alter plant competition. Overall, however, most data have only examined single insect–host plant interactions in response to increasing CO_2, and a more complete assessment of insect herbivory within plant communities is lacking.

There are also a number of recognized CO_2-induced changes that could alter the susceptibility of plants to disease. For example, improved WUE and leaf water content could promote sporulation by foliar fungi (e.g. Thompson & Drake, 1994), while increases in leaf carbohydrate could promote the growth and reproduction of pathogens following infection (Hibberd *et al.*, 1996). Alternatively, reductions in leaf nitrogen content could reduce pathogen load and disease severity (Thompson *et al.*, 1993) and CO_2-induced changes in senescence could either increase or decrease exposure duration to pathogens (e.g. Malmstrom & Field, 1997). At the community level, reductions in evaporation or transpiration could reduce canopy humidity with consequent effects on the growth and sporulation of most fungi (Chakraborty *et al.*, 2000; Chakraborty & Data, 2003), while increases in productivity in high [CO_2] could increase plant residues, with potentially greater pathogenic overwintering (Manning & Tiedemann, 1995). In addition, increased root production and/or changes in root exudation would increase the proportion of host tissue available for pathogenic infection (Manning & Tiedemann, 1995). Overall, however, the extremely limited attention given to this fiel of study precludes any ability to make generalized predictions with confidence We are left with the rather general prediction that 'diseases may increase, decrease, or show no change' (Coakley, 1995).

2.9 Global and evolutionary scales

2.9.1 Rising CO_2 as a selection factor

In using Figure 2.2 as a guide to examine how rising CO_2 alters plant biological function over time and space, we have not considered limits along either axis. For example, it seems unlikely that plant or community responses to CO_2 will remain stable over time; yet little attention has been paid to the consequences of increasing CO_2 on evolutionary timescales. As with light, nutrients, and water, there is considerable genetic variation in response to CO_2 (e.g. Curtis *et al.*, 1994; Bazzaz *et al.*, 1995), suggesting that plants have altered reproductive and evolutionary success. On a shorter timescale, evaluation of these selective changes may have pragmatic consequences, such as selection for high-yielding agronomic cultivars in managed plant systems (e.g. Ainsworth *et al.*, 2002) or the success of invasive plant species within a community (Smith *et al.*, 2000; Hattenschwiler & Körner, 2003). At present, long-term evaluations of species success, changes in biodiversity, or community selection at higher trophic levels in response to CO_2 per se are unavailable.

2.9.2 Global impacts

If evolution represents a long-term temporal change, then global estimates regarding the impact of rising CO_2 on ecosystem function reflec a very large spatial scale. Alterations on such a scale cannot be addressed experimentally, but only by means of global modeling, and a number of general circulation models as well as regional climate assessments are available that include atmosphere–biosphere exchanges (e.g. Boer *et al.*, 2000). Such models serve to integrate and synthesize existing information regarding CO_2 impacts and to project this information to global outcomes. As such, modeling efforts are useful as potential projections of global consequences, and highlight areas where additional enquiry is needed.

However, given the large adjustment in scale, uncertainties at the ecosystem level are magnifie considerably in global assessments. For example, there is a great deal of interest in quantifying the role of plant communities in sequestering additional carbon as a potential means to mitigate the rate of increase in atmospheric CO_2 (e.g. Gurney *et al.*, 2002). At the community/ecosystem level, there is some question as to the ability of forest systems to act as long-term carbon sinks (e.g. Schlesinger & Lichter, 2001), emphasizing the need for a better understanding of carbon/nitrogen cycling in forest soils. Indeed, the role of soil nitrogen pools appears crucial in understanding global carbon sequestration and remains the subject of much discussion (e.g. Norby & Cotrufo, 1998; Zak *et al.*, 2003; Hungate *et al.*, 2004, see also Chapters 8 and 9). Yet, global modeling estimates of climate and CO_2-induced increases in NPP (and subsequent changes in carbon sequestration) (e.g. Nemani *et al.*, 2003) may not consider such subtleties. Unfortunately, policymakers often view climate change models as a final authoritative evaluation, and not as works in progress.

2.10 Uncertainties and limitations

Carbon dioxide is one of four necessary abiotic inputs for plant biology and the recent and projected changes in its concentration have already impacted, and will continue to impact, on plant function. Although the primary physiological effects of CO_2 are directly related to carbon uptake/loss and water use, it is clear that these changes alter plant function at every organizational level (Figure 2.3). It is also clear that an experimental or conceptual understanding at one organizational level may not necessarily serve as a reliable guide to predicting the functional behavior at different levels. A thorough grasp of leaf-level processes, for example, only provides limited insight into ecosystem responses. Yet much of what is known regarding the impact of CO_2 on plant biology remains descriptive and not mechanistic; focused on single plant responses, and non-integrative. Overall, in evaluating the response of plants to CO_2, there is a clear imperative for researchers to 'scale-up' their findings

Which organizational levels require greater experimental study? While there are numerous evaluations that have examined the photosynthetic and growth response of individual plants to a 'doubling' of [CO_2], there are relatively fewer reports regarding the impact of rising CO_2 on spatial or temporal extremes. For example, we know little about specifi CO_2-induced changes in genetic expression, or how these changes would be influence in an evolutionary sense; similarly, we know relatively little about the impact of CO_2 on long-term ecosystem function and the interactions between carbon, water, and nutrient cycles (e.g. Pan *et al.*, 1998).

But ultimately, ecosystems are integrators of genetic, structural, plant–plant interactions, higher trophic levels, and evolutionary responses over time and space. In addition, those impacts, resulting from global increases in atmospheric [CO_2], whether economic or environmental, are most likely to be evidenced in ecosystem function. For managed systems, the role of increasing CO_2 in the success of undesirable plants, particularly invasive or noxious species, deserves particular attention; in unmanaged systems, differential species response, and the resulting implications for species selection, diversity, and higher trophic levels, remains a crucial area of

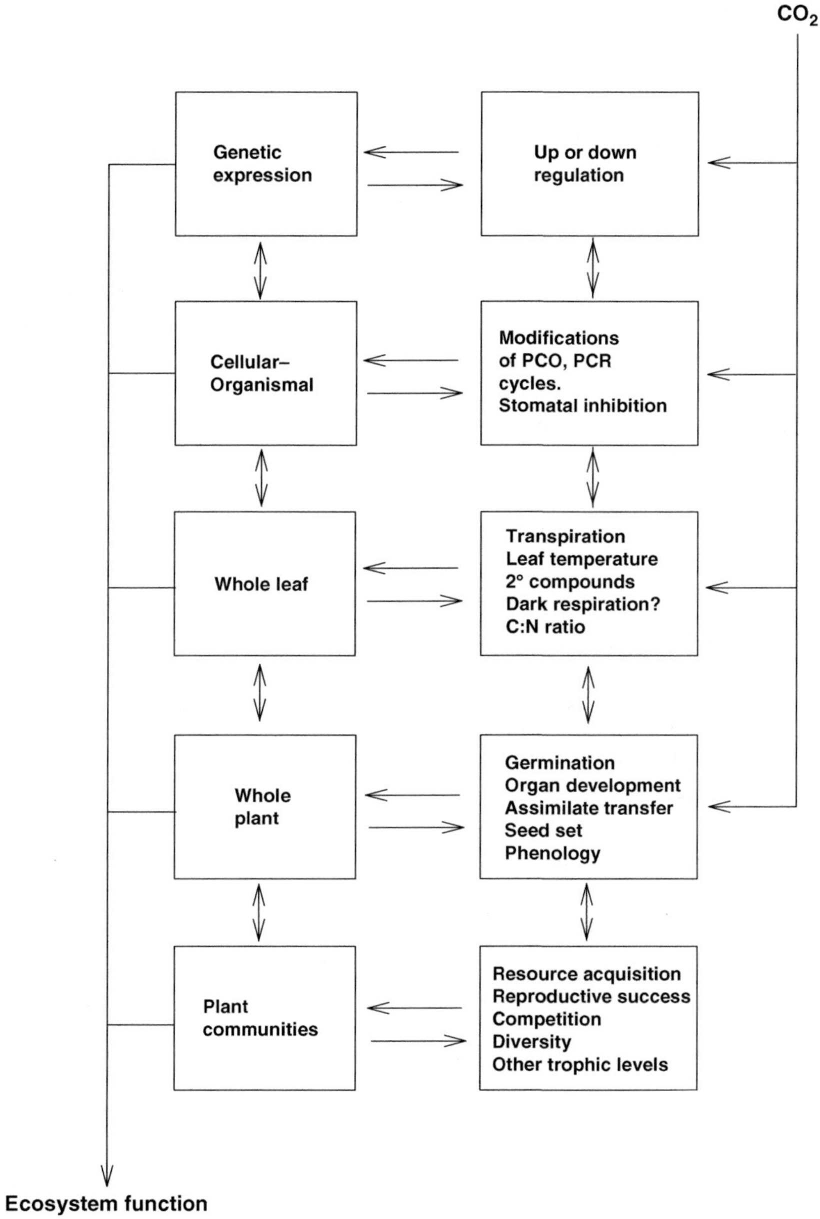

Figure 2.3 Potential impact of rising atmospheric CO_2 for the different organizational levels shown in Figure 2.2. Additional details are given in text.

study. There are obvious experimental challenges to studying ecosystems (e.g. abiotic vicissitude and year-to-year variation in primary productivity, quantificatio of below-ground processes particularly nutrient cycling and carbon storage, potential changes in herbivory, pathogen load, etc.); however, those hypotheses that consider multifactor responses, particularly at the ecosystem level, are necessary if we are to explicitly recognize and adapt to CO_2-induced changes in plant systems.

There is one other fundamental challenge: the need to recognize that global increases in CO_2 are only one aspect of unprecedented anthropogenic change. With a population of 6 billion, humans are significantl altering rates of nitrogen deposition (e.g. Wedin & Tilman, 1996), the extent of tropospheric ozone (e.g. Krupa & Manning, 1988), and land use patterns (Pielke *et al.*, 2002). Any experimental approach that focuses on ecosystem dynamics, therefore, should take not only CO_2 into account but also other rapidly changing abiotic variables, whenever possible. It is hoped that a multifactor approach integrating ecosystem function can also be used to increase the predictive capacity of existing global change models.

References

Acock, B. & Allen, L.H., Jr (1985) Crop responses to elevated carbon dioxide concentrations. In: *Direct Effects of Increasing Carbon Dioxide on Vegetation* (eds B.R. Strain & J.D. Cure), pp. 53–98. U.S. Department of Energy, DOE/ER-0238, Washington, D.C.

Adam, N.R., Wall, G.W., Kimball, B.A., Pinter, P.J., Jr, LaMorte, R.L., Hunsaker, D.J., Adamsen, F.J., Thompson, T., Matthias, A.D., Leavitt, S.W. & Webber, A.N. (2000) Acclimation response of spring wheat in a free-air CO_2 enrichment (FACE) atmosphere with variable soil nitrogen regimes. 1. Leaf position and phenology determine acclimation response. *Photosynth. Res.*, **66**, 65–77.

Ainsworth, E.A., Davey, P.A., Bernacchi, C.J., Dermody, O.C., Heaton, E.A., Moore. D.J., Morgan, P.B., Naidu, S.L., Yoo, H.-S., Zhu, X.G., Curtis, P.S. & Long, S.P. (2002) A meta-analysis of elevated [CO_2] effects on soybean (*Glycine max*) physiology, growth and yield. *Global Change Biol.*, **8**, 695–709.

Ainsworth, E.A., Davey, P.A., Hymus, G.J., Osborne, C.P., Rogers, A., Blum, H., Nosberger, J. & Long, S.P. (2003) Is stimulation of leaf photosynthesis by elevated carbon dioxide concentration maintained in the long term? A test with *Lolium perenne* grown for 10 years at two nitrogen fertilization levels under free air CO_2 enrichment (FACE). *Plant Cell Environ.*, **26**, 705–714.

Alberto, A.M., Ziska, L.H., Cervancia, C.R. & Manalo, P.A. (1996) The influenc of increasing carbon dioxide and temperature on competitive interactions between a C_3 crop, rice (*Oryza sativa*), and a C_4 weed (*Echinochloa glabrescens*). *Aust. J. Plant Physiol.*, **23**, 795–802.

Amthor, J.S. (1994) Scaling CO_2-photosynthesis relationships from the leaf to the canopy. *Photosynth. Res.*, **39**, 321–350.

Amthor, J.S., Koch, G.W., Williams, J.R. & Layzell, D.B. (2001) Leaf O_2 uptake in the dark is independent of coincident CO_2 partial pressure. *J. Exp. Bot.*, **52**, 2235–2238.

Arp, W.J. (1991) Effect of source-sink relations on photosynthetic acclimation to elevated CO_2. *Plant Cell Environ.*, **14**, 869–875.

Assman, S.M. (1999) The cellular basis of guard cell sensing of rising CO_2. *Plant Cell Environ.*, **22**, 629–638.

Bae, H. & Sicher, R. (2004) Changes of soluble protein expression and leaf metabolite levels in *Arabidopsis thaliana* grown in elevated atmospheric carbon dioxide. *Field Crops Res.*, **90**, 61–73.

Bazzaz, F.A. (1996) *Plants in Changing Environments: Linking Physiological, Population and Community Ecology*. Cambridge University Press, Cambridge, UK.

Bazzaz, F.A., Jasienski, M., Thomas, S.C. & Wayne, P. (1995) Micro-evolutionary responses in experimental populations of plants to CO_2-enriched environments: parallel results from two model systems. *Proc. Natl. Acad. Sci.* USA, **92**, 8161–8165.

Bazzaz, F.A. & McConnaughay, K.D.M. (1992) Plant-plant interactions in elevated CO_2 environments. *Aust. J. Bot.*, **40**, 547–563.

Billings, S.A., Zitzer, S.F., Weatherly, H., Schaeffer, S.M., Charlet, T., Arnone, J.A., III & Evans, R.D. (2003) Effects of elevated CO_2 on leaf litter quality in a Mojave desert ecosystem. *Global Change Biol.*, **9**, 729–735.

Boer, G.J., Flato, G.M., Reader, M.C. & Ramsden, D. (2000) A transient climate change simulation with historical and projected greenhouse gas and aerosol forcing: experimental design and comparison with the instrumental record for the 20th century. *Clim. Dyn.*, **16**, 405–425.

Bowes, G. (1996) Photosynthetic responses to changing atmospheric carbon dioxide concentration. In: *Photosynthesis and the Environment* (ed. N.R. Baker), pp. 387–407. Kluwer Publishing, Dordrecht, Netherlands.

Bunce, J.A. (1993) Growth, survival, competition, and canopy carbon dioxide and water vapor exchange of firs year alfalfa at an elevated CO_2 concentration. *Photosynthetica*, **29**, 557–565.

Bunce, J.A. (1994) Responses of respiration to increasing atmospheric carbon dioxide concentrations. *Physiol. Plant.*, **90**, 427–430.

Bunce, J.A. (1995a) Effects of elevated carbon dioxide concentration in the dark on the growth of soybean seedlings. *Ann. Bot.*, **75**, 365–368.

Bunce, J.A. (1995b) Long-term growth of alfalfa and orchard grass plots at elevated carbon dioxide. *J. Biogeochem.*, **22**, 341–348.

Bunce, J.A. (2000) Acclimation of photosynthesis to temperature in eight cool and warm climate herbaceous C_3 species: temperature dependence of parameters of a biochemical photosynthesis model. *Photosynth. Res.*, **63**, 59–67.

Bunce, J.A. (2001) Direct and acclimatory responses of stomatal conductance to elevated carbon dioxide in four herbaceous crop species in the field *Global Change Biol.*, **7**, 323–331.

Bunce, J.A. (2004a) Carbon dioxide effects on stomatal responses to the environment and water use by crops under fiel conditions. *Oecologia*, **140**, 1–10.

Bunce, J.A. (2004b) A comparison of the effects of carbon dioxide concentration and temperature on respiration, translocation and nitrate reduction in darkened soybean leaves. *Ann. Bot.*, **93**, 665–669.

Bunce, J.A. & Sicher, R.C. (2001) Water stress and day-to-day variation in apparent photosynthetic acclimation of field-gr wn soybeans to elevated carbon dioxide concentrations. *Photosynthetica*, **39**, 95–101.

Bunce, J.A. & Sicher, R.C. (2003) Daily irradiance and feedback inhibition of photosynthesis at elevated carbon dioxide concentration in *Brassica oleracea*. *Photosynthetica*, **41**, 481–488.

Carter, D.R. & Peterson, K.M. (1983) Effects of a CO_2-enriched atmosphere on the growth and competitive interaction of a C_3 and C_4 grass. *Oecologia*, **58**, 188–193.

Chakraborty, S. & Data, S. (2003) How will plant pathogens adapt to host plant resistance at elevated CO_2 under a changing climate? *New Phytol.*, **159**, 733–745.

Chakraborty, S., Pangga, I.B., Lupton, J., Hart, L., Room, P.M. & Yates, D. (2000) Production and dispersal of *Colletotrichum gloeosporioides* spores on *Stylosanthes scabra* under elevated CO_2. *Environ. Pollut.*, **108**, 381–387.

Cheng, S.-H., Moore, B.D. & Seemann, J.R. (1998) Effects of short- and long-term elevated CO_2 on the expression of Ribulose-1,5-bisphosphate carboxylase/oxygenase genes and carbohydrate accumulation in leaves of *Arabidopsis thaliana* (L.) Heynh. *Plant Physiol.*, **116**, 715–723.

Christ, R.A. & Körner, C. (1995) Responses of shoot and root gas exchange, leaf blade expansion and biomass production to pulses of elevated CO_2 in hydroponic wheat. *J. Exp. Bot.*, **46**, 1661–1667.

Chu, C.C., Coleman, J.S. & Mooney, H.A. (1992) Control of biomass partitioning between roots and shoot: atmospheric carbon dioxide enrichment and the acquisition and allocation of carbon and nitrogen in wild radish. *Oecologia*, **89**, 580–587.

Coakley, S.M. (1995) Biospheric change: will it matter in plant pathology? *Can. J. Plant Pathol.*, **17**, 147–165.

Cousins, A.B., Adam, N.R., Wall, G.W., Kimball, B.A., Pinter, P.J., Jr, Ottman, M.J., Leavitt, S.W. & Weber, A.N. (2003) Development of C_4 photosynthesis in sorghum leaves grown under free-air CO_2 enrichment (FACE). *J. Exp. Bot.*, **54**, 1969–1975.

Curtis, P.S., Drake, B.G., Leadley, P.W., Arp, W.J. & Whigham, D.F. (1989) Growth and senescence in plant communities exposed to elevated CO_2 concentrations on an estuarine marsh. *Oecologia*, **78**, 20–26.

Curtis, P.S., Snow, A.A. & Miller, A.S. (1994) Genotype-specifi effects of elevated CO_2 on fecundity in wild radish (*Raphanus raphanistrum*). *Oecologia*, **97**, 100–105.

Curtis, P.S. & Wang, X. (1998) A meta-analysis of elevated CO_2 effects on woody plant mass, form and physiology. *Oecologia*, **113**, 299–313.

Davey, P.A., Hunt, S., Hymus, G.J., DeLucia, E.H., Drake, B.G., Karnosky, D.F. & Long, S.P. (2004) Respiratory oxygen uptake is not decreased by an instantaneous elevation of CO_2, but is increased with long-term growth in the fiel at elevated CO_2. *Plant Physiol.*, **134**, 520–527.

DeLucia, E.H., Hamilton J.G., Shawna L.N., Thomas R.B., Andrews J.A., Finzi A., Lavine M., Matamala, R., Mohan, J.E., Hendrey, G.R. & Schlesinger, W.H. (1999) Net primary production of a forest ecosystem with experimental CO_2 enrichment. *Science*, **284**, 1177–1179.

Drake, B.G., Azcon-Bieto, J., Berry, J.A., Bunce, J.A., Dijkstra, P., Farrar, J., Koch, G.W., Lambers, H., Siedow, J. & Wullschleger, S. (1999) Does elevated CO_2 inhibit plant mitochondrial respiration in green plants? *Plant Cell Environ.*, **22**, 649–657.

Drake, B.G., Gonzalez-Meler, M. & Long, S.P. (1997) More efficien plants: a consequence of rising CO_2? *Ann. Rev. Plant Physiol. Plant Mol. Biol.*, **48**, 609–639.

Drake, B.G. & Leadley, P.W. (1991) Canopy photosynthesis of crops and native plant communities exposed to long term elevated CO_2. *Plant Cell Environ.*, **14**, 853–860.

Drennan, P.M. & Nobel, P.S. (2000) Responses of CAM species to increasing atmospheric CO_2 concentrations. *Plant Cell Environ.*, **23**, 767–781.

Ehleringer, J.R. & Monson, R.K. (1993) Evolutionary and ecological aspects of photosynthetic pathway variation. *Ann. Rev. Ecol. Syst.*, **24**, 411–439.

Esashi, Y., Abe, Y. & Ashino, H. (1989) Germination of cocklebur seed and growth of their axial and cotyledonary tissues in response to ethylene, carbon dioxide, and/or oxygen under water stress. *Plant Cell Environ.*, **12**, 183–190.

Esashi, Y., Hase, S. & Kojima, K. (1987) Light actions in the germination of cocklebur seeds: V. Effectsof ethylene, carbon dioxide, oxygen on germination in relation to light. *J. Exp. Bot.*, **38**, 702–710.

Ferrario-Mery, S., Thidaud, M.-C., Betsche, T., Baladier, M.-H. & Foyer, C.H. (1997) Modulation of carbon and nitrogen metabolism, and of nitrate reductase, in untransformed *Nicotiana plumbaginifolia* during CO_2 enrichment of plants grown in pots and hydroponic culture. *Planta*, **202**, 510–521.

Garbutt, K. & Bazzaz, F.A. (1984) The effects of elevated CO_2 on plants. III. Flower, fruit and seed production and abortion. *New Phytol.*, **98**, 433–446.

Garbutt, K., Williams, W.E. & Bazzaz, F.A. (1990) Analysis of the differential response of fi e annuals to elevated carbon dioxide during growth. *Ecology*, **71**, 1185–1194.

Geiger, M., Haake, V., Ludewig, F., Sonnewald, U. & Stitt, M. (1999) The nitrate and ammonium nitrate supply have a major influenc on the response of photosynthesis, carbon metabolism, nitrogen metabolism and growth to elevated carbon dioxide in tobacco. *Plant Cell Environ.*, **22**, 1177–1199.

Geiger, M., Walch-Piu, L., Harnecker, J., Schulze, E.-D., Ludewig, F., Sonnewald, U., Scheible, W.-R. & Stitt, M. (1998). Enhanced carbon dioxide leads to a modifie diurnal rhythm of nitrate reductase activity in older plants, and a large simulation of nitrate reductase activity and higher levels of amino acid in higher plants. *Plant Cell Environ.*, **21**, 253–268.

Ghannoum, O., von Caemmerer, S., Ziska, L.H. & Conroy, J.P. (2000) The growth response of C_4 plants to rising atmospheric CO_2 partial pressure: a reassessment. *Plant Cell Environ.*, **23**, 931–942.

Gonzalez-Meler, M.A., Ribas-Carbo, M., Siedow, J.N. & Drake, B.G. (1996) Direct inhibition of plant mitochondrial respiration by elevated CO_2. *Plant Physiol.*, **112**, 1349–1355.

Gonzalez-Meler, M.A., Taneva, L & Trueman, R.J. (2004) Plant respiration and elevated atmospheric CO_2 concentration: cellular responses and global significance *Ann. Bot.*, **94**, 647–656.

Granados, J. & Körner, C. (2002) In deep shade, elevated CO_2 increases the vigor of tropical climbing plants. *Global Change Biol.*, **8**, 1109–1117.

Grulke, N.E., Riechers, G.H., Oechel, W.C., Hjelm, U. & Jaeger, C. (1990) Carbon balance in tussock tundra under ambient and elevated atmospheric CO_2. *Oecologia*, **83**, 485–494.

Grunzweig, J. & Körner, C. (2000) Growth and reproductive responses to elevated CO_2 in wild cereals of the northern Negev of Israel. *Global Change Biol.*, **6**, 631–638.

Gurney, K.R., Law, R.M., Denning, A.S., Rayner, P.J., Baker, D., Bousquet, P., Bruhwiler, L., Chen, Y.-H., Ciais, P., Fan, S., Fung, I.Y., Gloor, M., Heimann, M., Higuchi, K., John, J., Maki, T., Maksyutov, S., Masarie, K., Peylin, P., Prather, M., Pak, B.C., Randerson, J., Sarmiento, J., Taguchi, S., Takahashi, T. & Yuen, C.-W. (2002) Towards robust regional estimates of CO_2 sources and sinks using atmospheric transport models. *Nature*, **415**, 626–630.

Hardy, R.W.F. & Havelka, U.D.K. (1975) Symbiotic N_2 fixation multi-fold enhancement by CO_2-enrichment of field-groun soybeans. *Plant Physiol.*, **48**, 35.

Hattenschwiler, S. & Körner, C. (2003) Does elevated CO_2 facilitate naturalization of the non-indigenous *Prunus laurocerasus* in Swiss temperate forests? *Funct. Ecol.*, **17**, 778–785.

Hattenschwiler, S. & Schafellner, C. (2004) Gypsy moth feeding in the canopy of a CO_2 enriched mature forest. *Global Change Biol.*, **10**, 1899–1908.

Hendrey, G.R. & Kimball, B.A. (1994) The FACE program. *Agric. For. Meteorol.*, **70**, 3–14.

Hibberd, J.M., Whitbread, R. & Farrar, J.F. (1996) Effect of elevated concentrations of CO_2 on infection of barley by *Erysiphe graminis*. *Physiol. Mol. Plant Pathol.*, **48**, 37–49.

Holtum, J.A.M. & Winter, K. (2003) Photosynthetic CO_2 uptake in seedlings of two tropical tree species exposed to oscillating elevated concentrations of CO_2. *Planta*, **218**, 152–158.

Houghton, J.T., Ding, Y., Griggs, D.J., Noguer, M., Van der Linden, P.J. & Xiaosu, D. (eds) (2001) *Climate Change 2001: The Scientific Basis*. Cambridge University Press, Cambridge, UK.

Hughes, L. & Bazzaz, F.A. (1997) Effect of elevated CO_2 on interactions between the western fl wer thrips, *Frankliniella occidentalis* (Thysanoptera: Thripidae) and the common milkweed, *Ascelias syriaca*. *Oecologia*, **109**, 286–292.

Hungate, B.A., Stiling, P.D., Dijkstra, P., Johnson, D.W., Ketterer, M.E., Hymus, G.J., Hinkle, C.R. & Drake, B.G. (2004) CO_2 elicits long-term decline in nitrogen fixation *Science*, **304**, 1291.

Idso, C.D., Idso, S.B. & Balling, R.C., Jr (1998) The urban CO_2 dome of Phoenix, Arizona. *Phys. Geog.*, **19**, 95–99.

Jablonski, L.M., Wang, X. & Curtis, P.S. (2002) Plant reproduction under elevated CO_2 conditions: a meta-analysis of reports on 79 crop and wild species. *New Phytol.*, **156**, 9–26.

Jach, M.E. & Ceulemans, R. (1999) Effects of elevated atmospheric CO_2 on phenology, growth and crown structure of Scots pine (*Pinus sylvestris*) seedlings after two years of exposure in the field *Tree Physiol.*, **19**, 289–300.

Jahnke, S. & Krewitt, M. (2002) Atmospheric CO_2 concentration may directly affect leaf respiration measurement in tobacco, but not respiration itself. *Plant Cell Environ.*, **25**, 641–651.

Janssens, I.A., Mousseau, M. & Ceulemans, R. (2000) Crop ecosystem responses to climatic change: tree crops. In: *Climate Change and Global Crop Productivity* (eds K.R. Reddy & H.F. Hodges), pp. 245–263. CABI Publishing, New York.

Jarvis, P.G. & McNaughton, K.G. (1986) Stomatal control of transpiration: scaling up from leaf to region. *Adv. Ecol. Res.*, **15**, 1–49.

Jokela, E.J., Dougherty, P.M. & Martin, T.A. (2004) Production dynamics of intensively managed loblolly pine stands in the southern United States: a synthesis of seven long-term experiments. *For. Ecol. Manage.*, **192**, 117–130.

Jones, H.G. (1998) Stomatal control of photosynthesis and transpiration. *J. Exp. Bot.*, **49**, 387–398.

Kaplan, A., Gale, J. & Poljakoff-Mayber, A. (1977) Effect of O_2 and CO_2 concentrations on gross dark CO_2 fixatio and dark respirations in *Bryophyllum daigrmontianum. Aust. J. Plant Physiol.*, **4**, 745–752.

Karnosky, D.F., Ceulemans, R., Scarascia-Mugnozza, G.E. & Innes, J.L. (eds) (2001) *The Impact of Carbon Dioxide and Other Greenhouse Gases on Forest Ecosystems*. CABI Publishing, Wallingford, UK.

Kerslake, J.E., Woodin, S.J. & Hartley, S.E. (1998) Effects of carbon dioxide and nitrogen enrichment on a plant-insect interaction: the quality of *Calluna vulgaris* as a host for *Operophtera brumata. New Phytol.*, **140**, 43–49.

Kimball, B.A. (1993) Effects of increasing atmospheric CO_2 on vegetation. *Vegetatio*, **104/105**, 65–83.

Kimball, B.A. & Mauney, J.R. (1993) Response of cotton to varying CO_2, irrigation and nitrogen: yield and growth. *Agron. J.*, **85**, 706–712.

Kimball, B.A., Pinter, P.J., Jr, Garcia, R.L., LaMorte, R.L., Wall, G.W., Hunsaker, D.J., Wechsung, G., Wechsung, F. & Kartschall, T. (1995) Productivity and water use of wheat under fee-air CO_2 enrichment. *Global Change Biol.*, **1**, 429–442.

Koch, G.W. & Mooney, H.A. (1996) *Carbon Dioxide and Terrestrial Ecosystems*. Academic Press, San Diego, CA.

Koch, K.E. (1996) Carbohydrate-modulated gene expression in plants. *Ann. Rev. Plant Physiol. Plant Mol. Biol.*, **47**, 509–540.

Körner, C. (1997) The responses of alpine grassland to four seasons of CO_2-enrichment: a synthesis. *Acta Oecol.*, **18**, 165–175.

Krapp, A., Hofmann, B., Schafer, C. & Stitt, M. (1993) Regulation of the expression of rbcS and other photosynthetic genes by carbohydrates: a mechanism for the 'sink' regulation of photosynthesis? *Plant J.*, **3**, 817–828.

Krupa, S.V. & Manning, W.J. (1988) Atmospheric ozone: formation and effects on vegetation. *Environ. Pollut.*, **50**, 101–137.

Lavola, A. & Julkunen-Titto, R. (1994) The effect of elevated carbon dioxide and fertilization on primary and secondary metabolites in birch, *Betula pendula* (Roth). *Oecologia*, **99**, 315–322.

Lewis, J.D., Wang, X.Z., Griffin K.L. & Tissue, D.T. (2002) Effects of age and ontogeny on photosynthetic responses of a determinate annual plant to elevated CO_2 concentrations. *Plant Cell Environ.*, **25**, 359–368.

Lin, W., Ziska, L.H., Namuco, O.S. & Bai, K. (1997) The interaction of high temperature and elevated CO_2 on photosynthetic acclimation of single leaves of rice in situ. *Physiol. Plant.*, **99**, 178–184.

Lincoln, D.E. & Couvet, D. (1989) The effect of carbon supply on allocation to allelochemicals and caterpillar consumption of peppermint. *Oecologia*, **78**, 112–120.

Lindroth, R.L. & Kinney, K.K. (1998) Consequences of enriched atmospheric CO_2 and defoliation for foliar chemistry and gypsy moth performance. *J. Chem. Ecol.*, **24**, 1677–1689.

Lindroth, R.L., Kinney, K.K. & Platz, C.L. (1993) Responses of deciduous trees to elevated atmospheric CO_2: productivity, phytochemistry, and insect performance. *Ecology*, **74**, 763–772.

Lloyd, J. & Farquhar, G.D. (1996) The CO_2 dependence of photosynthesis, plant growth responses to elevated atmospheric CO_2 concentrations and their interaction with soil nutrient status. I. General principles and forest ecosystems. *Funct. Ecol.*, **10**, 4–32.

Long, S.P. (1991) Modificatio of the response of photosynthetic productivity to rising temperature by atmospheric CO_2 concentrations: has its importance been underestimated? *Plant Cell Environ.*, **14**, 729–739.

Ludewig, F. & Sonnewald, U. (2000) High CO_2-mediated down-regulation of photosynthetic gene transcripts is caused by accelerated leaf senescence rather than sugar accumulation. *FEBS Lett.*, **479**, 19–24.

Luo, Y. & Mooney, H.A. (eds) (1999) *Carbon Dioxide and Environmental Stress*. Academic Press, San Diego, CA.

Makino, A., Nakano, H., Mae, T., Shimada, T. & Yamamoto, N. (2000) Photosynthesis, plant growth and N allocation in transgenic rice plants with decreased rubisco under CO_2 enrichment. *J. Exp. Bot.*, **50**, 383–389.

Malmstrom, C.M. & Field, C.B. (1997) Virus-induced differences in the response of oat plants to elevated carbon dioxide. *Plant Cell Environ.*, **20**, 178–185.

Manning, W.J. & Tiedemann, A.V. (1995) Climate Change: potential effects of increased atmospheric carbon dioxide (CO_2), ozone (O_3) and ultraviolet-B (UV-B) radiation on plant diseases. *Environ. Pollut.*, **88**, 219–234.

Masle, J. (2000) The effects of elevated CO_2 concentrations on cell division rates, growth patterns, and blade anatomy in young wheat plants are modulated by factors related to leaf position, vernalization and genotype. *Plant Physiol.*, **122**, 1399–1416.

Matsui, T., Namuco, O.S., Ziska, L.H. & Horie, T. (1997) Effects of high temperature and CO_2 concentration on spikelet sterility in Indica rice. *Field Crops Res.*, **51**, 213–221.

Medlyn, B.E., Rey, A., Barton, C.V.M. & Forstreuter, M. (2001) Above-ground growth responses of forest trees to elevated atmospheric CO_2 concentrations. In: *The Impact of Carbon Dioxide and Other Greenhouse Gases on Forest Ecosystems* (eds D.F. Karnosky, R. Ceulemans, G.E. Scarascia-Mugnozza & J.L. Innes), pp. 127–146. CABI Publishing, New York.

Miyazaki, S., Fredricksen, M., Hollis, K.C., Proyko, V., Shepley, D., Galbraith, D.W., Long, S.P. & Bohnert, H.J. (2004) Transcript expression profile of *Arabidopsis thaliana* grown under controlled conditions and open-air elevated concentrations of CO_2 and O_3. *Field Crops Res.*, **90**, 47–59.

Mohan, J.E., Clark, J.S. & Schlesinger, W.H. (2004) Genetic variation in germination, growth and survivorship of red maple in response to subambient through elevated atmospheric CO_2. *Global Change Biol.*, **10**, 233–247.

Moore, B.D., Cheng, S.-H., Sims, D. & Seemann, J.R. (1999) The biochemical and molecular basis for photosynthetic acclimation to elevated atmospheric CO_2. *Plant Cell Environ.*, **22**, 567–582.

Morgan, J.A., Hunt, H.W., Monz, C.A. & LeCain, D.R. (1994) Consequences of growth at two carbon dioxide concentrations and two temperatures for leaf gas exchange in *Pascopyrum smithii* (C_3) and *Bouteloua gracilis* (C_4). *Plant Cell Environ.*, **17**, 1023–1033.

Morgan, J.A., Pataki, D.E., Körner, C., Clark, H., Del Grosso, S.J., Grunzweig, J.M., Knapp, A.K., Mosier, A.R., Newton, P.C.D., Niklaus, P.A., Nippert, J.B., Nowak, R.S., Parton, W.J., Polley, H.W. & Shaw, M.R. (2004) Water relations in grassland and desert ecosystems exposed to elevated atmospheric CO_2. *Oecologia*, **140**, 11–25.

Morison, J.I.L. (1998) Stomatal response to increased CO_2 concentration. *J. Exp. Bot.*, **49**, 443–452.

Murray, D.R. (1997) *Carbon Dioxide and Plant Responses*. John Wiley and Sons, New York.

Nemani, R.R., Keeling, C.D., Hashimoto, H., Jolly, W.M., Piper, S.C., Tucker, C.J., Myneni, R.B. & Running, S.W. (2003) Climate-driven increases in global terrestrial net primary production from 1982 to 1999. *Science*, **300**, 1560–1562.

Newton, P.C.D., Clark, H., Bell, C.C. & Glasgow, E.M. (1996) Interaction of soil moisture and elevated CO_2 on the above-ground growth rate, root length density and gas exchange of turves from temperate pasture. *J. Exp. Bot.*, **47**, 771–779.

Norby, R.J. & Cotrufo, M.F. (1998) A question of litter quality. *Nature*, **396**, 17–18.

Obrist, D., Arnone, J.A., III & Körner, C.H. (2001) In situ effects of elevated atmospheric CO_2 on leaf freezing resistance and carbohydrates in a native temperate grassland. *Ann. Bot.*, **87**, 839–844.

Olesniewicz, K.S. & Thomas, R.B. (1999) Effects of mycorrhizal colonization on biomass production and nitrogen fixatio of black locust (*Robinia pseudoacacia*) seedlings grown under elevated atmospheric carbon dioxide. *New Phytol.*, **142**, 133–140.

Ottman, M.J., Kimball, B.A., Pinter, P.J., Wall, G.W., Vanderlip, R.L., Leavitt, S.W., LaMorte, R.L., Matthias, A.D. & Brooks, T.J. (2001) Elevated CO_2 increases sorghum biomass under drought conditions. *New Phytol.*, **150**, 261–273.

Owensby, C.E., Coyne P.I., Ham, J.M., Auen, L.M., Knapp, A.K. (1993) Biomass production in a tallgrass prairie ecosystem exposed to ambient and elevated CO_2. *Ecol. Appl.*, **3**, 644–653.

Palta, J.A. & Nobel, P.S. (1989) Influenc of soil O_2 and CO_2 on root respiration for *Agave deserti*. *Physiol. Plant.*, **76**, 187–192.

Pan, Y., Melillo, J.M., McGuire, A.D., Kicklighter, D.W., Pitelka, L.F., Hibbard, K., Pierce, L.L., Running, S.W., Ojima, D.S., Parton, W.J. & Schimel, D.S. (1998) Modeled responses of terrestrial ecosystems to elevated atmospheric CO_2: a comparison of simulations by the biogeochemistry models of the vegetation/ecosystem modeling and analysis project (VEMAP). *Oecologia*, **114**, 389–404.

Patterson, D.T. (1986) Responses of soybean (*Glycine max*) and three C_4 grass weeds to CO_2 enrichment during drought. *Weed Sci.*, **34**, 203–210.

Patterson, D.T., Flint, E.P. & Beyers, J.L. (1984) Effects of CO_2 enrichment on competition between a C_4 weed and a C_3 crop. *Weed Sci.*, **32**, 101–105.

Phillips, O.L., Martinez, R.V., Arroyo, L., Baker, T.R., Killeen, T., Lewis, S.L. Malhi, Y., Mendoza, A.M., Neill, D., Vargas, P.N., Alexiades, M., Ceron, C., Di Flore, A., Erwin, T., Jardim, A., Palacios, W., Saidlas, M. & Vinceti, B. (2002) Increasing dominance of large llianas in Amazonian forests. *Nature*, **418**, 770–774.

Pielke, R.A., Marland, G., Betts, R.A., Chase, T.N., Eastman, J.L., Niles, J.O., Niyogi, D.S. & Running, S.W. (2002) The influenc of land-use change and landscape dynamics on the climate system: relevance to climate-change policy beyond the radiative effect of greenhouse gases. *Philos. Trans. R. Soc. Lond.*, **360**, 1705–1719.

Pinelli, P & Loreto, F. (2003) $^{12}CO_2$ emission from different metabolic pathways measured in illuminated and darkened C_3 and C_4 leaves at low, atmospheric and elevated CO_2 concentration. *J. Exp. Bot.*, **54**, 1761–1769.

Pons, T.L. & Welschen, R.A.M. (2002) Overestimation of respiration rates in commercially available clamp-on leaf chambers. Complications with the measurement of net photosynthesis. *Plant Cell Environ.*, **25**, 1367–1372.

Poorter, H. (1998) Do slow-growing species and nutrient-stressed plants respond relatively strongly to elevated CO_2? *Global Change Biol.*, **4**, 693–697.

Poorter, H. & Navas, M.-L. (2003) Plant growth and competition at elevated CO_2: on winners, losers and functional groups. *New Phytol.*, **157**, 175–198.

Poorter, H. & Perez-Soba, M. (2001) The growth response of plants to elevated CO_2 under non-optimal conditions. *Oecologia*, **129**, 1–20.

Potvin, C. & Vasseur, L. (1997) Long-term CO_2 enrichment of a pasture community. *Ecology*, **78**, 666–677.

Rachmilevitch, S., Cousins, A.B. & Bloom, A.J. (2004) Nitrate assimilation in plant shoots depends on photorespiration. *Proc. Natl. Acad. Sci USA.*, **101**, 11506–11510.

Rasse, D.P., Peresta, G. & Drake, B.G. (2005) Seventeen years of elevated CO_2 exposure in a Chesapeake Bay Wetland: sustained but contrasting responses of plant growth and CO_2 uptake. *Global Change Biol.*, **11**, 369–377.

Reddy, K.R. & Hodges, H.F. (eds) (2000) *Climate Change and Global Crop Productivity*. CABI Publishing, New York.

Reekie, E.G. & Bazzaz, F.A. (1991) Phenology and growth in four annual species grown in ambient and elevated CO_2. *Can. J. Bot.*, **69**, 2475–2481.

Reekie, J.Y.C., Hicklenton, P.R. & Reekie, E.G. (1997) The interactive effects of carbon dioxide enrichment and daylength on growth and development in *Petunia hybrida*. *Ann. Bot.*, **80**, 57–66.

Reich, P.B., Knops, J., Tilman, D., Craine, J., Ellsworth, D., Tjoelker, M., Lee, T., Wedin, D., Naeem, S., Bahauddin, D., Hendrey, G., Jose, S., Wrage, K., Goth, J. & Bengston, W. (2001) Plant diversity enhances ecosystem responses to elevated CO_2 and nitrogen deposition. *Nature*, **410**, 809–812.

Reuveni, J. & Gale, J. (1985) The effect of high levels of carbon dioxide on dark respiration and growth of plants. *Plant Cell Environ.*, **8**, 623–628.

Reuveni, J., Gale, J. & Mayer, A.M. (1993a) Photosynthesis, Respiration and dry matter growth of *Lemna gibba*, as affected by day/night [CO_2] regimes. In: *Photosynthetic Responses to the Environment* (eds H.Y. Yamamoto & C.M Smith), Vol. 8, pp. 201–206. American Society of Plant Physiologists Series, Rockville, MD.

Reuveni, J., Gale, J., & Mayer, A.M. (1993b) Reduction of respiration by high ambient CO_2 and the resulting error in measurements of respiration made with O_2 electrodes. *Ann. Bot.*, **72**, 129–131.

Rogers, H.H., Peterson, C.M., McCrimmon, J.N. & Cure, J.D. (1992) Response of plant roots to elevated atmospheric carbon dioxide. *Plant Cell Environ.*, **15**, 749–752.

Rogers, H.H., Runion, G.B. & Krupa, S.V. (1994) Plant responses to CO_2 enrichment with emphasis on roots and the rhizosphere. *Environ. Pollut.*, **83**, 155–189.

Seneweera, S., Ghannoum, O. & Conroy, J.P. (2001) Root and shoot factors contribute to the effect of drought on photosynthesis and growth of the C_4 grass *Panicum coloratum* at elevated CO_2 partial pressures. *Aust. J. Plant Physiol.*, **28**, 451–460.

Schafer, K.V.R., Oren, R., Lai, C.-T. & Katul, G.G. (2002) Hydrologic balance in an intact temperate forest ecosystem under ambient and elevated atmospheric CO_2 concentration. *Global Change Biol.*, **8**, 895–911.

Schlesinger, W.H. & Lichter, J. (2001) Limited carbon storage in soil and litter of experimental forest plots under increased atmospheric CO_2. *Nature*, **411**, 466–469.

Sharkey, T.D. (1985) O_2-insensitive photosynthesis in C_3 plants. Its occurrence and a possible explanation. *Plant Physiol.*, **78**, 71–75.

Sheen, J. (1994) Feedback control of gene expression. *Photosynth. Res.*, **39**, 427–438.

Sicher, R.C. (1998) Yellowing and photosynthetic decline of barley primary leaves in response to atmospheric CO_2 enrichment. *Physiol. Plant.*, **103**, 193–199.

Sims, D.A., Cheng, W., Luo, Y. & Seemann, J.R. (1999) Photosynthetic acclimation to elevated CO_2 in a sunfl wer canopy. *J. Exp. Bot.*, **50**, 645–653.

Sionit, N. & Patterson, D.T. (1984) Responses of C_4 grasses to atmospheric CO_2 enrichment. I. Effect of irradiance. *Oecologia*, **65**, 30–34.

Smith, S.D., Huxman, T.E., Zitzer, S.F., Charlet, T.N., Housman, D.C., Coleman, J.S., Fenstermaker, L.K., Seemann, J.R. & Nowak, R.S. (2000) Elevated CO_2 increases productivity and invasive species success in an arid ecosystem. *Nature*, **408**, 79–82.

Socias, F.X., Medrano, H. & Sharkey, T.D. (1993) Feedback limitation of photosynthesis of *Phaseolus vulgaris* L. grown in elevated CO_2. *Plant Cell Environ.*, **16**, 81–86.

Stiling, P., Cattell, M., Moon, D.C., Rossi, A., Hungate, B.A., Hymus, G. & Drake, B.G. (2002) Elevated atmospheric CO_2 lowers herbivore abundance, but increases leaf abscission rates. *Global Change Biol.*, **8**, 658–667.

Stitt, M. (1991) Rising CO_2 levels and their potential significanc for carbon low in photosynthetic cells. *Plant Cell Environ.*, **14**, 741–762.

Stitt, M. & Krapp, A. (1999) The interaction between elevated carbon dioxide and nitrogen nutrition: the physiological and molecular background. *Plant Cell Environ.*, **22**, 583–621.

St. Omer, L. & Horvath, S.M. (1983) Elevated carbon dioxide concentrations and whole plant senescence. *Ecology*, **64**, 1311–1314.

Takeuchi, Y., Kubiske, M.E., Isebrands, J.G., Pregtizer, K.S., Hendrey, G. & Karnosky, D.F. (2001) Photosynthesis, light and nitrogen relationships in a young deciduous forest canopy under open-air CO_2 enrichment. *Plant Cell Environ.*, **24**, 1257–1268.

Temperton, V.M., Grayston, S.J., Jackson, G., Barton, C.V.M., Millard, P. & Jarvis, P.G. (2003) Effects of elevated carbon dioxide concentration on growth and nitrogen fixatio in *Alnus glutinosa* in a long-term fiel experiment. *Tree Physiol.*, **23**, 1051–1059.

Thomas, R.B. & Strain, B.R. (1991) Root restriction as a factor in photosynthetic acclimation of cotton seedlings grown in elevated CO_2. *Plant Physiol.*, **96**, 627–634.

Thompson, G.B., Brown, J.K.M. & Woodward, F.I. (1993) The effects of host carbon dioxide, nitrogen and water supply on the infection of wheat by powdery mildew and aphids. *Plant Cell Environ.*, **16**, 687–702.

Thompson, G.B. & Drake, B.G. (1994) Insects and fungi on a C_3 sedge and a C_4 grass exposed to elevated atmospheric CO_2 concentrations in open-top chambers in the field *Plant Cell Environ.*, **17**, 1161–1172.

Tissue, D.T., Megonigal, J.P. & Thomas, R.B. (1997) Nitrogenase activity and N_2 fixatio are stimulated by elevated CO_2 in a tropical N_2-fixin tree. *Oecologia*, **109**, 28–33.

Tyree, M.T. & Alexander, J.D. (1993) Plant water relations and the effects of elevated CO_2: a review and suggestions for future research. *Vegetatio*, **104/105**, 47–62.

van Oosten, J.-J. & Besford, R.T. (1995) Some relationships between the gas exchange, biochemistry and molecular biology of photosynthesis during leaf development of tomato plants after transfer to different carbon dioxide concentrations. *Plant Cell Environ.*, **18**, 1253–1266.

Ward, J.K. & Strain, B.R. (1997) Effects of low and elevated CO_2 partial pressure on growth and reproduction of *Arabidopsis thaliana* from different elevations. *Plant Cell Environ.*, **20**, 254–260.

Ward, J.K. & Strain, B.R. (1999) Elevated CO_2 studies: past, present and future. *Tree Physiol.*, **19**, 211–220.

Watling, J.R., Press, M.C. & Quick, W.P. (2000) Elevated CO_2 induces biochemical and ultrastructural changes in leaves of the C_4 cereal, sorghum. *Plant Physiol.*, **123**, 1143–1152.

Wedin, D.A. & Tilman, D. (1996) Influenc of nitrogen loading and species composition on the carbon balance of grasslands. *Science*, **274**, 1720–1723.

Weerakoon, W.M., Olszyk, D.M. & Moss, D.N. (1999) Effects of nitrogen nutrition on responses of rice seedlings to carbon dioxide. *Agric. Ecosyst. Environ.*, **72**, 1–8.

Wilson, K.B., Carlson, T.N. & Bunce, J.A. (1999) Feedback significantl influence the simulated effect of CO_2 on seasonal evapotranspiration from two agricultural species. *Global Change Biol.*, **5**, 903–917.

Wullschleger, S.D., Ziska, L.H. & Bunce, H.A. (1994) Respiratory response of higher plants to atmospheric CO_2 enrichment. *Physiol. Plant.*, **90**, 221–229.

Zak, D.R., Holmes, W.E., Finzi, A.C., Norby, R.J. & Schlesinger, W.H. (2003) Soil nitrogen cycling under elevated CO_2: a synthesis of forest FACE experiments. *Ecol. Appl.*, **13**, 1508–1514.

Ziska, L.H. (2000) The impact of elevated carbon dioxide on yield loss from a C_3 and C_4 weed in fiel grown soybean. *Global Change Biol.*, **6**, 899–905.

Ziska, L.H. (2001a) Acclimation to growth temperature alters the temperature dependent stimulation of photosynthesis to elevated carbon dioxide in *Albutilon theophrasti*. *Physiol. Plant.*, **111**, 322–328.

Ziska, L.H. (2001b) Changes in competitive ability between a C_4 crop and a C_3 weed with elevated carbon dioxide. *Weed Sci.*, **49**, 622–627.

Ziska, L.H. (2003a) Evaluation of the growth response of six invasive species to past, present and future carbon dioxide concentrations. *J. Exp. Bot.*, **54**, 395–401.

Ziska, L.H. (2003b) Evaluation of yield loss in fiel sorghum from a C_3 and C_4 weed with increasing CO_2. *Weed Sci.*, **51**, 914–918.

Ziska, L.H. & Bunce, J.A. (1993) The influenc of elevated carbon dioxide and temperature on seed germination and emergence from soil. *Field Crops Res.*, **34**, 147–157.

Ziska, L.H. & Bunce, J.A. (1995) Growth and photosynthetic response of three soybean cultivars to simultaneous increases in growth temperature and CO_2. *Physiol. Plant.*, **94**, 575–584.

Ziska, L.H. & Bunce, J.A. (1997a) Influenc of increasing carbon dioxide concentration on the photosynthetic and growth stimulation of selected C_4 crops and weeds. *Photosynth. Res.*, **54**, 199–208.

Ziska, L.H. & Bunce, J.A. (1997b) The role of temperature in determining the stimulation of CO_2 assimilation at elevated carbon dioxide concentration in soybean seedlings. *Physiol. Plant.*, **100**, 126–132.

Ziska, L.H. & Bunce, J.A. (1998) The influenc of increasing growth temperature and CO_2 concentration on the ratio of respiration to photosynthesis in soybean seedlings. *Global Change Biol.*, **4**, 637–643.

Ziska, L.H. & Caulfield F.A. (2000) Rising carbon dioxide and pollen production of common ragweed, a known allergy-inducing species: implications for public health. *Aust. J. Plant Physiol.*, **27**, 893–900.

Ziska, L.H., Gebhard, D.E., Frenz, D.A., Faulkner, S.S., Singer, B.D. & Straka, J.G. (2003) Cities as harbingers of climate change: common ragweed, urbanization, and public health. *J. Allergy Clin. Immunol.*, **111**, 290–295.

Ziska, L.H., Ghannoum, O., Baker, J.T., Conroy, J., Bunce, J.A., Kobayashi, K & Okada, M. (2001) A global perspective of ground level, 'ambient' carbon dioxide for assessing the response of plants to atmospheric CO_2. *Global Change Biol.*, **7**, 789–796.

Ziska, L.H., Morris, C.F. & Goins, E.W. (2004) Quantitative and qualitative evaluation of selected wheat varieties released since 1903 to increasing atmospheric carbon dioxide: can yield sensitivity to carbon dioxide be a factor in wheat performance? *Global Change Biol.*, **10**, 1810–1819.

Ziska, L.H., Sicher, R.C. & Bunce J.A. (1999) The impact of elevated carbon dioxide on the growth and gas exchange of three C_4 species differing in CO_2 leak rates. *Physiol. Plant.*, **105**, 74–80.

Ziska, L.H., Sicher, R.C. & Kremer, D.F. (1995) Reversibility of photosynthetic acclimation of swiss chard and sugarbeet grown at elevated concentrations of CO_2. *Physiol. Plant.*, **95**, 355–364.

Ziska, L.H., Weerakoon, W., Namuco, O.S. & Pamplona, R. (1996) The influenc of nitrogen on the elevated CO_2 response in field-gr wn rice. *Aust. J. Plant Physiol.*, **23**, 45–52.

3 Significance of temperature in plant life

Christian Körner

3.1 Two paradoxes

Life is inevitably tied to certain temperature conditions that facilitate metabolism. Since plants are poikilothermic organisms (i.e. organisms whose body temperature varies with the temperature of their immediate environment) and since they commonly cannot move, except through reproduction, they have to cope with whatever the environment offers. In this overview, I will firs revisit classical responses of plant metabolism to temperature (T) and will then explore the significanc of such responses for plant life in the 'real world'. In doing so, some paradoxes will become apparent, the most significan of which I will place at the beginning of the chapter to give the reader a f avour of how difficul it is to bridge from well-understood and established physiological knowledge to things like ecosystem productivity or soil CO_2 emission.

3.1.1 Paradox 1

Across the globe's humid biota there is a well-known productivity gradient from high latitude or high altitude low-temperature environments (e.g. alpine grassland with 0.4 kg m^{-2} a^{-1}, 2-month growing season) to equatorial forests with their annual productivity of around 2.5 kg m^{-2} a^{-1} (12-month growing season). If one divides the annual productivity by the number of months available for growth, it comes perhaps as a surprise that the monthly productivity is approximately 0.2 kg m^{-2} in both biomes and the annual productivity differences emerge as a pure time effect, with the actual mean growing season air temperatures still being 8°C in one case and 28°C in the other, i.e., 20 K different (I will use K for T differences throughout). Although there are large variations around these approximate means (data compiled in Körner, 2003a), the basic message is that the mean productivity of native vegetation is insensitive to the global amplitude of temperature in places which permit full ground cover, provided water is available.

3.1.2 Paradox 2

Just like plant metabolism, soil microbial (including fungal) metabolism is very sensitive to temperature. A measure of the overall microbial metabolism in soils is soil CO_2-efflu (often called soil respiration), which commonly accounts for half of the total efflux the other half is root respiration. Most of this metabolism is associated with litter decomposition. One would expect respiratory efflu rates to be far greater in the tropics than in the arctic. Raich and Nadelhoffer (1989) explored this and noted with surprise that the monthly rates of soil CO_2 evolution during the growing season (see above) do not differ and the annual efflu was well explained by total litter input (i.e., is substrate driven and not temperature driven) which, following from paradox 1, is a function of time only.

Both these examples draw on data from natural ecosystems, which are in a long-term steady state and carry a vegetation that had been selected for meeting the regional environmental demands (both abiotic and biotic). These systems are fully coupled to their soil biota and natural soil resources. The situation may be very different in crops, which are not in any steady state and which still carry an evolutionary memory to the often warmer areas where they originated, compared to those where they are currently grown. Furthermore, crops are managed in a way that they become independent from natural soil mineral resources, and hence, microbial biomass recycling; i.e. they are to a large extent decoupled from soil processes.

What we learn from the two examples is that large differences in temperature (fi e times a 3–4 K IPCC warming scenario) may have no net effect on some key plant and ecosystem processes, provided these mean differences occurred for a long enough period of time. For how long, we do not know. It is a different issue how plants and ecosystems respond to day-by-day changes in temperature or to a rapid change of means over a couple of decades. Plant growth outside the tropics is always sensitive to temperature as evidenced by tree rings or crop yields, but it seems that these are variations around a mean, which is in large controlled by factors other than the direct influenc of temperature, at least when natural vegetation is considered.

Temperature-driven seasonality (similar to moisture-driven seasonality) comes in as an indirect influence which truncates the period during which plant growth is possible. It is important to separate such time effects from direct temperature effects on metabolism. The following sections will now return to shorter timescales, to the 'day-to-day business' of plant life, where temperature matters. But it is important to bear in mind that these temperature effects must diminish as we scale up in time, given the above paradoxes.

3.2 Baseline responses of plant metabolism to temperature

In the following, I will firs briefl recall instantaneous T responses; i.e., responses seen when plant tissue is experimentally exposed to a series of temperatures over no more than a few hours experimental duration. All classical textbook temperature response curves refer to such test conditions, although often without mentioning this. As a result, the literature is full of problematic long-term extrapolations based on such curves, which will be discussed later.

I will mention only two cardinal types of gradual responses to temperature, namely that for net photosynthesis, A (for CO_2 assimilation), and that for dark respiration, R. The latter also stands for the response type of many other metabolic

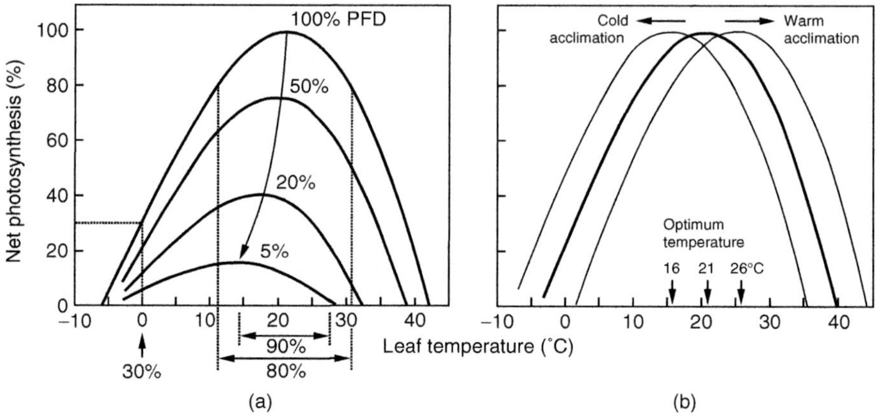

Figure 3.1 The 'classical' responses of net photosynthesis of leaves (A) to temperature (cf. Larcher, 1969, 2003). (a) Typical response curves for a temperate plant species measured at different light intensities (PFD, photon flu densities). Note the shift of the temperature, optimum to lower temperatures as light supply is diminished. The range at which 80 and 90% of maximum A (photosynthetic capacity) is reached under light saturation is indicated. (b) Thermal acclimation of A to different growth/habitat temperatures. Lefthand curve, cold habitat; middle, mild habitat; right, very warm habitat.

reactions. The two responses differ fundamentally, because the photosynthetic response is in fact a net response of two opposite processes: the rate of the so-called dark reaction of photosynthesis, i.e., CO_2 fixatio by rubisco, which increases with temperature, and two types of concurrent CO_2 release processes, a partly suppressed R (see below) and photorespiration, which also increase with temperature. Beyond a certain temperature (the optimum temperature) the balance between CO_2 fixatio and CO_2 release shifts in favour of release, because the affinit of rubisco to CO_2 declines and because the solubility of CO_2 in water declines more rapidly than that of oxygen as temperatures increase. The net result is the well-known bell-shaped curve (Figure 3.1a). In contrast, dark respiration (mitochondrial respiration) steadily (exponentially) increases with temperature until the rate collapses near the lethal heat limit (Larcher, 1969, 2003; Figure 3.2a).

3.2.1 Photosynthesis

At firs glance, these curves suggest a very high temperature sensitivity of both processes. However, for A this is a wrong impression for three reasons: (1) Over a wide range of temperatures (in this example, 12 K), a typical leaf with this response characteristic achieves approximately 80% of maximum A. (2) This T response interacts with light availability. As light intensity goes down and photosynthesis diminishes, the whole curve shifts to the left and gets flatte , so reaches maximum CO_2 fixatio at a lower temperature and the 80% capacity range for each of these curves becomes even wider (Figure 3.1a). The lower the light, the lower is the temperature sensitivity of photosynthesis. Because temperatures are commonly also

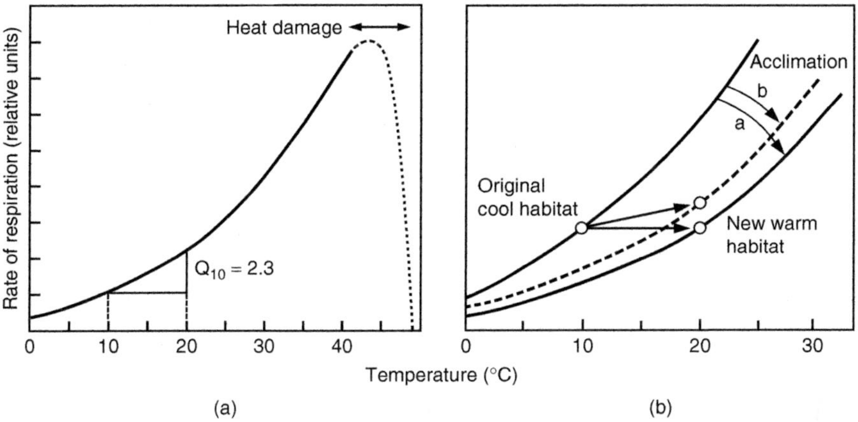

Figure 3.2 The instantaneous response of dark respiration (R) to temperature (T). (a) A response for a typical temperate species, with sub-zero activity and exponential increase with temperature following the mean Q_{10} of 2.3 (R at 20:10°C). When temperatures reach a damaging range, R collapses and reaches zero at the heat death of the tissue. The shape of this transition varies with species and is drawn here only schematically. (b) Acclimation of R to prevailing growth/habitat conditions. The arrow indicates the effect of shift from a cool to a warm habitat. A species that shows full acclimation is shown (no long-term change in R despite an increase in T). The dashed line illustrates a more common case of partial acclimation.

cooler as radiation declines, this response removes part of the T effect one would predict from a high-light T response alone. A surprising net result of this T-response characteristic of A is that even in cold alpine climates the ‘missed’ CO_2 uptake compared to a theoretical maximum reached, if temperatures were always optimal for any given light level, is only approximately 7%, and is even smaller in warmer climates. (3) The whole curve shifts with the mean growth temperature, hence it peaks at lower temperature in plants from cool habitats (e.g. 16°C in treeline trees) and at higher temperatures in warm habitats (e.g. 27°C in tropical plants). Photosynthesis is particularly robust to low temperatures in adapted and acclimated species, and A becomes zero only when tissues freeze. In many cold-adapted plants, A reaches 30% of the maximum at 0°C, so photosynthesis is largely light driven and temperature plays only a marginal role (Körner, 2003a). This is not so in dark respiration.

3.2.2 Dark respiration

Dark respiration, R, is very small at 0°C, < 10% of that, for instance, measured at 25°C. For leaves, the mean increase of R for any 10 K increase of T (the Q_{10}) is 2.3 (Larigauderie & Körner, 1995). During light hours, much of the R activity in leaves is shut down (Atkin *et al.*, 1998), because metabolism is directly supplied with energy through photosynthesis (this is of course not true for all non-green tissue, roots in particular). Hence the common practice to add rates of R in leaves to those of A to calculate a ‘gross photosynthesis’ is simply wrong. In leaves in the light, R is already accounted for in A as measured with common equipment. It is very hard to separate the concurrent R term from A. Another common mistake leading to exaggerated rates of R is darkening a leaf during the day instead of measuring R at night. 'Black cloth' (measuring chamber darkened) rates of R can be twice as high as R at night at the same temperature, largely independent of how long the leaf is darkened when it is daytime.

These are the instantaneous responses to temperature, commonly measured at rates of change in temperature (in order to arrive at a complete curve) far greater than anything likely to occur in nature. However, such curves are valuable physiological fingerprint of the momentary tuning of the metabolism. Under no condition should such curves be used to make predictions about the effects of changing temperatures in the field longer term changes in particular, because these functions are not fi ed.

3.3 Thermal acclimation of metabolism

Medium and long-term exposure (one to several days, up to a full season) to a new temperature regime leads to acclimative adjustments of metabolism, causing the baseline responses discussed above to shift (Larcher, 1969). Any of these curves shown in Figures 3.1 and 3.2a already reflect the temperature conditions plants had experienced before the study. For instance, douglas-fi seedlings adjust their photosynthesis response to a new thermal regime within 10 days (Sorensen & Ferrell, 1972). Because the T dependency of photosynthesis plays such a minor role as compared to the dominant dependency on light conditions in the field it is sufficien to remember that the bell-shaped T-response curve can move by several K up and down also in a single leaf when temperature regimes change, further minimising photosynthetic T-constraints (Figure 3.1b). In contrast, temperature exerts strong instantaneous influence on respiration and growth, but before discussing acclimation of these processes, I will firs comment on a few common misconceptions about dark respiration.

1. Mitochondria are often seen as machines that run independently, with rates of R inevitably tied to T. In reality R has to meet a demand. When demand for metabolic energy goes down for whatever reason, R should and does go down as well. Demand can decline because a tissue matured and what is often termed *growth respiration* (cf. Lambers, 1985) is not needed anymore. Demand can go down because a plant becomes dormant or because it enters a new ontogenetic stage. Demand can go down because nutrients become readily available and the often-termed *nutrient uptake respiration* can be turned down. Demand can go down when temperatures go up, for instance because development was completed earlier and storage compartments had been filled hence also less phloem-loading energy is needed, and certain chemical reactions may run more efficientl when it is warmer and require less biochemical energy. The number of mitochondria and their activity is not 'a given', but can vary with demand. For instance, plants in cold

climates tend to have more mitochondria (Miroslavov & Kravkina, 1991). So R is demand driven, and demand is often controlled by factors other than temperature (Amthor & Baldocchi, 2001; Kurimoto *et al.*, 2004).

2. R is commonly studied at and reported for standard temperatures (e.g. 20°C) 'to be readily comparable' across plant growth conditions, which is in fact the opposite to comparability. R should correctly be compared at the temperatures under which plants actually grow. This problem led to the frequently published false notion that plants of cold climates respire more. This may be true when cold climate plants, which would hardly ever experience a 20°C night, are tested at 20°C, whereas the comparison warm climate plants grown at approximately 20°C are also measured at 20°C. In reality, cold climate plants may experience 5°C at night, and their leaves actually respire much less at night than leaves in plants in warm habitats. So, a rate of R measured in temperature regimes that a plant or a tissue is not experiencing in nature has little meaning.
3. R, as any other metabolic processes, needs a reference against which rates are expressed (dry weight, fresh weight, volume, area, water content, chlorophyll content, protein content etc.). This is all but trivial, because growth conditions selected for the study of R may also have influence the reference (e.g. Mitchell *et al.* (1999) showed that specifi leaf area, i.e., leaf area per unit leaf dry matter, had a great influenc on conclusions drawn from respiration measurements in 18 Appalachian tree species). Thus, differences in R may, in reality, reflec thicker cell walls, increase in protein, etc., depending on the reference chosen. Unfortunately, there is not a single best reference. For convenience, R is commonly dry matter based, although tissue density is known to change with growth conditions. Any comparison of plants from warmer and cooler sites should therefore include a suite of reference parameters to check for such bias.
4. Finally, opposite to what is often believed, the correct measurement of R is far more delicate and the responses are far more sensitive to the plant's growth conditions than are photosynthesis responses. Measurements of A can be restricted to leaves, whereas R measurements need to account for all tissue/organ types to gain comparable weight, and the crucial plant part is roots, which cannot be studied without massive intervention, i.e., decoupling them from the microbial rhizosphere community and mycorrhizal fungi, by interrupting nutrient uptake and transport, including downloading of sugar from phloem sap. If one accepts that R is not a self-fulfillin activity of mitochondria but has a purpose, i.e., is meeting a demand (see point 1), the removal of the functions that are inevitably tied to demand must affect R.

With these four precautions in mind, it is key to acknowledge that the T-response function of R is not static, but shows an *acclimatory response* (Figure 3.2b), and it is this reaction which needs to be known to draw meaningful conclusions about, for instance, the consequences of a warmer climate. However, the acclimation potential varies a lot across taxa and biogeographic origin. Some highly specialised cold climate species show almost no acclimation to temperatures warmer than the ones they come from (*Saxifraga* spp., *Ranunculus glacialis*; Larigauderie & Körner, 1995). Warming their habitat must have fatal metabolic consequences. They commonly die rapidly in lowland rock gardens. Other species show almost perfect acclimation so that their rate of respiration stays constant over a 10-K shift in growth temperature (e.g. in some wheat cultivars; Kurimoto *et al.*, 2004). Most species perform partial acclimation. The increased engagement of alternative respiratory pathways (no ATP production) in cold-adapted plants (e.g. McNulty *et al.*, 1988) may have to do with mitigating overshooting metabolism in the case of canopy overheating under extreme solar radiation in otherwise cold-adapted plants.

In order to account for acclimation, one needs to know LTR_{10}, the long-term temperature response of respiration (Larigauderie & Körner, 1995). When $Q_{10} = 2.3$ (the instantaneous 2.3-fold increase of R for a 10-K warming) and $LTR_{10} = 2.3$ (the increase of R after a long period of living at a 10-K warmer climate), then there is no acclimation; when $LTR_{10} = 1$, acclimation is complete (homoeostatic response). LTR_{10} data are very rare in the literature, but from greenhouse acclimation studies it seems that values are commonly bigger than 1 and are smaller than 2. With long-term growth in increased temperature, Q_{10} declines nearly linearly (Atkin & Tjoelker, 2003). Criddle *et al.* (1994) also showed that plants that experience broader ranges of temperatures during growth in their native habitat have a smaller temperature coefficien of respiration. It is about 70 years since the German ecophysiologist Otto Stocker (1935) noted with surprise that leaves of tropical trees in Java respired at about the same rate as leaves of willows in Greenland, when both were measured in situ, at their natural habitat temperatures. It is time for a wide acknowledgement that R does not follow long-term trends in temperature in the way indicated by short-term T-response curves, at least not in a straightforward manner. Such predictions need to account for LTR_{10}, with Q_{10} only driving relative responses around absolute rates set by LTR_{10}.

However, even perfect LTR_{10} data cannot solve the ultimate dilemma: plants may accelerate their development (e.g. earlier fl wering, senescence etc.), and, thus, the lifelong net C balance at higher temperatures compared to lower temperatures has little in common with the carbon balance measured at one point in time. It is not the actual rate of R that matters for understanding the carbon balance, but the integrated response over the lifespan of an organ or plant, and these integrated losses in CO_2 need to be balanced by the concurrent gain in carbon. Given that the difference between uptake and loss of C in a growing plant represents biomass production, it is often more informative and safe, and also much easier, to explore this net effect, i.e., the T responses of growth, instead of the opposing, delicate metabolic processes involved. Not knowing LTR_{10} responses, a best firs approximation often is that R follows growth rates (demand), which in turn control rates of whole plant photosyntheses (except in light-limited conditions, where it is the opposite situation). R and A often correlate well (Gifford, 1995) because both are driven by the demand of the same active carbon sinks. In addition to temperature, sink activity (growth) depends on nutrient and water availability and developmental stage, each with its own temperature dependency, explaining why measured short-term respiratory temperature responses do not normally scale to long-term responses as would be desirable for modelling.

3.4 Growth response to temperature

Growth measurements 'suffer' from their being simple: they require no expensive equipment but are often tedious, so lack academic appeal. This is sad, because the amount of good time resolution growth data that permit direct linking of the growth process with temperature is scarce. In terms of understanding temperature effects on plant life, growth data are far more informative than photosynthesis data, simply because actual photosynthetic carbon gain exhibits little sensitivity to temperature, whereas growth exhibits high sensitivity to temperature. The cooler the temperatures, the more the growth response lags behind the photosynthetic machinery's capacity to provide new assimilates (Figure 3.3; Körner, 2003b). As an example, Ford *et al.* (1987) found that the extension growth of sitka spruce shoots was fi e times more sensitive to temperature than sensitivity to changes in solar radiation, which had a large effect on photosynthesis, but little impact on growth.

There are many reasons why leaf photosynthesis data, which have been well explored, relate so poorly to growth. Most important are reasons related to tissue density, tissue duration and overall plant allometry. This fiel had been illuminated

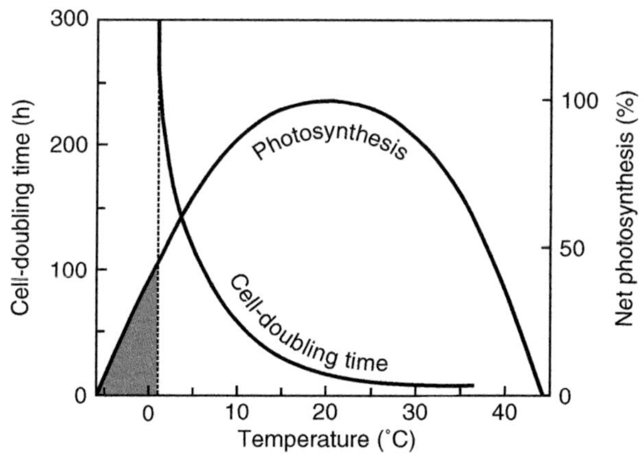

Figure 3.3 The temperature dependency of leaf net photosynthesis versus the temperature dependency of cell cycle duration (the rate at which new cells are formed). Note the large discrepancy at temperatures close to zero degrees. (From Körner, 2003b.)

by functional growth analysis (e.g. Lambers *et al.*, 1989), which does account for biomass allocation to organs and 'costs' of organs (e.g. their aerial or volume density, the N concentration) and their amortisation over time. Yield-oriented crop breeding had therefore not succeeded by selecting for leaf photosynthesis traits (Evans & Dunstone, 1970; Biscoe & Gallagher, 1977; Woolhouse, 1981; Saugier, 1983; Wardlaw, 1990). This is still not widely acknowledged in the scientifi community, but it is very important for developing scenarios for plant growth under changing atmospheric conditions. Woolhouse, in his plea for a change in paradigm, quoted Monteith and Elston (1971, cited in Woolhouse, 1981) by stating that 'limitation of growth under the cool conditions reside primarily with the capacity for cell division and expansion rather than with photosynthesis', and his concern that we know almost nothing about the nature of the rate-limiting steps to growth at low temperature is still true. As an example for the type of studies needed at the tissue level, I refer to Creber *et al.* (1993) who explored genotypic variation in cell division in *Dactylis glomerata*, showing that plants can compensate for the slowing of the cell cycle at low temperatures by greater numbers of cycling cells. Understanding effects of warming will require an understanding of such processes. Improved yields of cereals, in essence, have largely resulted from increase in harvest index rather than from increased leaf-level assimilation, but, as Monteith and Elston (1983) state, the ratio of papers that refer to growth versus photosynthesis in a climate context is 1:3.

By growth, I mean the formation of new plant tissue. In terms of mass accretion, this is in essence cell wall construction; in terms of metabolic infrastructure, it is the build up of the protoplast's inventory. Of the three steps, cell division, cell enlargement and cell differentiation, it appears to be the last step where thermal limitations come into play, but the three steps inevitably are tightly coupled (see Dale & Milthorpe, 1983; Gallagher, 1985 for further reading). As mentioned previously, photosynthesis may reach a third of full capacity at 0°C but no plant can grow at 0°C. The cell cycle duration (the full time it takes a cell to double) may be 10 h at 25°C but approaches infinit a few degrees above zero. It is at low temperatures where small amounts of warming can have immediate and strong effects by activating meristems (sinks).

Even most cold-adapted plants, including winter cereals, show negligible growth at 2–3°C (for wheat; e.g. Gallagher *et al.*, 1979; Hay & Wilson, 1982) and significan rates may only be found at >6°C. Indeed, 6°C is a well-known threshold for crop growth as reflecte by officia farmer recommendations, dating back to the nineteenth century as for instance: 'the advent of spring may properly be considered as taking place at the advent of a 6–7°C isotherm' (Harrington, 1894; for the UK). The US Department of Agriculture recommended 6°C as the zero point of 'vital temperature' (Smith, 1920). De Candolle (1855) had already noted that 'there seems to be a 6°C threshold temperature for plant development', and similar comments can be found in Hoffmann (1859; references collected by Gensler, 1946).

The significanc of 6°C has acquired an interesting new dimension with the result of a global survey of cold limits of tree growth, which yielded an average mean 6.7 ± 0.8°C seasonal mean temperature for nearly 40 treeline locations worldwide

(Körner & Paulsen, 2004). This mean is for the growing season as define by a critical daily mean soil temperature of >3.2°C at 10-cm soil depth, roughly corresponding to a mean air temperature of zero degrees, irrespective of the actual length of the season (12 months at the equator and 2.5 months at sub-polar latitudes). A surprising aspect of this analysis was that neither thermal sums nor a median temperature yielded a better global fit and that season length had very little influenc on the treeline position. Means are slightly lower at tropical treelines (5–6°C) compared to higher latitudes (6–7°C), but even the Betula treeline in northern Fennoscandia (68°N) is at a mean 6.5°C temperature. A 5–7°C threshold for any significan growth to occur is also well known for cold-adapted trees (e.g. James *et al.*, 1994; Vapaavuori *et al.*, 1992). Knowledge of such thresholds is critically important for modelling.

Taken together, there seems to be an absolute limit for any growth activity between 0 and 2°C, but growth rates really become measurable only at around 6°C irrespective of plant life form or taxon among the cold-tolerant taxa. The reason why upright trees fin a lower elevation/latitude limit than low-stature plants has nothing to do with the physiology of growth but is related to tree morphology, which couples tree crowns closer to air temperature, as will be discussed later (see Colour Plate 4). I presume that even the most cold-adapted alpine and arctic species are tied to this threshold, but they need much shorter time to pass through the seasonal growth cycle and they profi from solar heat, periodically accumulating near the ground. It is an interesting perspective that there might be one common lower thermal threshold for the basic processes involved in tissue formation of higher plants such as winter wheat, treeline trees and alpine buttercups. The cellular processes responsible are not really understood; the only thing which is certain is that this limitation has very little to do with the availability of photoassimilates (Körner & Pelaez Menendez-Riedl, 1989; Hoch *et al.*, 2002; Körner, 2003a,b).

I am not aware of growth studies that would yield data similar to those as shown for R in Figure 3.2b. What would be needed would be the growth rates at a define growth stage (e.g. a herbaceous plant growing from the 8-leaf to the 10-leaf stage) at different growth temperatures and in plants that had experienced different temperatures while growing to the 8-leaf stage. Of course, there are lots of biomass data for plants grown at different temperatures, but such data cannot reveal thermal acclimation.

There have been many studies on the genetic (ecotypic) adaptation of growth to habitat temperature, using common garden or greenhouse conditions (e.g. Clements *et al.*, 1950; Lyr & Garbe, 1995; Oleksyn *et al.*, 1998). As an example, Figure 3.4 illustrates T-response functions for leaf growth of *Poa* species that are native to cold and warm habitats. In contrast to the R response (Figure 3.2), growth has a discrete end at certain low, positive threshold temperatures, and the slope of the response is steeper in the warm habitat species and flatte in the cold habitat species, opposite to the responses known for R. Hence, the cold-adapted species are more able to grow at lower temperatures than the warm-adapted species, but at higher temperature they reach only half the growth rate as exhibited by the warm-adapted species. It appears as if there were a trade-off between the ability to grow at low temperatures and the maximum rate of growth achieved when conditions are favourable. The cold-adapted

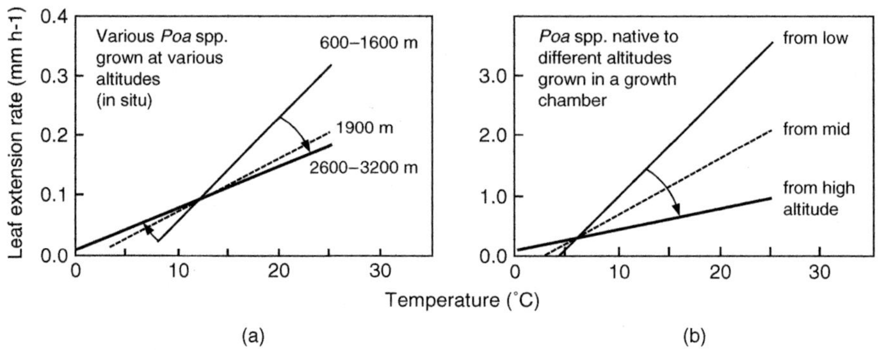

Figure 3.4 The in situ temperature response of leaf extension growth in grasses (*Poa* spp.; recorded with an electromagnetic displacement recorder) from thermally different habitats. Note the different low-temperature thresholds and slopes. (From Körner & Woodward, 1987; Körner, 2003a.)

species would primarily profi from an extension of favourable periods. The warm-adapted species would take additional advantages from higher temperatures that permit relatively greater acceleration of the rate of growth. In agreement with the available information for R, the Q_{10} of growth declines with habitat temperature.

3.5 Temperature extremes and temperature thresholds

In addition to the gradual responses of life processes to temperature (or other climatic factors) discussed above, threshold phenomena are in fact the overarching filte by which the presence and absence of taxa in a given region is determined. Low-temperature extremes are far more significant and plant sensitivity to low temperatures varies to much greater extent than is the case for high-temperature extremes. All plants are killed somewhere between 46 and 56°C (mostly around 48–50°C). Such temperatures commonly occur only at unshaded soil surfaces, hence may affect plant establishment and require some facilitative initial shading to allow plants to establish, as is common in semiarid regions. However, critically low (damaging) temperatures vary from +7°C in chilling-sensitive tropical species such as coffee or cacao to −70°C in the most frost-tolerant taxa of the continental boreal forest, and these thresholds vary with season (acclimation), tissue type, plant age and other environmental factors such as water and nutrients (Sakai & Larcher, 1987; Larcher, 2003). The important point is that critically low temperatures need to hit a population of a species only once in the course of many years to become decisive and eliminate a species from an area unless there is recruitment from the soil seedbank or resprouting from rootstock.

The other side of the coin is that taxa that have passed this centennial filte are not at danger; hence frost has no serious influenc on their life. In fact, it is safeguarding the frost-resistant taxa against the invasion of less robust taxa. In this respect, resistant species need regular near-critical frost events to keep the habitat free of competing invaders. So-called (low temperature) *stress-dominated habitats* are thus inhabited by plants for which severe frost is a vital requirement rather than a constraint. The mitigation of frost severity is anything but a relief; it is like a breaking dam, which opens the arena for a 'flood of non-resistant taxa, with fatal consequences for the native species. It is one of the common misconceptions that plants from cold habitats are cold stressed. They become stressed once temperatures rise (Körner, 2003c).

This does not mean that plants native to cold habitats are not impacted by extreme events. In the case of low-temperature extremes, native vegetation may well be hit by late spring frost and lose a leaf cohort or all fl wers in a given year, but this damage is not fatal. The dangerous periods are not the coldest periods, but the transition periods, when plants are already dehardened or not yet fully hardened when a freezing event occurs (Taschler & Neuner, 2004). In the tropics, plants never harden, hence may be hit at any time at certain elevations or marginal latitudes. Furthermore, extreme events (e.g. low minimum air temperature) are not necessarily tied to mean temperature trends. The climate may get warmer, but the likelihood of polar air masses to reach lower latitudes once every 30 years may actually increase, e.g. because of a new arrangement of atmospheric pressure systems. The climate may also be relatively cool, but also frost free, permitting the growth of tropical species, as is the case on some temperate ocean islands (e.g. sub-tropical plants growing in gardens in Southern Ireland or the coastal flor of southwestern New Zealand). Hence, annual means have little meaning for assessing the probability of frost damage. On a shorter timescale, it is the minimum night-time temperature and not the daily mean temperature that matters. Given that radiative cooling on clear nights may reduce plant temperatures by 4 K below ambient, night-time cloudiness may significantl modify plant temperatures compared to actual meteorological records of air temperature.

The mechanisms by which plants can cope with low-temperature extremes, in brief, are escape or resistance, and the latter is achieved through either avoidance or tolerance. *Escape* means not being present with sensitive tissue when extremes prevail, which can be reached by leaf shedding, overwintering by below ground organs or by seeds, or selecting habitats with secure snow cover (e.g. 'snow beds' in mountains). *Avoidance* means being exposed to sub-freezing temperatures, but avoiding ice formation in the leaf by either freezing point depression through solutes (a very inefficien mechanism, because it requires the doubling of common osmotic pressures to arrive at a 1.5–2 K freezing point depression) or so-called supercooling. Super cooling means retaining water in a gel-type stage by avoiding ice nucleation. This is a risky strategy, adopted by some tropical alpine species and more commonly (and less risky) in the xylem of trees; risky, because once a critical temperature is surpassed (in leaves commonly −12°C), the tissue will freeze immediately and will be killed. The most common way of coping with freezing temperatures is *tolerance* of ice formation outside the protoplast, which means extracellular accumulation of ice (largely in the intercellular space), which gradually dehydrates the protoplast, and thus prevents it from freezing (which would be lethal). This mechanism requires an intact and flui plasmalemma membrane, which permits orderly efflu of water out of the protoplast at low temperatures, and some protective compounds (certain sugars, proteins) that safeguard the membranes in the shrinking, dehydrating protoplast. Frost resistance by tolerance requires biochemical adjustments of membranes when it gets cold, a key process in thermal acclimation. If temperatures drop too rapidly so that the rate of efflu of water cannot cope, the protoplast will freeze and die (Sakai & Larcher, 1987; Larcher, 2005).

In the climate change context, it is important to distinguish between cold acclimation, a reversible process, induced by environmental conditions, and the evolutionary (genetic) adaptation to life in cold climates. The latter sets the ultimate limit, the former depends on developmental state, temperature history and photoperiod (see below). There is no absolute thermal limit one can defin for a plant – frost resistance is a context dependent variable.

3.6 The temperatures experienced by plants

It is often assumed that plant tissue temperatures correspond to the temperatures measured in the air surrounding the plant. However, in the real world, plants air-condition their organs and their micro-environment, and to some degree can escape certain thermal constraints or build up new ones (in the case of heat). Any body exposed to solar radiation will inevitably warm, and any body vaporising water will inevitably cool, and the net balance between the two processes controls body temperature during the day. How much these two processes will cause an object to depart from surrounding air temperature depends on the rate at which heat is exchanged between surrounding air and this body, which depends on wind speed, humidity and aerodynamic properties of the body. Plants that are short of water have to close their stomata, and thus lose the cooling power of transpiration and leaves will warm up under solar irradiance. By their leaf size and whole morphology (architecture), plants can be well- or poorly coupled with atmospheric conditions. Highly coupled plant types are tall, with an open canopy of narrow leaves (e.g. trees of the genus *Casuarina* on tropical islands or some *Pinus* species in higher latitudes); poorly coupled structures are of low stature and form rosettes, dense mats or cushions.

A classical example for the mismatch between climate station data and the temperatures experienced by plants are low-stature alpine biota, which ‘collect’ substantial solar heat, in contrast to upright trees, which are constrained by their architecture, which thus explains their high elevation limit (the treeline). By their stature, trees track air temperature and prevent solar heating of their soils by shading the ground, once the canopy has closed. It then does not come as a surprise that treeless alpine vegetation several hundred meters above treeline experiences a warmer season than treeline trees (Körner & Paulsen, 2004) and the alpine vegetation has a temperature of 22°C optimum for photosynthesis, similar to lowland grassland plants (Körner, 2003a). Depending on growth form, temperatures in prostrate vegetation may be

3–20 K above air temperature during sunshine periods. Soil heat flu is high under such conditions, also leading to warmer night-time temperatures for the predominant sub-surface meristems (Körner & Cochrane, 1983; Grace *et al.*, 1989; Körner *et al.*, 2003). There is no physiological evidence that trees are less capable of handling low temperatures than grasses, herbs and dwarf shrubs. Trees simply experience a colder world than grasses and dwarf shrubs, the life forms they are forced to yield place to when it gets too cold (Colour Plate 4).

Because of these different degrees of coupling to ambient conditions, trees will be affected to greater extent when temperatures change than low-stature vegetation. The temperature of low-stature plants also varies greatly with topography (orientation to the sun, slope, shelter), and microhabitats differing in temperature by several Kelvin may be found within a meter from each other. These small-scale thermal mosaics contrast with the common isotherm-oriented scenarios of those models that model vegetation as driven by climatic changes or which simulate climatic change effects on vegetation or with large-scale natural thermal gradients.

On a technical note it has become very simple to record the thermal characteristics of a habitat at very little cost and effort. Robust, waterproof data loggers of the size of a coin are available for less than €100. If buried in the meristematic zone of grassland plants or exposed in full shade of tall vegetation, year round temperatures can be recorded at high temporal resolution (examples for the usefulness of such records are in Körner & Paulsen, 2004). There is only one precaution: the sun must never hit such devices. The true leaf surface temperature will thus remain unknown. The best devices to obtain such surface temperatures are high-resolution digital thermal cameras as the one used for producing Colour Plate 4 (for a review of such techniques consult Jones *et al.*, 2003; Jones & Leinonen, 2003).

3.7 Temperature and plant development

Temperature controls rates of plant development, but not necessarily critical ontogenetic phase changes such as induction of bud-break, fl wering or leaf senescence, which may be determined by photoperiod. Commonly, it is the speed at which plants and their organs pass through developmental phases, which depends on temperature. For instance, a higher temperature may shorten the period of grain fillin in wheat (Wheeler *et al.*, 1996). So, temperature effects interact with other environmental and internal drivers of development. Temperatures, low ones in particular, may serve as a signal which alone or together with photoperiod set the receptivity of plants to the gradual direct influence of temperature on metabolism and growth as described above.

Temperature as a signal is best known under terms like *vernalisation* or *chilling requirement*. The firs one specificall refers to the induction of fl wer buds, the second refers to general growth activity. Most plants from higher latitudes require the experience of a certain degree of cold weather (a sum of hours or days below a certain threshold temperature) before they resume growth in spring (Cannell & Smith, 1986). This is one way of detecting that winter is over, but not a very secure one, when winters are mild or rapidly get milder as we have seen in the recent past. These effects are well understood in so-called winter and spring cereals. 'Winter varieties' will not set ears unless they experienced a cold winter as it naturally occurs in their steppe-type original habitats. 'Spring cereals', cultivars sown in spring, have very little or no chilling requirement for initiating the reproductive phase. A winter variety sown in a climate with a warm winter will thus fail to produce a harvestable crop, but remain trapped in the vegetative life phase (a green meadow, Colour Plate 5).

There is a rich literature on the influenc of temperature and chilling requirements in tree development, dating back to the beginning of the last century (a rather complete account is given by Klebs (1914) for European trees). The theme is complicated, because tree species and provenances differ not only in their chilling requirement but also in the sum of heat required after the chilling requirement is met, before they start to flush There is a negative interaction between the degree of chilling and the heat sum needed for flushin (the less chilling, the more heat is needed to break dormancy). For instance, the thermal time (e.g. number of days with T > 5°C since 1 January) remains high in *Fagus sylvatica,* the late flushin European beech, irrespective of a warmer climate, because the less chill it receives, the longer it takes to bud burst. In contrast, a species with a small thermal time and chilling requirement like *Crataegus monogyna* flushe much earlier in a simulated warmer spring (Murray *et al.*, 1989).

Because temperature is often an unreliable marker of seasonality, most long-lived plant species native to areas outside the tropics have evolved a second line of safeguarding them against 'misleading' temperature conditions: photoperiodism. The significanc of photoperiodism increases with latitude, not only because the annual variation of the photoperiod becomes more pronounced, but also because of its biological function. There are two major roles of photoperiodism: (1) synchronisation of fl wering in populations and thus ensuring reproductive success and (2) preventing phenology from following temperature as a risky environmental signal for development. Although the two functions are linked, the second is the one most relevant here. It is an insurance for plants against temperature-induced break of dormancy too early in the season, and induction of dormancy too late in the season. Thus, photoperiodism constrains the influenc of temperature on development to 'safe periods'.

Taken together, chilling requirement and photoperiodism represent a dual induction system, which in combination with actual temperatures (e.g. heat sums) determine development (Hay, 1990). Needless to say that this complicates predictions of phenology in a warmer climate; even more so, because species greatly differ in their chilling and photoperiod requirements. The problem becomes worse, because these interactions between photoperiod and temperature are not firm but are in part substitutive, which means particularly warm temperatures can override photoperiod controls and particularly long days can override chilling requirements (Heide, 1993a). A common pattern is that photoperiod and/or a sufficien chill dose releases/induces a developmental step, and follow-up temperatures determine the speed of progress (Figure 3.5A). There are domestic tree species that have an evolutionary history where photoperiodism did not play a significan role in spring, as for instance,

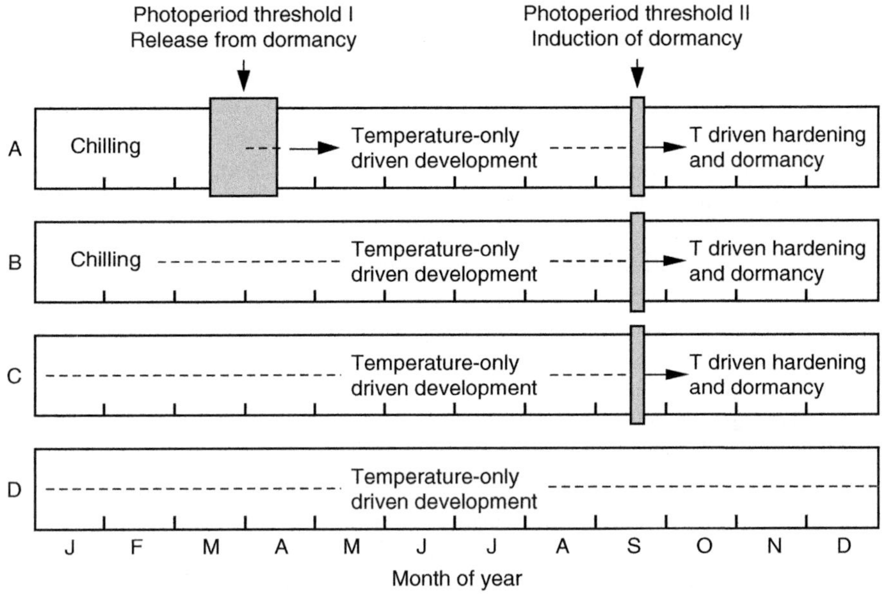

Figure 3.5 A schematic representation of the interaction of temperature and photoperiodism in photoperiod-sensitive species from cool temperate climates. Boxes illustrate the photoperiod-driven windows that permit development, the speed of which is controlled by the actual temperature. A depicts a triple control of bud burst, B a double control (no spring photoperiod effect), C an opportunistic behaviour (only actual temperature matters), with A–C still adopting a photoperiod control of timely senescence or dormancy induction in a seasonal climate. D represents a tropical ecotype with no regular threshold controls of phenology (but there may be other triggers).

apricot and cherry from central Asia, but also horse chestnut. Once they have experienced some (limited) frost, they fl wer whenever temperature permits (even in mid winter) – and then commonly lose all fl wers or young fruit when winter returns with freezing temperatures (Figure 3.5B). Apple and pear, however, cannot be 'tricked' to the same extent by warm spells, and photoperiodism protects them from premature bud-break. However, some cultivars are at great risk of frost damage in a warmer climate (Cannell & Smith, 1986). Most trees in the wild operate with a combined triple system (Figure 3.5C), with (1) chilling requirements and (2) photoperiodism preventing warm temperatures from becoming effective too early in the season, but both interacting with (3) the heat sum, required to flus (Heide, 1993b). For a number of tree species, Heide showed that the thermal time required to bud burst decreases nonlinearly with increasing duration of previous chilling, with different slopes for each species. The more photoperiod controlled a species is, the less likely will its budbreak occur too early with respect to the risk of freezing damage. Species with no day length control are *Sorbus aucuparia*, *Rubus* spp., *Carpinus betulus* (Heide, 1993a,b).

In contrast to the more fl xible onset of seasonal growth, the late season induction of dormancy (e.g. bud maturation, leaf senescence) is more tightly photoperiod-controlled in climates with a cold season. This does not necessarily include the discolouration of senesced leaves (which may still depend on cold nights), but photoperiod sets the internal physiological state and guarantees bud ripening irrespective of temperature. If induction of dormancy was delayed until the onset of the cold period, plants would fail to produce the necessary structures and make the biochemical adjustments required in time. The autumnal transition to dormancy (and full frost resistance) starts with a photoperiod signal, is enhanced by cool nights and reaches its full strength after exposure to frost (Larcher, 2003).

I want to close this section by pointing out three potential problems, when vegetation is photoperiod- and chilling-requirement controlled and climatic conditions become warmer at a rate exceeding that of evolutionary adjustments.

The firs problem is related to soils. Free-living soil organisms are commonly opportunistic and become active whenever temperatures and soil moisture permit. One could envisage warmer winters with high microbial activity and release of nutrients by the decomposer food web, but plants, with their evolutionary 'memory', are still constrained to use these resources because of photoperiod- and chilling-controlled dormancy. These free nutrients need to be either stored in microbial biomass or become tied to charged surfaces (ion exchange) in the substrate or else become washed out by winter rains. It could well be that the ion exchange capacity of soils will determine whether such genotype controls of plant dormancy will lead to nutrient losses (leaching) of the system.

The second problem relates to predictions of future season length and the related plant activities by using current trends in plant phenology and climate. Numerous phenological observations, both direct and by remote sensing (see Chapter 4) have documented that the warming trends observed during the last century were associated with earlier greening/fl wering and later senescence of plants. However, what might have been seen so far in the majority of the tree taxa is (a) the response for species with weak photoperiodism or chill control, or (b) a phenology that was pushed by temperatures to the far end of phenology 'windows' controlled by genetically controlled phenology. In the latter case, we should not see a further extension of these trends, or the slowing of the trends should not be confused with a slowing of warming (which may have other biological effects, see problem one). In the firs case, we should see community effects, with the photoperiod insensitive taxa taking an advantage. Exotic taxa, as commonly grown in cities may track the climate, whereas the native vegetation may not.

The third problem relates to a paradox that mild winters may either (1) delay spring development because of insufficien winter chill and thus higher heat sum required to bud burst, or (2) may lead to earlier bud-break in photoperiod insensitive taxa with low chill requirement, with an enhanced risk of frost damage (Cannell & Smith, 1986; Myking & Heide, 1995). A number of alpine plant species are unresponsive to temperature, but will not even start leafin in a warmer climate unless photoperiod requirements are met (Keller & Körner, 2003; Colour Plate 6).

The key message of this section is that plant development will not necessarily track a given trend in temperature. Annual plants may rapidly evolve new photoperiodism genotypes (Heide, 2001), and in the case of trees, this is not a reasonable scenario. In case of rapid climatic warming, the given diversity of genotypic phenology responses will affect intraspecifi competition and may change species composition, at least in long-lived plant taxa that have no time to evolve new genotypes.

3.8 The challenge of testing plant responses to temperature

There are four principal empirical ways to assess plant responses to temperature: (1) looking into the past, using historical trends of temperature and growth, in essence restricted to dendrology, (2) studying current growth processes across thermal gradients, or (3) studying current growth in response to the natural temporal variation in temperature and (4) manipulating temperatures around plants and testing their responses under controlled conditions (both indoors and in the field) Each of these approaches has some advantages and disadvantages. While type 4 tests are best controlled in terms of environmental influences they are limited in time and space and are commonly confine to very artificia growth conditions and obviously restricted to very young ages in the case of trees. The other three options are commonly less 'precise' in the sense of isolating temperature effects from other effects and good replication, but they are closer to real world conditions. It is the challenge of empirical sciences to make maximum use of all these options, but there is a great need to complement the predominance of type 4 studies with more type 1–3 studies (Körner, 2001). The area second best explored is tree rings that, for instance, allowed the demonstration of clear warming effects in treeline trees in recent decades (Rolland *et al.*, 1998; Paulsen *et al.*, 2000) in some regions but not in others (Kirchhefer, 2005). I would like to argue for greater attention to type 2 and 3 studies.

Using either temporal or spatial patterns of temperature and concurrent growth processes in established plants has a number of advantages. Plants growing along thermal gradients have had time to adjust, grow in undisturbed soil and under a natural variation of temperature. The dichotomy of (a) studying plants of one species across the thermal range of that species versus (b) studying plants in the centre of the range of species restricted to different thermal ranges offers the study of contrasting evolutionary history and likely genetic adaptation (Körner, 2003a). The inclusion of invasive species permits tracking rapid evolutionary processes. On the other hand, the in situ study of the influenc of short-term natural variation in temperature on metabolism and growth provides information on instantaneous response characteristics in a natural situation. Comparing data for plants that have experienced different thermal prehistory also permits exploring acclimative trends. These classical approaches (e.g. Gallagher *et al.*, 1979; Ford *et al.*, 1987; Körner & Woodward, 1987; James *et al.*, 1994) are under-represented tools in experimental biology.

On the other hand, caution should be exercised with the interpretation of experimental warming trials in the field There is no way one can warm plants in a given environment without affecting humidity. So any warming test will inevitably be confounded with changed evaporative conditions. It is also very difficul to simulate a warmer atmosphere with point sources of heat such as blowers without affecting aerodynamics. Radiative heaters, which are in widespread use, do not simulate convective (diffuse) warming, but exert a directional heat with vertical gradients, unlike that of a warmer climate, even if mean temperatures may match. The same applies to soil warming, which, for physical reasons, induces water diffusion away from the heat source. In addition, step increases of temperature in soils represent a major disturbance which may take years to lead to a new steady state, with initial responses in essence documenting the disturbance of the rather delicate balance between plant roots, fungi, microbes and the soil fauna associated with it. Given these intrinsic constraints, it is far safer to build upon short-distance natural topographic or narrow elevational gradients which easily can be found to offer, e.g., 2K warmer condition under otherwise similar overall test conditions (soils, flora precipitation). An alternative is the use of soil monoliths at least in grassland. These can be transplanted or transferred in the fiel to controlled environments, although the 'step change' problem cannot be overcome. There are several psychological barriers to the use of these elegant tools that nature offers to the experimentalist, who often prefers to interfere with some technological glamour rather than capitalise on these free-of-charge test conditions.

Whenever possible, the various techniques should be combined to capitalise on the advantage of each. I emphasise the simpler, often overlooked tools for biological temperature research offered in situ, because the lack of high-tech facilities is often seen to preclude upfront research. Controlling life conditions in closed research units is and will remain a key tool for understanding plant temperature responses. However, such data are not necessarily more 'accurate' or relevant than those obtained in the field although this assumption is the tradition that I and many of my age class grew up with. I want to encourage the next generation to be more open to the alternative approaches with much greater 'experimental noise' incurred, but this may become manageable with high replication and with the modern statistical and computational tools.

References

Amthor, J.S. & Baldocchi, D.D. (2001) Terrestrial higher plant respiration and net primary production. In: *Terrestrial Global Productivity* (eds J. Roy, B. Saugier & H.A. Mooney), pp. 33–59. Academic Press, San Diego, CA.

Atkin, O.K., Evans, J.R., Ball, M.C., Siebke, K., Pons, T.L. & Lambers, H. (1998) Light inhibition of leaf respiration: the role of irradiance and temperature. In: *Plant Mitochondria: From Gene to Function* (eds I.M. Møller, P. Gardeström, K. Glimelius & E. Glaser), pp. 567–574. Backhuys, Leiden, The Netherlands.

Atkin, O.K. & Tjoelker, M.G. (2003) Thermal acclimation and the dynamic response of plant respiration to temperature. *Trends Plant Sci.*, **8**, 343–351.

Biscoe, P.V. & Gallagher, J.N. (1977) Weather, dry matter production and yield. In: *Environmental Effects on Crop Physiology* (eds J.J. Landsberg & C.V. Cutting), pp. 75–100. Academic Press, New York.

Cannell, M.G.R. & Smith, R.I. (1986) Climatic warming, spring budburst and frost damage in trees. *J. Appl. Ecol.*, **23**, 177–191.

Clements, F.E., Martin, E.V. & Long, F.L. (1950) *Adaptation and Origin in the Plant World. The Role of Environment in Evolution*. Waltham, MA.

Creber, H.M.C., Davies, M.S. & Francis, D. (1993) Effects of temperature on cell division in root meristems of natural populations of Dactylis glomerata of contrasting latitudinal origins. *Environ. Exp. Bot.*, **33**, 433–442.

Criddle, R.S., Hopkin, M.S., McArthur, E.D. & Hansen, L.D. (1994) Plant distribution and the temperature coefficien of metabolism. *Plant Cell Environ.*, **17**, 233–243.

Dale, J.E. & Milthorpe, F.L. (1983) *The Growth and Functioning of Leaves*. Cambridge University Press, Cambridge, UK.

De Candolle, M.A. (1855) *Géographie Botanique Raissonnée*. V. Masson, Paris.

Evans, L.T. & Dunstone, R.L. (1970) Some physiological aspects of evolution in wheat. *Aust. J. Biol. Sci.*, **23**, 725–741.

Ford, E.D., Milne, R. & Deans, J.D. (1987) Shoot extension in *Picea sitchensis* II. Analysis of weather influence on daily growth rate. *Ann. Bot.*, **60**, 543–552.

Gallagher, J.N. (1985) The way ahead: a crop physiologist's viewpoint. In: *Control of Leaf Growth* (eds N.R. Baker, W.J. Davies & C.K. Ong), pp. 319–343. Cambridge University Press, Cambridge, UK.

Gallagher, J.N., Biscoe, P.V. & Wallace, J.S. (1979) Field studies of cereal leaf growth. IV. Winter wheat leaf extension in relation to temperature and leaf water status. *J. Exp. Bot.*, **30**, 657–668.

Gensler, G.A. (1946) *Der Begriff der Vegetationszeit*. Engadin Press, Samedan, Switzerland.

Gifford, R.M. (1995) Whole plant respiration and photosynthesis of wheat under increased CO_2 concentration and temperature: long-term vs. short-term distinctions for modelling. *Global Change Biol.*, **1**, 385–396.

Grace, J., Allen, S.J. & Wilson, C. (1989) Climate and the meristem temperatures of plant communities near the tree-lines. *Oecologia*, **79**, 198–204.

Harrington, M.W. (1894) The advent of spring. *Harper's New Monthly Mag.*, **27** (European edition), 874–879.

Hay, R.K.M. (1990) Tansley review no. 26. The influenc of photoperiod on the drymatter production of grasses and cereals. *New Phytol.*, **116**, 233–254.

Hay, R.K.M. & Wilson, G.T. (1982) Leaf appearance and extension in field-gr wn winter wheat plants: the importance of soil temperature during vegetative growth. *J. Agric. Sci.*, **99**, 403–410.

Heide, O.M. (1993a) Dormancy release in beech buds (*Fagus sylvatica*) requires both chilling and long days. *Physiol. Plant.*, **89**, 187–191.

Heide, O.M. (1993b) Daylength and thermal time response of budburst during dormancy release in some northern deciduous trees. *Physiol. Plant.*, **88**, 531–540.

Heide, O.M. (2001) Flowering responses of contrasting ecotypes of *Poa annua* and their putative ancestors *Poa infirma* and *Poa supina*. *Ann. Bot.*, **87**, 795–804.

Hoch, G., Popp, M. & Körner, Ch. (2002) Altitudinal increase of mobile carbon pools in *Pinus cembra* suggests sink limitation of growth at the Swiss treeline. *Oikos*, **98**, 361–374.

Hoffmann, H. (1859) Ueber den klimatischen Coefficiente der Vegetation. *Bot. Zeitg.*, **17**, 85–88.

James, J.C., Grace, J. & Hoad, S.P. (1994) Growth and photosynthesis of *Pinus sylvestris* at its altitudinal limit in Scotland. *J. Ecol.*, **82**, 297–306.

Jones, H.G., Archer, N., Rotenberg, E. & Casa, R. (2003) Radiation measurement for plant ecophysiology. *J. Exp. Bot.*, **54**, 879–889.

Jones, H.G. & Leinonen, I. (2003) Thermal imaging for the study of plant water relations. *J. Agric. Meteorol.*, **59**, 205–217.

Keller, F. & Körner, Ch. (2003) The role of photoperiodism in alpine plant development. *Arct. Antarct. Alp. Res.*, **35**, 361–368.

Kirchhefer, A.J. (2005) A discontinuous tree-ring record AD 320–1994 from Dividalen, Norway: inferences on climate and treeline history. In: *Mountain Ecosystems. Studies in Treeline Ecology* (eds G. Broll & B. Keplin), pp. 219–235. Springer-Verlag, Berlin.

Klebs, G. (1914) *Über das Treiben der einheimischen Bäume speziell der Buche*. Carl Winters Universitätsbuchhandlung, Heidelberg.

Körner, Ch. (2001) Experimental plant ecology: some lessons from global change research. In: *Ecology: Achievement and Challenge* (eds M.C. Press, N.J. Huntly & S. Levin), pp. 227–247. Blackwell Science, Oxford.

Körner, Ch. (2003a) *Alpine Plant Life*, 2nd edn. Springer-Verlag, Berlin.

Körner, Ch. (2003b) Carbon limitation in trees. *J. Ecol.*, **91**, 4–17.

Körner, Ch. (2003c) Limitation and stress – always or never? *J. Veg. Sci.*, **14**, 141–143.

Körner, Ch. & Cochrane, P. (1983) Influenc of plant physiognomy on leaf temperature on clear mid-summer days in the Snowy Mountains, south-eastern Australia. *Acta Oecol.* (*Oecol. Plant.*), **4**, 117–124.

Körner, Ch. & Paulsen, J. (2004) A world-wide study of high altitude treeline temperatures. *J. Biogeogr.*, **31**, 713–732.

Körner, Ch., Paulsen, J. & Pelaez-Riedl, S. (2003) A bioclimatic characterisation of Europe's alpine areas. In: *Alpine Biodiversity in Europe.* Ecological Studies 167 (eds L. Nagy, G. Grabherr, Ch. Körner & D.B.A. Thompson), pp. 13–28. Springer-Verlag, Berlin.

Körner, Ch. & Pelaez Menendez-Riedl, S. (1989) The significanc of developmental aspects in plant growth analysis. In: *Causes and Consequences of Variation in Growth Rate and Productivity of Higher Plants* (eds H. Lambers, M.L. Cambridge, H. Konings & T.L. Pons), pp. 141–157. SPB Academic Publisher, The Hague, The Netherlands.

Körner, Ch. & Woodward, F.I. (1987) The dynamics of leaf extension in plants with diverse altitudinal ranges. 2. Field studies in *Poa* species between 600 and 3200 m altitude. *Oecologia*, **72**, 279–283.

Kurimoto, K., Day, D.A., Lambers, H. & Noguchi, K. (2004) Effect of respiratory homeostasis on plant growth in cultivars of wheat and rice. *Plant Cell Environ.*, **27**, 853–862.

Lambers, H. (1985) Respiration in intact plants and tissues: its regulation and dependence on environmental factors, metabolism and invaded organisms. In: *Encyclopedia of Plant Physiology, Ns 18, Higher Plant Cell Respiration* (eds R. Douce & D.A. Day), pp. 418–473. Springer-Verlag, New York.

Lambers, H., Cambridge, M.L., Konings, H. & Pons, T.L. (1989) *Causes and Consequences of Variation in Growth Rate and Productivity of Higher Plants*. SPB Academic Publisher, The Hague, The Netherlands.

Larcher, W. (1969) The effect of environmental and physiological variables on the carbon dioxide gas exchange of trees. *Photosynthetica*, **3**, 167–198.

Larcher, W. (2003) *Physiological Plant Ecology*, 4th edn. Springer-Verlag, Berlin.

Larcher, W. (2005) Climatic constraints drive the evolution of low temperature resistance in woody plants. *J. Agr. Meteorol.*, **61**, 189–202.

Larigauderie, A. & Körner, Ch. (1995) Acclimation of leaf dark respiration to temperature in alpine and lowland plant species. *Ann. Bot.*, **76**, 245–252.

Lyr, H. & Garbe, V. (1995) Influenc of root temperature on growth of *Pinus sylvestris*, *Fagus sylvatica*, *Tilia cordata* and *Quercus robur*. *Trees*, **9**, 220–223.

McNulty, A.K., Cummins, W.R. & Pellizzar, A. (1988) A fiel survey of respiration rates in leaves of arctic plants. *Arctic*, **41**, 1–5.

Miroslavov, E.A. & Kravkina, I.M. (1991) Comparative analysis of chloroplasts and mitochondria in leaf chlorenchyma from mountain plants grown at different altitudes. *Ann. Bot.*, **68**, 195–200.

Mitchell, K.A., Bolstad, P.V. & Vose, J.M. (1999) Interspecifi and environmentally induced variation in foliar dark respiration among eighteen southeastern deciduous tree species. *Tree Physiol.*, **19**, 861–870.

Monteith, J.L. & Elston, J. (1983) Performance and productivity of foliage in the field In: *The Growth and Functioning of Leaves* (eds J.E. Dale & F.L. Milthorpe), pp. 499–518. Cambridge University Press, Cambridge, UK.

Murray, M.B., Cannell, M.G.R. & Smith, R.I. (1989) Date of budburst of fiftee tree species in Britain following climatic warming. *J. Appl. Ecol.*, **26**, 693–700.

Myking, T. & Heide, O.M. (1995) Dormancy release and chilling requirement of buds of latitudinal ecotypes of *Betula pendula* and *B. pubescens*. *Tree Physiol.*, **15**, 697–704.

Oleksyn, J., Modrzynski, J., Tjoelker, M.G., Zytkowiak, R., Reich, P.B. & Karolewski, P. (1998) Growth and physiology of *Picea abies* populations from elevational transects: common garden evidence for altitudinal ecotypes and cold adaptation. *Funct. Ecol.*, **12**, 573–590.

Paulsen, J., Weber, U.M. & Körner, Ch. (2000) Tree growth near treeline: abrupt or gradual reduction with altitude? *Arctic. Antarct. Alp. Res.*, **32**, 14–20.

Raich, J.W. & Nadelhoffer, K.J. (1989) Belowground carbon allocation in forest ecosystems: global trends. *Ecology*, **70**, 1346–1354.

Rolland, C., Petitcolas, V. & Michalet, R. (1998) Changes in radial tree growth for *Picea abies*, *Larix decidua*, *Pinus cembra* and *Pinus uncinata* near the alpine timberline since 1750. *Trees Struct. Funct.*, **13**, 40–53.

Sakai, A. & Larcher, W. (1987) *Frost Survival of Plants. Responses and Adaptation to Freezing Stress.* Ecological Studies 62. Springer-Verlag, Berlin.

Saugier, B. (1983) Plant growth and its limitations in crops and natural communities. In: *Disturbance and Ecosystems: Components of Response.* Ecological Studies 44 (eds H.A. Mooney & M. Godron), pp. 159–174. Springer- Verlag, Berlin.

Smith, J.W. (1920) *Agricultural Meteorology, the Effect of Weather on Crops*. Rural Text Book Series. H. Bailey, New York.

Sorensen, F.C. & Ferrell, W.K. (1972) Photosynthesis and growth of Douglas-fi seedlings when grown in different environments. *Can. J. Bot.*, **51**, 1689–1698.

Stocker, O. (1935) Assimilation und Atmung westjavanischer Tropenbaume. *Planta*, **24**, 402–445.

Taschler, D. & Neuner, G. (2004) Summer frost resistance and freezing patterns measured in situ in leaves of major alpine plant growth forms in relation to their upper distribution boundary. *Plant Cell Environ.*, **27**, 737–746.

Vapaavuori, E.M., Rikala, R. & Ryyppö, A. (1992) Effects of root temperature on growth and photosynthesis in conifer seedlings during shoot elongation. *Tree Physiol.*, **10**, 217–230.

Wardlaw, I.F. (1990) Tansley review no. 27. The control of carbon partitioning in plants. *New Phytol.*, **116**, 341–381.

Wheeler, T.R., Hong, T.D., Ellis, R.H., Batts, G.R., Morison, J.I.L. & Hadley, P. (1996) The duration and rate of grain growth, and harvest index, of wheat (*Triticum aestivum* L.) in response to temperature and CO_2. *J. Exp. Bot.*, **47**, 623–630.

Woolhouse, H.W. (1981) Crop physiology in relation to agricultural production: the genetic link. In: *Physiological Processes Limiting Plant Productivity* (ed. C.B. Johnson), pp. 1–21. Butterworths, London.

4 Temperature and plant development: phenology and seasonality

Annette Menzel and Tim Sparks

4.1 The origins of phenology

The recording of the timing of life-cycle events has only recently been considered as an area of climate impacts research. For a much longer period, phenology has been recorded by those with an interest in natural history, by those engaged in agriculture and horticulture and where traditional local festivals have been associated with plant phases.

Some plant species and some phases are more apparent than others. Hence the brilliant displays of cherry fl wering at the Royal Court in the former Japanese capital of Kyoto or of peach fl wering in Shanghai are very obvious and are associated with local festivals. Flowering of forsythia, for example, is much more obvious than that of beech trees. In Europe, religion and folklore may associate some plants with specifi calendar dates: for example, daffodil fl wering with St David's Day (March 1), snowdrop fl wering with Candlemas (February 2) and the Devil spitting on blackberries on the night of October 10. Flowering of other species is of considerable importance for tourism, such as of fruit trees in south-eastern Norway, of crocuses at Husum, Germany, and of tulips in the Netherlands. Given these facts, it is not surprising that the emphasis in traditional plant phenology is biased towards trees and towards plants with obvious fl wers, and may have a different emphasis in different countries. At a later date, the importance of phenology to assess environmental conditions for annual and perennial crop cultivation was recognised.

The life cycles of most deciduous plants go through recognisable phases, e.g., leafing fl wering, fruiting, leaf colouration, leaf fall, bare. For some species it is possible to sub-divide these broad categories, for example firs fl wering, 50% fl wering and end of fl wering. However, phenology has traditionally been involved with easy-to-record events where fewer opportunities exist for individual interpretation. As a consequence, events such as firs leafin and firs fl wering dates are by far the most popular. This does not mean that there is no room for inconsistency as the time at which complex leaves and fl wers open may be subject to interpretation. Species do differ in their dates of phenological phases and the order in which these events occur. For example, in Table 4.1 it is obvious that ash fl wers before leafing and loses its leaves early, in contrast to oak.

The oldest known phenological series is that of the Kyoto cherry (*Prunus jamasakura*) f owering, mentioned above (Menzel, 2002b). Data on this series stretch back to the year 705 AD. Some Chinese series stretch back to the sixteenth

Table 4.1 Average dates of phenological phases of ash (*Fraxinus excelsior*) and oak (*Quercus robur*) in Worcestershire, UK, in the early twentieth century*

	Ash	Oak
First leafin	May 11	April 29
Full leafin	May 24	May 11
First fl wer	April 19	April 30
First tint	September 16	September 18
Full tint	October 15	November 4
Fruit ripe	September 27	October 4
Bare	October 29	November 26

*Recorded by F. Lowe.

century or earlier (Hameed & Gong, 1993). Within Europe, data sets exist from the eighteenth century onwards, with a few also from the fifteent century. Two examples of this are the very long record of wheat harvest dates in Sussex from 1769 to 1910 (Figure 4.1) and grapevine harvest in France, Switzerland and Rhineland (Figure 4.2a). A slightly later series on horse chestnut leafin dates in Geneva commenced in 1808 and continues to the current day (Defil & Clot, 2001). The latter has shown considerable variation in timing of 110 days with a steady advance from the beginning of the twentieth century to the current day. The mean date of leafin around 1900 was early April and is currently the end of February. With a series such as this, it is inevitable that heat and light pollution in the city will have had some impact (Rötzer *et al.*, 2000) on leafin dates over and above that which would occur in the countryside. Considerable variation in vegetation development can also be seen in photographs of plants taken on fi ed calendar dates, for example by Willis (1944).

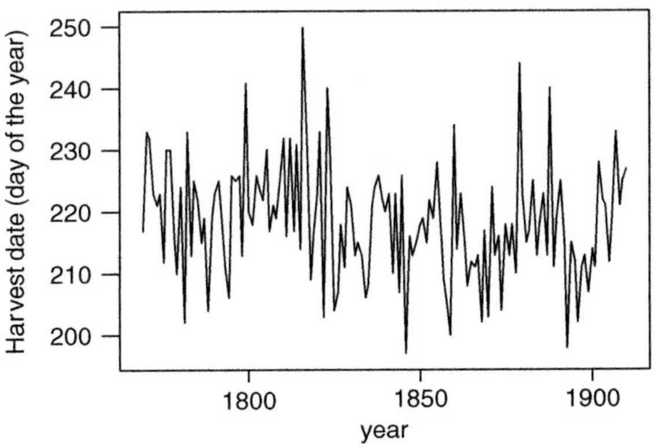

Figure 4.1 Wheat harvest dates in Sussex, England, from 1769 to 1910 (based on data in Russell, 1921). Considerable variation in harvest date (~50 days) exists in a period when no rapid agricultural advances, for example plant breeding, were taking place. Year-to-year variations in harvest date correlate very closely with mean May–July temperatures ($r = -0.710$, $P < 0.001$).

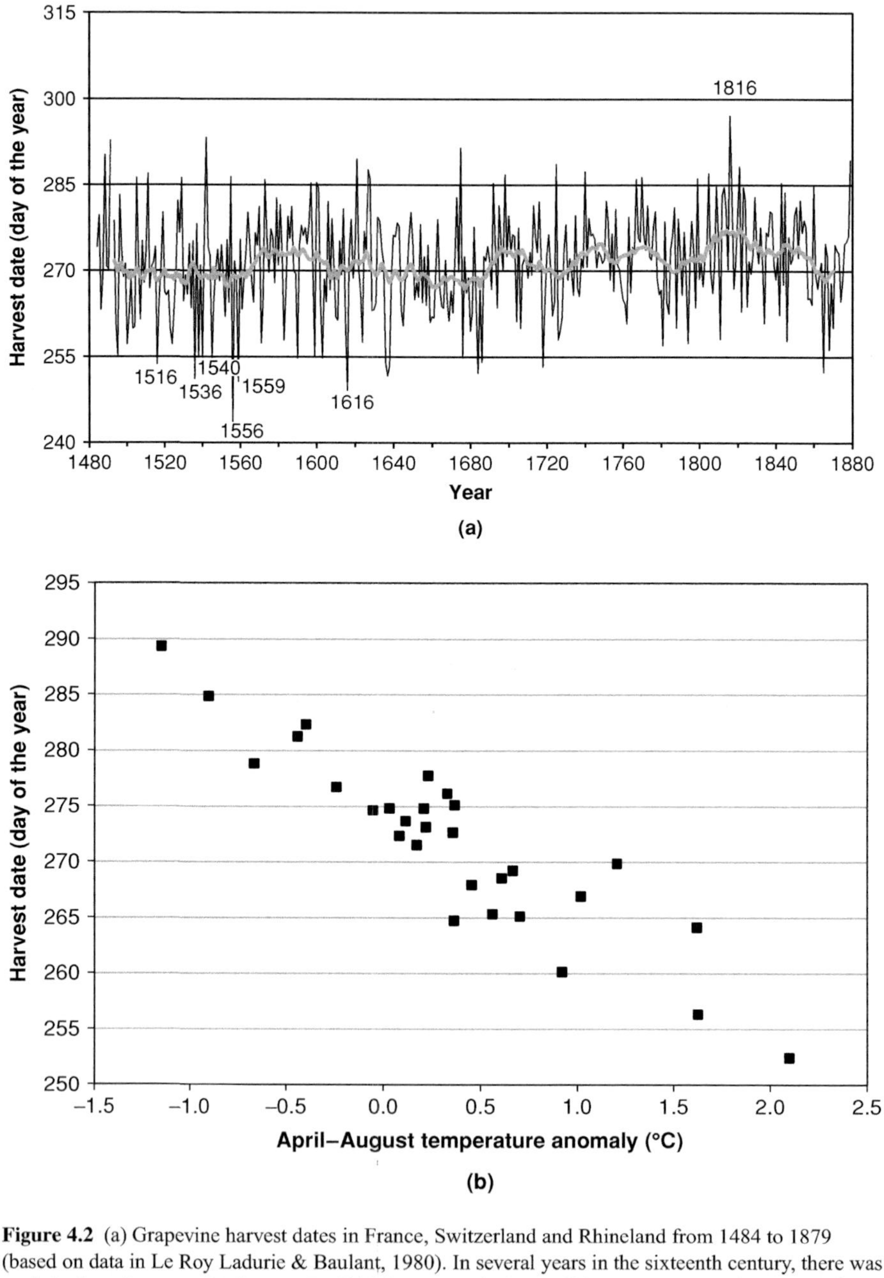

Figure 4.2 (a) Grapevine harvest dates in France, Switzerland and Rhineland from 1484 to 1879 (based on data in Le Roy Ladurie & Baulant, 1980). In several years in the sixteenth century, there was a relatively early grapevine harvest. In 1816, known as the 'year without a summer' after the volcanic eruption of Tambora, Indonesia, the harvest started very late. (b) Year-to-year variations in grapevine harvest date correspond very closely to mean April–August growing season temperatures (averaged for 0–10°E and 45–50°N (France, Switzerland, southwest Germany), Jones *et al.*, www.cru.uea.ac.uk/cru/data/temperature/, 1851–1879, $R^2 = 83.9\%$, $P < 0.0001$; Menzel, 2005).

From the late nineteenth century, phenological recording became more systematic and more organised. In the United Kingdom, the Royal Meteorological Society started a phenological network in 1875 that was to last until 1947. This scheme specificall requested firs fl wering dates of a range of plant species from hazel to ivy, right through the season. The scheme expanded to include other plant and animal events as time passed. Some tree leafin dates were included but were less frequently recorded than fl wering. The British Naturalists' Association began a scheme among its members in 1905, which continues on a small scale until the current time.

In Germany, Professors Hoffmann and Ihne began to coordinate records from across Europe in 1882 (Hoffmann & Ihne, 1882), which was to last through to 1941 (Ihne, 1883–1841). This pre-computer, pre-email collaboration is a perfect and lasting example of both meticulous coordination and ideal international collaboration. In addition to all these sources of data a large number of individuals have maintained records of events that are of specifi interest to them and also to the current phenological recording initiatives.

Undoubtedly further important sources of phenological data exist in obscure books, and more ephemeral diaries and manuscripts. These are not recorded on electronic catalogues and painstaking detective work is required to identify them. These historic data can be very important in many ways. They can provide a baseline against which to assess current phenology, and they allow us to examine the historical reaction of species to temperature and other climatic variables at a time in history when many other environmental factors were relatively stable. An example of such an obscure source is a manuscript summary of the fruit ripening dates of strawberry (among other events) held by the Linnean Society of London in its library (Figure 4.3). As mentioned above, records such as these were taken for personal

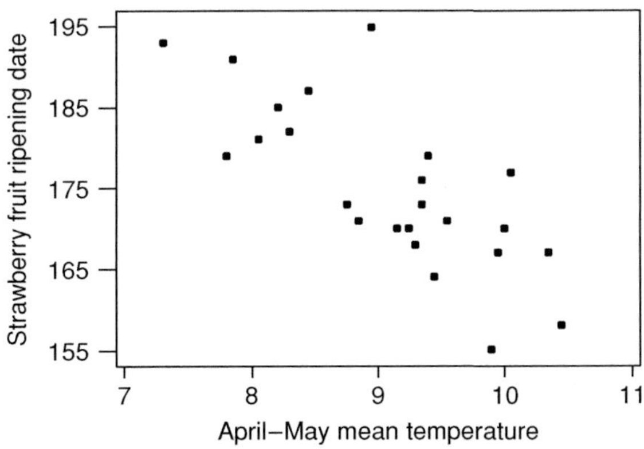

Figure 4.3 The fruit ripening dates of strawberry recorded by J. Roger Dutton (1874–1902) in relation to mean April–May central England temperature. (Data source: manuscript in library of the Linnean Society of London.)

interest, or for the calculation of 'calendars' of gardening or natural history. Only in the last two decades has it become apparent that they provide some of the best documented evidence of response to global warming. The example of grapevine harvest dates provides an even better correlation of harvest dates with growing season temperatures (Figure 4.2b).

In the current recording schemes, operated by national weather services (for example in central and eastern European countries) as well as by newly installed or revived networks (for example United Kingdom, the Netherlands), it is mostly native wild species, crops and fruit trees that are observed (for a more detailed review, see Menzel, 2003b). Sometimes animal phenology is also included. The discernable stages in the life cycle of plants, so-called phenophases, comprise, e.g., bud burst, beginning of fl wering, full fl wering, leaf unfolding, fruit ripening, leaf colouring and leaf fall. In order to reduce possible subjectivity in the observations, phenological manuals describe the procedure and defin phenophases, with graphs or pictures often accompanying the text. For agricultural and fruit tree species, there exists the problem of genotypic changes due to plant breeding, and in short-lived wild plants adaptation might possibly confound observed temporal and seasonal changes. Thus, long-lived tree species are especially useful as they will not show genotypic change during at least medium term time series.

The following attempts have been undertaken to improve phenological monitoring, especially in regard to their recent use in climate change studies:

1. The homogenisation of recording by further development of common guidelines (e.g. by the World Meteorological Organisation) (www.cost725.org)
2. Exact definition of plant growth stages (e.g. by the BBCH code, which is a system for uniform coding of phenologically similar growth stages of all monocotyledonous and dicotyledonous plant species, Federal Biological Research Centre for Agriculture and Forestry, 1997)
3. Define location of observations (e.g. in phenological gardens) in conjunction with:
4. The use of cloned plant material instead of native plants (e.g. IPG, International Phenological Gardens, which were founded by Schnelle & Volkert, 1957).

4.2 Recent changes in phenology

Recently published summaries of climate impacts (Walther *et al.*, 2002; Parmesan & Yohe, 2003; Root *et al.*, 2003) have emphasised the importance of phenology in demonstrating the impacts of climate warming on the natural world. These impacts have been, and continue to be, considered in the evidence-gathering activities of the Intergovernmental Panel on Climate Change (IPCC; McCarthy *et al.*, 2001). It has been acknowledged that there is a bias in the availability of results towards the

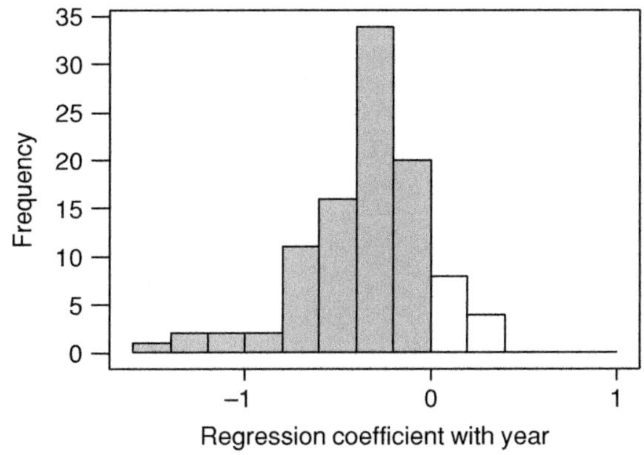

Figure 4.4 A histogram of the regression coefficient in fl wering date of the 100 plant species reported by Abu-Asab *et al.* (2001) for the period 1970–1999. Shaded bars indicate negative trends, i.e. towards earlier fl wering.

Northern Hemisphere and particularly to Europe and North America. Attempts are being made to rectify this imbalance.

Evidence exists that change in phenology is happening in both cultivated (e.g. Menzel, 2000b; Chmielewski *et al.*, 2004) and native plants (e.g. Fitter & Fitter, 2002), in Europe (e.g. Menzel & Fabian, 1999), North America (e.g. Abu-Asab *et al.*, 2001) and Japan (e.g. Matsumoto *et al.*, 2003). Several published studies incorporate results from many species and all sites in networks, and thus are not selective and not biased towards results that indicate an advance in phenology. These papers are of especial importance because they give a broader view of the overall change.

Data from Abu-Asab *et al.* (2001) allow us to look in detail at the overwhelming change towards earlier fl wering (Figure 4.4). The ratio of negative-to-positive change is much greater than would be expected by chance (sign test $P < 0.001$). The situation is similar, but not as marked for the fl wering data reported by Fitter and Fitter (2002) where 69% of the 385 plant species show a negative trend, i.e. earlier fl wering. This is still much greater than would be expected by chance (sign test $P < 0.001$).

We can look at the Abu-Asab *et al.* (2001) data in more detail. Figure 4.5 shows the regression coefficient converted to estimated changes in fl wering dates over the 30-year period, plotted against mean fl wering date. It is evident from this that, despite the greater variability in earlier events (see Figure 4.5), earlier fl wering species have changed more than later fl wering species. The regression line plotted on this graph is significan at $P = 0.019$.

Temperature is the most likely factor to be driving these changes; an initial assessment suggests that in the Fitter and Fitter (2002) data set about 83% of the 385 species exhibit a significan relationship with mean monthly air temperature.

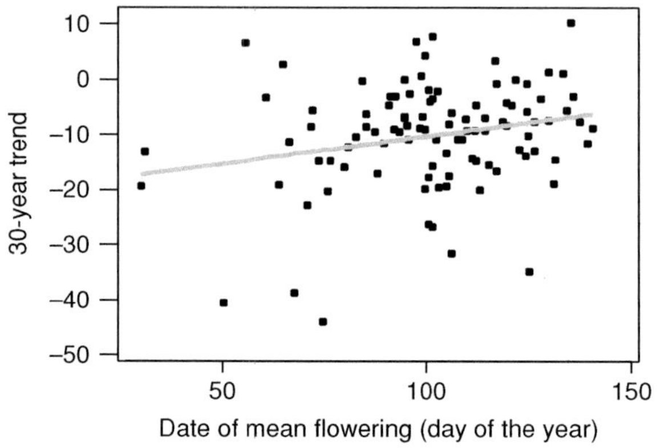

Figure 4.5 30-year trends in the data set reported by Abu-Asab *et al.* (2001) for 100 species, plotted against mean fl wering date.

The most comprehensive assessment of leafin dates across Europe was reported by Menzel and Fabian (1999) based on cloned trees grown in the IPG network over 30 years. From 616 spring time series, an average advance of 6 days over the 30 years was apparent (Figure 4.6a).

In northeast Spain, Peñuelas *et al.* (2002) reported an overwhelming trend towards earlier leafing significan for 24 of 25 examined species (Figure 4.7a), advancing by an average of 20 days over 48 years. Advances in fl wering date (not summarised here) were also reported.

For autumn events fewer data are available, but fruiting tends to be advanced and leaf colouration and leaf fall is most likely delayed by increasing temperatures. Menzel and Fabian (1999) reported that, of 178 autumn series, a delay of 5 days over the 30 years of data was apparent (see Figure 4.6b). Peñuelas *et al.* (2002) reported a consistent trend towards earlier fruiting in 27 species in northeast Spain in the period 1952–2000, averaging 8 days (Figure 4.7b), and a consistent trend towards later leaf fall (Figure 4.8), averaging 13 days later.

From this summary of reported changes it is clear that there has been a marked advance in leafing fl wering and fruiting dates of plants in the last half-century and a delay in leaf fall. One of the consequences of this is an increase in the length of the growing season. This affects not only native species as outlined here but also forestry and agriculture (e.g. Chmielewski *et al.*, 2004; Williams & Abberton, 2004).

What is readily apparent is that the trends reported are not homogeneous in terms of both location and species. It is the latter that is of most concern. It is quite clear that at a single location there have been a range of responses to a warming climate; some quite dramatic and some negligible and some even opposite to that expected. Fitter and Fitter (2002) reported that annual species responded more than perennials, insect-pollinated species responded more than wind-pollinated species, and those close to the centre of their distributional range responded more than range edge

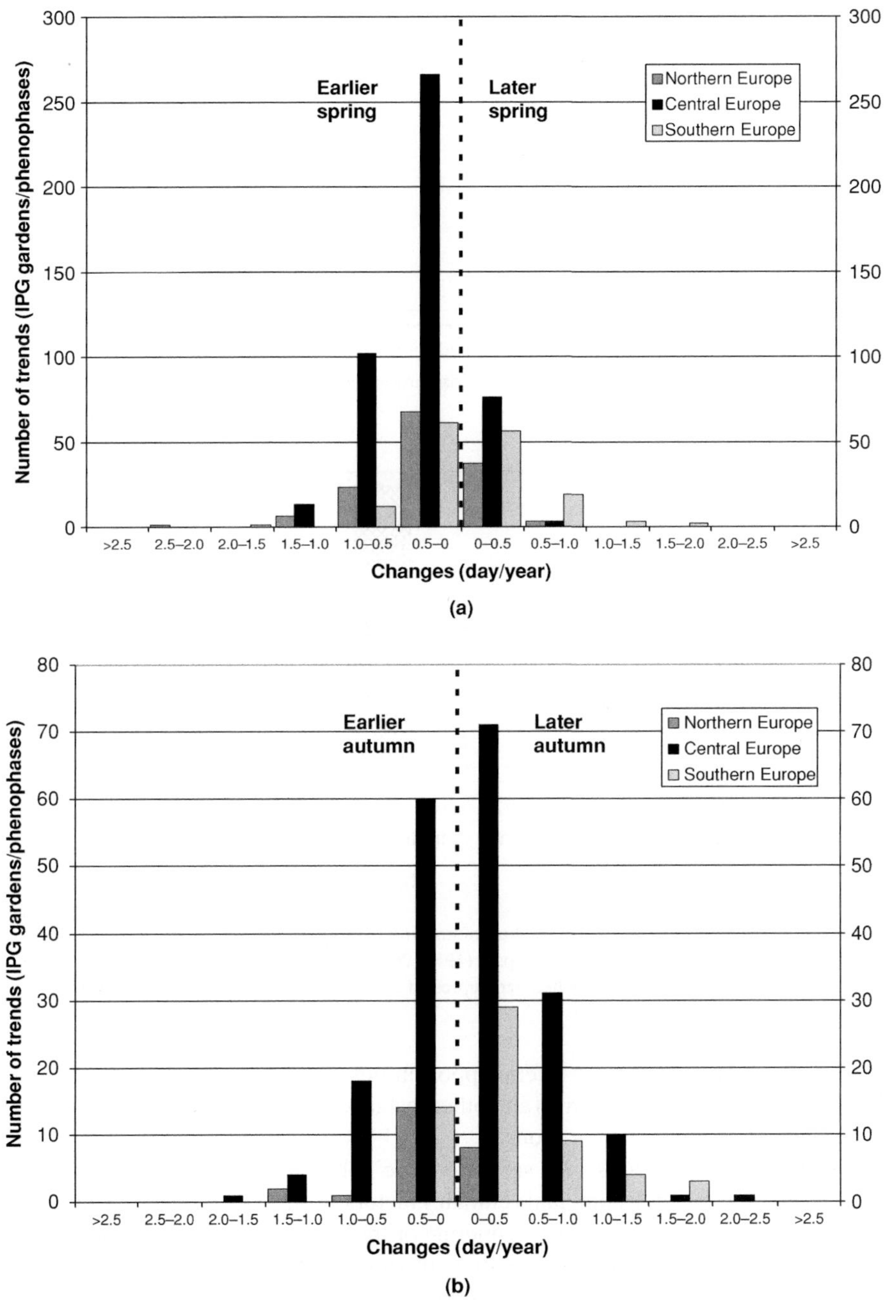

Figure 4.6 20+-year trends in the IPG data (1959–1998) for (a) spring and (b) autumn, updated following Menzel and Fabian (1999) and Menzel (2000a).

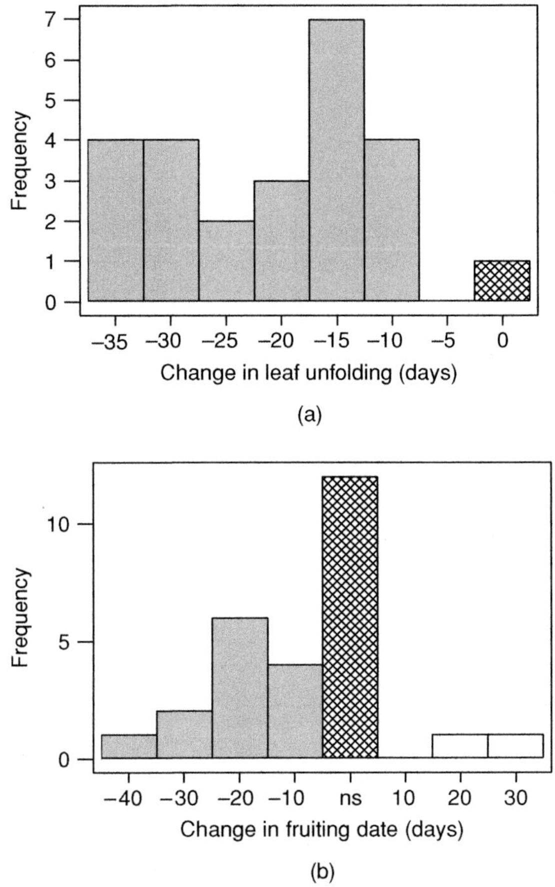

Figure 4.7 A summary of change in (a) leaf unfolding and (b) fruiting dates as reported by Peñuelas *et al.* (2002), based on data from northeast Spain (1952–2000). The shaded bars represent species with significan advances and the cross-hatched bars represent species where no significan trend was detected.

species. They also pointed out several opportunities where hybridising species were separating in their fl wering times and other instances where potentially hybridising species were becoming synchronous.

Among all phenophases, we separate 'true' phenological phases, which are mainly triggered by environmental (climate) factors, from 'false' phases, which are under the influenc of humans for economic or traditional reasons, e.g. sowing and harvesting of agricultural crops. The most obvious example here is grapevine harvest (see Figure 4.2a), in former times taking place on a preferential day of the week (Pfiste *et al.*, 1999) or crop harvest dates, which depend in recent times on the availability of rented harvesters or the correct ground conditions. Figure 4.9 shows the mean onset of phases of winter wheat (*Triticum aestivum*) in Germany (1951–1998). The beginning of rapid growth in height (mean May 8, trend −0.26 days/year,

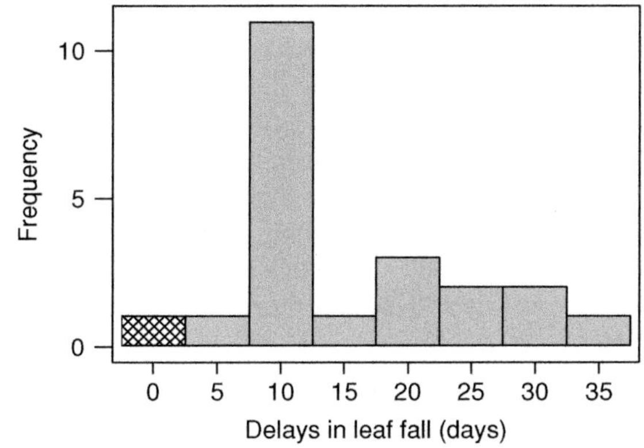

Figure 4.8 A summary of change in leaf fall dates as reported by Peñuelas *et al.* (2002) based on data from northeast Spain (1952–2000). The shaded areas represent significan delays in leaf fall dates and the cross-hatched areas represent where no significan trend was detected.

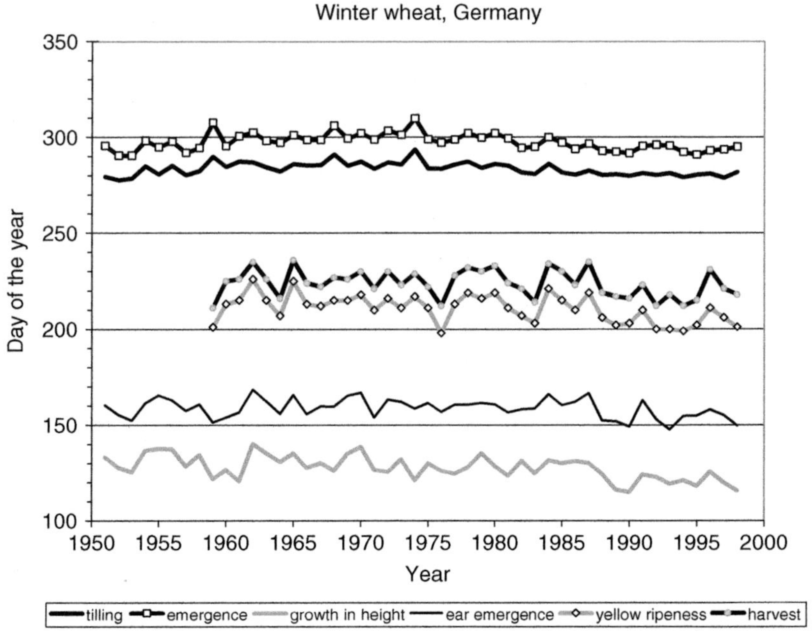

Figure 4.9 Mean onset of phenophases of winter wheat observed in Germany (1951–1998) (after Menzel, 2000b).

$P < 0.01$), beginning of ear emergence (mean June 9, −0.10 days/year, $P < 0.05$) and beginning of yellow ripeness (mean August 1, −0.30 days/year, $P < 0.01$) in spring and summer, as true phases which react to temperature changes, have clearly advanced. In contrast, the beginning of harvest (mean August 14, −0.16 days/year, $P < 0.10$), tilling/sowing in autumn (mean October 13, −0.07 days/year, $P < 0.07$) and the beginning of emergence (mean October 28, −0.07 days/year, $P < 0.11$) have advanced by less and not significantl so (Menzel, 2000b). Thus, in general, true phases can be used as bioindicators of climate change; many of the false phases, however, may also be of importance as they identify human-induced adaptation processes.

Network data have shown spatial variability, with differences between sites apparent. At particular sites, the response of different species is distinct, often combined with a strong seasonal variation (highest advances in early spring to almost no response in summer and early autumn). Comparatively few studies have been performed with autumn phenological phases, such as leaf colouring and leaf fall, but in this season temporal changes seem to be less pronounced and show a more heterogeneous pattern.

Large-scale studies reveal regional differences, e.g. in Europe the phenological shifts are more pronounced in the western maritime areas than in the eastern continental ones (Ahas *et al.*, 2002; Menzel *et al.*, 2005).

4.3 Attribution of temporal changes

Our interests in this book lie in the relationship between phenology and climate. The use of phenology as a biological indicator of climate change presupposes precise quantitative analysis of changes in phenological time series and a known relationship with temperature. We will thus ignore general modelling applications of phenology, for example in the timing of agricultural cultivation, pest control or pollen warning forecasts. To this end we can restrict ourselves to examining following three questions:

1. How can we properly detect phenological changes?
2. How can we attribute year-to-year changes in phenology to temperature and other factors?
3. Are there other confounding factors?

There will inevitably be other questions that will arise as a consequence of intensive phenological study. There are several ways to approach any problem, and this is plainly true in this section.

4.3.1 Detection of phenological change

The most common approach to detection of phenological change is a simple regression of phenological phase on year, followed by an examination of the resulting regression coefficien for statistical significance Negative coefficient imply an advance in the phenological phase and positive ones imply a retardation. Regression such as this relies on a number of assumptions that include independence in data points and residuals that follow the normal distribution. If these assumptions are not met, then regression coefficient will be unaffected but significanc levels may be inflated In our experience, autocorrelation (correlation between successive data points) is not a serious problem and can usually be ignored. Alternative analyses include a whole raft of time series methods (to accommodate autocorrelation) and non-parametric regression methods.

The detection of significanc is affected by three factors: the strength of any true trend, the number of data points examined and the background variability in the data. We (Sparks & Menzel, 2002) have recommended 20 years as an appropriate length of series to detect effects, and Sparks and Tryjanowski (2005) have given examples of the problems of start year, end year and series length on the conclusions that may be drawn. The background variability appears to be greater in early season species than in later season ones (Figure 4.10), which would imply that trend detection would be harder in early species. Figure 4.11 demonstrates this change in the fl wering date of daffodil. There is a clear change in fl wering date, which may be more of a step change than the straight-line fi suggests, but linear regression is still one of the helpful tools in detecting significan change.

A new method for the analysis of long-term phenological time series, based on Bayesian concepts, was recently introduced by Dose and Menzel (2004). Compared to traditional trend analysis by linear regression, phenological time series are analysed by Bayesian non-parametric function estimation, which allows a quantifie comparison of different models to describe their functional behaviour. A model with two linear segments with one change point (one change point model, example in Figure 4.12) is compared to less sophisticated alternatives; a zero trend in the data (constant model) and a constant trend over the data (linear model).

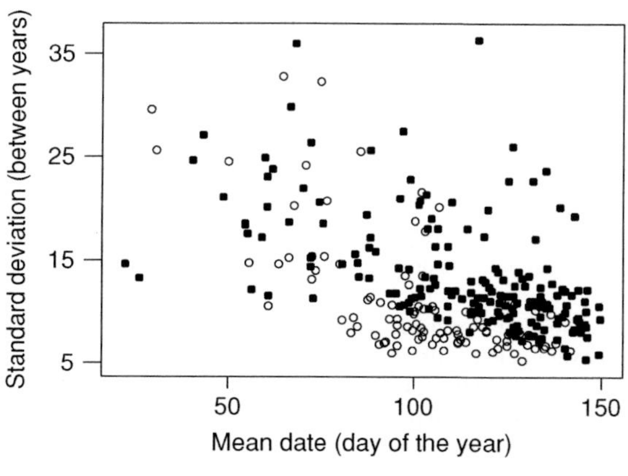

Figure 4.10 Relationship between year-to-year variability and mean fl wering date is apparent in examples of spring fl wering plants from Washington, DC (Abu-Asab *et al.*, 2001; open symbols) and Oxfordshire, UK (Fitter and Fitter, 2002; solid symbols).

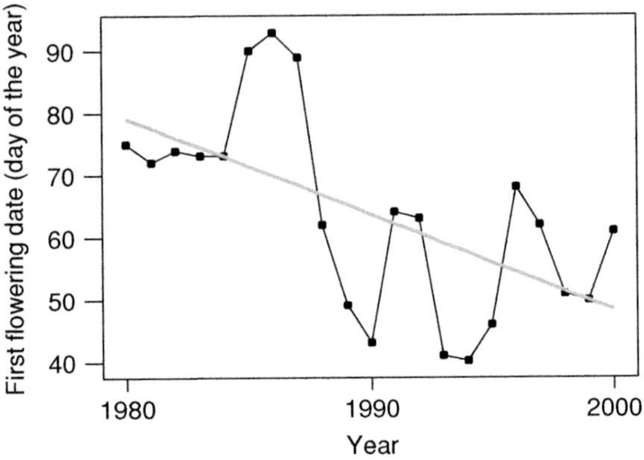

Figure 4.11 First fl wering dates of daffodil in Sussex, UK (1980–2000). The straight line represents the regression of date on year ($b = -1.54$) and is very significan ($P = 0.004$).

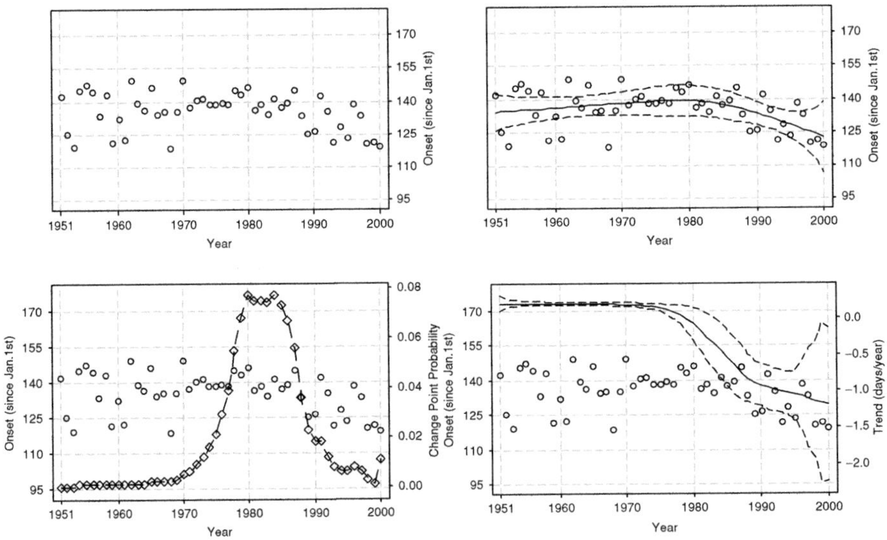

Figure 4.12 Bayesian analysis of fl wering of lilac (*Syringa vulgaris*) at Grünenplan, Germany (1951–2000). On the upper left plot, the original data series is plotted with a notable scatter of the data, and these data points are superimposed on the remaining graphs. On the lower left plot the change point probabilities (i.e. the likelihood of a gradient change in the data series), normalised to unit area, are displayed with a clear maximum in the firs half of the 1980s. The rate of change, determined by the one change point model (lower right plot), reaches −1.19 days/year; the confidenc ranges indicate advancing fl wering, clearly different from zero change. On the upper right plot, the average functional behaviour and its confidenc ranges estimated from the one change point model are plotted (Schleip, 2005).

Temporal changes are clearly detected by analysis of their development and respective change point probabilities. The most important aspects of the method are a rigorous treatment of uncertainties, e.g. of trend (days/year) or functional behaviour, and the possibility of prediction of missing and future data with associated uncertainties. Figure 4.12 displays the results for the analysis of a 50-year record of fl wering of lilac at Grünenplan, Germany. The one change point model is preferred by 97.9%, the linear model by 1.6% and the constant model by 0.5% likelihood. The analysis of change point probabilities for the one change point model reveals a clear maximum in the firs half of the 1980s. The average functional behaviour is a steady, modest delay of fl wering till the mid 1980s and then a sharp advance. The resulting trends reach −1.19 days/year, clearly different from zero.

For many parts of central Europe, this example may be typical of the temporal variability of changes, as over most of the last century the trend is almost zero; however, from the mid 1980s onwards, the rate of change is clearly negative, indicating a discontinuous shift towards earlier occurrence dates (see Dose & Menzel, 2004; Menzel & Dose, 2005).

4.3.2 Attribution of year-to-year changes in phenology to temperature and other factors

The earlier onset of spring and summer is closely related to change in temperature. This can be demonstrated experimentally, with the support of physiologically based models of plant development in spring or with simple statistical relationships.

The common approach here uses simple or multiple regressions to relate the date of the phenological phase to a number of potential explanatory variables. To the dangers mentioned in the section above, for example genotype, we must add a new danger: that we have so many variables that some achieve significanc by chance alone (Sparks & Tryjanowski, 2005). We would recommend the reduction of variables to those that are logical and relevant, and cover an appropriate timeframe. This sifting of variables will reduce the number of chance significances although it may also fail to identify some unforeseen influenc on phenology. The response to temperature is well understood and accepted since the onset of spring and summer events, and consequently the length of the growing season is very sensitive to climate and weather (Sparks *et al.*, 2001; Menzel, 2003a).

The phenological clock in Figure 4.13 does not only display the mean onset of phenological seasons (1985–2000) at Geisenheim (Germany), but also displays the schematic partitioning of the year into growth and non-growth periods. During endo-dormancy or (true) dormancy, the plants need cool or chilling temperatures to break this stage; during exo-dormancy or rest, warm temperatures – so-called forcing – lead to bud break. As long as the chilling requirement of the plants is fulfille and the dormancy is broken, an increase in winter and spring temperatures will lead to an advancement of spring (and summer) phases. Thus, the observed temperature increase during recent decades has resulted in advancing spring and

Figure 4.13 The phenological year in a clock. Selected phenological phases determine the start of phenological seasons (1–9, outer two rings; example is for the 1985–2000 period at Geisenheim, Germany). The activity of the plant is physiologically divided up in the period of dormancy (para-, endo- and exodormancy) and growth (1–4, second ring). Corresponding to this, the frost hardiness of plants changes (third ring).

summer phenophases. In many parts of Europe (mainly central Europe) the response of onset dates to mean monthly temperatures of the preceding months is almost linear; however, there is the question whether this might change in extreme years and result in a more sigmoidal (s-shaped) relationship. The retrospective analysis of observation dates and temperatures does not allow an assessment of the possible consequences of highly variable, abrupt and totally extreme changes, which can be tested only by experiments.

Data for explaining phenological change are becoming easier and easier to obtain and are available at higher temporal and spatial resolutions. The most commonly available data are mean monthly air temperature, total monthly rainfall and monthly or seasonal indices of the North Atlantic Oscillation (NAO). It is now easier to obtain minimum air, maximum air and soil temperatures, sunshine hours and other climatic variables. The availability of daily data has encouraged some researchers to break away from fi ed calendar monthly periods. In our experience the use of monthly mean air temperatures usually produces acceptable results such as those shown in Figure 4.14 ($R^2 = 73.1\%$) or Figure 4.2b ($R^2 = 83.9\%$).

In general, there is overwhelming evidence from numerous studies that spring and summer phases are to a very large extent influence by the temperature of the preceding 1–3 months, as once the plants have experienced sufficien cold to

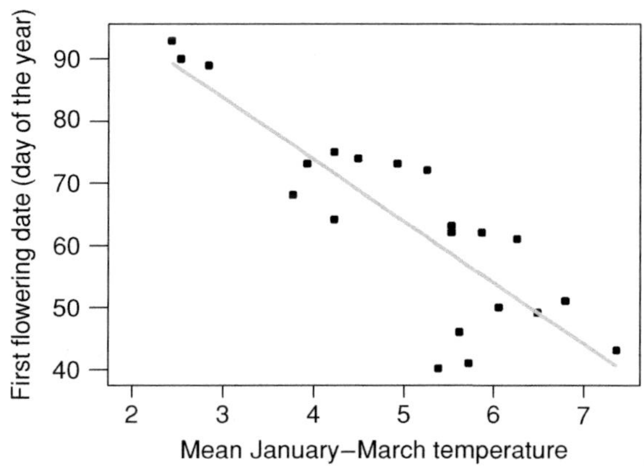

Figure 4.14 The daffodil data presented in Figure 4.11 plotted against mean January–March central England temperature. The regression coefficien suggests a 1°C increase in temperature would advance fl wering by 9.9 days and is highly significan ($P < 0.001$).

overcome their winter rest, warmer temperatures are all they need to start sprouting or fl wering. The resultant relationship with temperature is often almost linear (see Figures 4.14 and 4.2b). Spring temperatures also play a decisive part in determining the time at which the fruit ripens in summer and autumn, as well as the duration of the entire growing season. In order to prevent too early a break of dormancy or too late an induction of dormancy in autumn, photoperiodism (day length) plays an additional role in triggering phenological events. As discussed in Chapter 3, photoperiodism constrains the influenc of temperature on development to 'safe periods'. However, it seems that in many regions dormancy is broken by photoperiod and sufficien chilling, and only the subsequent warmer conditions trigger the onset dates of bud burst or other spring phases. Chilling requirements are quite modest relative to heat requirements so that most of the delay in phenology caused by reducing chilling under climate warming will likely be swamped by advanced phenology arising from spring warming.

The situation in autumn is much more complicated. There is no lack of general hypotheses about which environmental factors might trigger the autumnal phenophases, such as leaf colouring and leaf fall (Estrella & Menzel, 2006). Nevertheless, no physiologically based model is yet available that may simulate observations in this season. In addition, the attribution of observed changes to potential factors by statistical methods is more complex. There are some indications that warm weather in May and June will advance leaf colouring, whereas a warm late summer and early autumn in August and September delays leaf colouring (Menzel, 2003a). Other possible triggering factors, such as drought, soil moisture, atmospheric pollutants, constant growing season and critical day length, resist detection by statistical analysis.

New research, also by Bayesian analysis of time series, is investigating whether phenological and temperature records should be treated as coherent or incoherent (Dose & Menzel, in press).

Modelling the timing of early season plant phases, particularly in the agricultural sector, relies heavily on variants of accumulated heat units or growing degree days. Well-known examples in agriculture are 'Ontario units' in maize production (e.g. Easson & Fearnehough, 2003) and 'TSum200' in grass production. Most of these models accumulate daily mean or maximum and minimum temperatures above a threshold (which may be zero) from a starting date to predict a particular plant phase. The choice of the starting date and the threshold should be selected to optimise the relationship with the plant phase, i.e. to minimise the year-to-year variability in the accumulated heat units. In practice, the starting date may be selected for convenience and consistency between species (e.g. January 1) as may the threshold temperature (e.g. 0°C or 5°C).

Variants on these models include the need for accumulated chilling in autumn (again with a variable start date and threshold temperature) for vernalisation or the balance of chill and heat units. A summary of models is given in Chuine *et al.* (2003). In practice, temperatures are taken from a nearby met station and will be recorded in a screen at a given height above the ground. As such, they will at best approximate the temperatures the plant experiences. In woodland environments there will be considerable differences in temperature from the woodland floo to the canopy. Improvements to models may be achieved by using soil temperatures in some circumstances (e.g. Sparks *et al.*, 2005). Optimisation may reveal a range of starting dates and threshold temperatures with similar properties, and the selection of model parameters with wider applicability ('portability') should be sought. A pragmatic attitude to modelling is essential.

Some phases, such as the beginning and duration of the growing season in northern and central Europe, correlate very well with the NAO index. The NAO is an atmospheric circulation index describing a major driving force of the Northern Hemisphere climate system, which largely determines the interannual variation of winter temperatures in the northern Atlantic region (Hurrell, 1995). NAO reflect the weather conditions in winter, as such a positive index is linked to warm wet conditions in northern and central Europe. Phenological phases in these areas respond to higher NAO indices by earlier spring onset (advancing onset of leaf unfolding, fl wering and a lengthening of the growing season) (e.g. Chmielewski & Rötzer, 2001; Menzel, 2003a). Spatial studies for central (Scheifinge *et al.*, 2002) as well as central and eastern Europe (Aasa *et al.*, 2004) indicate a decreasing influenc of the NAO with increasing distance from the North Sea. Menzel *et al.* (2005) also reveal that the rate and the pattern of spring and summer progression through Europe is linked to the NAO. Figure 4.15 demonstrates this influenc of NAO index on the spring progression in Europe: the patterns of mean late spring onset in Europe for the 10 years with the highest NAO index and those for the 10 years with the lowest NAO index (November–March) are clearly different. Progress of spring phenological events run southwest to northeast in years with high NAO index and south to

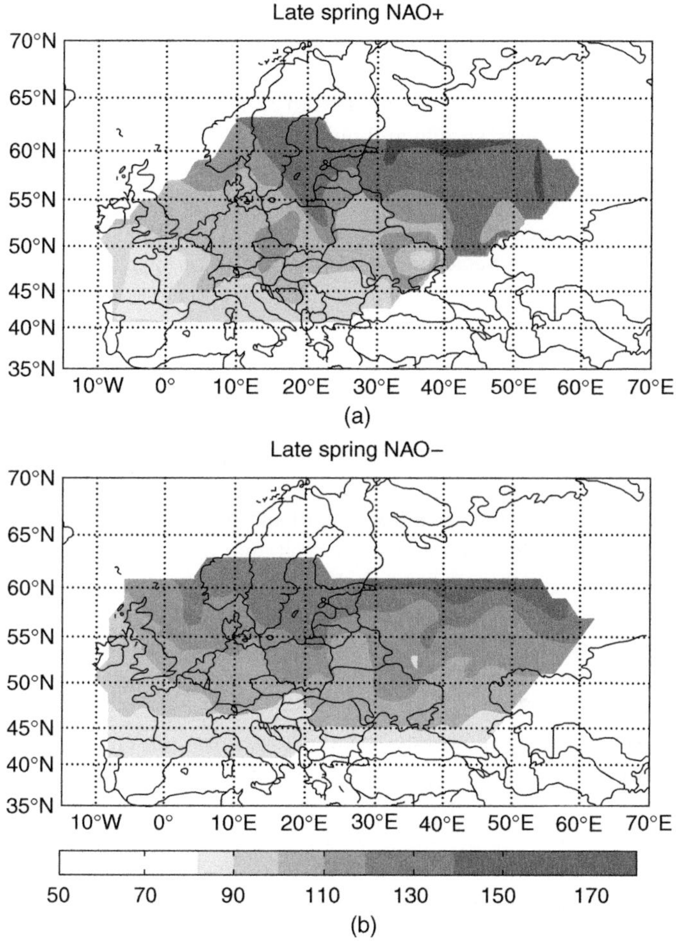

Figure 4.15 The mean onset of late spring (days of the year) in Europe for (a) the 10 years with the highest (1990, 1882, 1928, 1903, 1993, 1910, 1880, 1997, 1989, 1992, NAO+) and (b) the 10 years with the lowest (1969, 1936, 1900, 1996, 1960, 1932, 1886, 1924, 1941, 1895, NAO–) NAO winter and spring index (November–March) in the period 1879–1998 (after Menzel *et al.*, 2005).

north in years with low NAO index despite the known fact that the onset of spring phases in years with high NAO index is advanced.

4.3.3 Confounding factors

Analyses of phenological data series reveal evidence for climate change impacts: the times at which phenological events take place are very closely related to climate and weather conditions, especially in spring and summer (e.g. Sparks *et al.*, 2001; Menzel, 2003a). Unlike other observed changes in ecosystems, temperature is the factor of decisive influenc here with the result that phenology is probably the simplest way of detecting the effect of changes in temperature in temperate and boreal zones (Sparks & Menzel, 2002; Walther *et al.*, 2002).

Although attributing the observed changes to climate change for spring and summer conditions, is relatively simple and quite easy to understand, multiple forcing by numerous environmental factors, in particular in autumn, render the attribution tricky as plants are 'integrating measuring instruments' for all weather conditions. Thus, the plant responses may sometimes be non-homogeneous due to local microclimate conditions (see Section 4.2), natural variation, genetic differences (see Section 4.1) or other non-climatic factors. In addition to the current weather, weather conditions in the present and preceding growing season, as well as in the dormant season, a plant's phenological reaction can also be affected by the soil, nutrient application and availability, competition, genetics, pollutants and/or pests. Separating out the various potential causes of phenological variation can be problematic, often requiring data covering the full spectrum of conditions and by examination of partial regression coefficients

In many mid and higher latitude regions of the Northern Hemisphere the soils benefi from winter precipitation and snow melt, and thus are saturated with water in spring. However, in other regions such as the Mediterranean, phases may be triggered by drought. Peñuelas *et al.* (2002) found for the Mediterranean region that a relationship clearly exists between precipitation and the commencement dates of some species that are less resistant to drought, as well as farm crops that are not irrigated.

The influenc of rising atmospheric CO_2 concentrations on the phenology of plants can only be examined by experiments, as analyses of observational records do not allow a strict separation from other climatological factors. For example, Murray *et al.* (1994) found, in an experiment with Sitka spruce seedlings, that increased plant nutrient supply lengthened the growing season due to both earlier start and later end, whereas elevated CO_2 concentrations delayed bud burst in spring.

4.4 Evidence from continuous phenological measures

Besides phenological data, which are collected by observations on research plots or in phenological networks, there are also many indirect sources of phenological information, particularly satellite data, atmospheric CO_2 mixing ratios, climatological (e.g. frost-free season) as well as meteorological derived measures (e.g. by eddy covariance techniques or radiation measurements).

Compared to traditional, direct, 'individual' phenological observations, satellite data or analysis of CO_2 signals provide spatially and species-averaged information. Phenological observations mostly have daily temporal resolution whereas remote-sensed data are coarser (10 or 14 day MVC, maximum value composites). Remote sensing also scales up from specific site-restricted observations to coarse, area-averaged, regional scale information; from define plant species to pixels with more or less homogenous vegetation types. Precise phenophases, such as beginning of fl wering, are replaced by derived measures for seasonal phenology (e.g. start of a season).

The role of remote sensing in phenological studies is increasing as this allows the study and description of seasonal phenomena, such as start, duration and end of the growing season over larger areas. This information is critical in global vegetation models or coupled atmosphere vegetation models, especially for the timing of greening up in spring, as it determines the start of CO_2 assimilation and transpiration. Among various measures for greenness or vegetation indices (VIs) derived from remote-sensed data, the most commonly used is the NDVI (normalised difference vegetation index). As with many others, it is based on the red proportion of the radiation spectrum, where plants absorb red light for photosynthesis, and the near infrared part of the spectrum, which is reflecte by vegetation. Other VIs have been designed in order to reduce canopy background/soil effects and atmospheric contamination.

The longest time series exists from the AVHRR (advanced very high resolution radiometers) instruments on board the NOAA (National Oceanic and Atmospheric Administration) satellites, which were started in July 1981 and provide images with a 1 (to 8)-km resolution, covering the globe in a nearly daily repeat cycle. This data set is a standard one due to its availability and the relatively long time series. Newer moderate resolution satellite sensors, thus inevitably with shorter records, include SPOT Vegetation (1 km, since 1998), Envisat MERIS (300 m, since 2002) and MODIS (since 1999) (Reed *et al.*, 2003).

The NDVI mimics the photosynthetic capacity of the vegetation cover. In order to reduce inaccuracies due to clouds and other atmospheric effects, which express themselves by misleadingly low values, the VIs are constructed by taking the maximum value within a 10-day or 2-week compositing period. In general, further temporal smoothing allows an elimination of other false data. Figure 4.16 shows an example of an NDVI time series over mostly evergreen spruce (*Picea abies*) forest in southern Germany.

A variety of approaches and techniques are utilised to derive start and end of the growing season within these annual NDVI time series, including fi ed or site-specifi varying thresholds, inflectio points and moving average approaches. Each of these methods may result in fundamentally different phenomena of seasonality. Limitations (following Reed *et al.*, 2003) of the satellite-derived phenology include pixel size (spatial resolution), temporal resolution, general limitations of the VIs and confounding atmospheric or other environmental conditions, such as snow melt and soil moisture. Also some evergreen types of vegetation, or regions with no clear or multiple growing seasons, are more difficul to study.

Despite these limitations, remote sensing of phenology has been applied to study trends in the length of the growing season or vegetation production, and these analyses of longer time series have delivered important results. Myneni *et al.* (1997) revealed that both the length of the growing season and the photosynthetic activity of terrestrial vegetation in high northern latitudes increased from 1981 to 1991

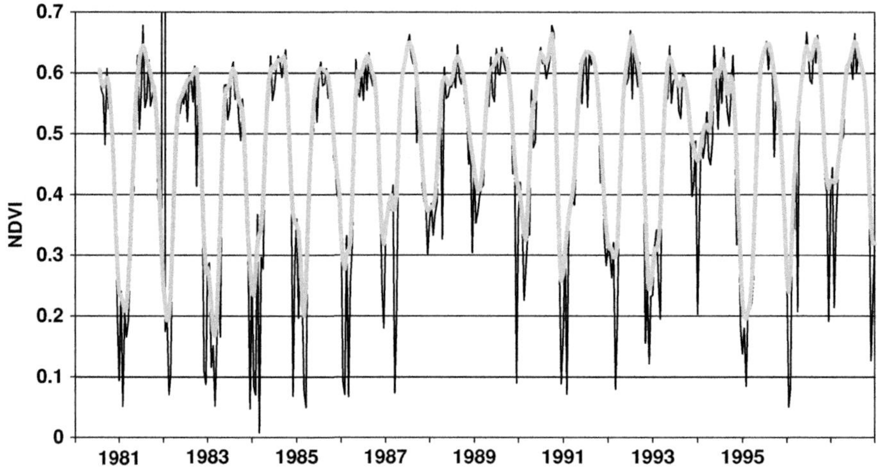

Figure 4.16 Time series NDVI over a mainly deciduous/evergreen forest land cover type (8 × 8 km pixels). Downward spikes due to clouds and other atmospheric perturbations may be corrected by various temporal smoothing techniques shown by the light grey curve (after Menzel, 2002a).

(growing season by 12 days due to an 8 ± 3 day earlier start and a 4 ± 2 day later end; photosynthetic activity by 7–14%, respectively). An increase in the May–September NDVI between 1982 and 1999 as well as an earlier start of the growing season was shown by Tucker *et al.* (2001) for the high latitudes (45–75°N). Zhou *et al.* (2001) also found a lengthening of the growing season by 18 days in Eurasia and 12 days in North America and a higher photosynthetic activity by 12% in Eurasia and 8% in North America (July 1981–December 1999). Analyses of NDVI records are consistent with an increase in the annual amplitude of CO_2 concentrations and an earlier onset of the spring downward crossing (Keeling *et al.*, 1996).

For Europe, trends in the greenness of the vegetation as well as trends in the start and the end of the growing seasons have been determined by Menzel (2002a). Figure 4.17 shows the average growing season (May–September) NDVI for the 1982–1999 period in Europe. Regions with the highest values can be found in central and eastern Europe; in southern Europe the growing season may be restricted by drought in summer, in northern Europe by low temperatures.

Various other definition of the length of the growing season comprise the frost-free period, the period when 5°C is permanently exceeded, the carbon uptake period or the days with fPAR (fraction of photosynthetically active radiation intercepted) > 0.5. For these measures, especially those having a strong relationship with temperature, there are indications that environmental conditions, which determine plant growth, have undergone equally substantial changes in the last few decades. The observed changes comprise, although differing in region and absolute extent, a lengthening of the frost-free season (e.g. Robeson, 2002; Menzel *et al.*, 2003), less frost days in total, reduced risk of late spring and early autumn frost damage (e.g.

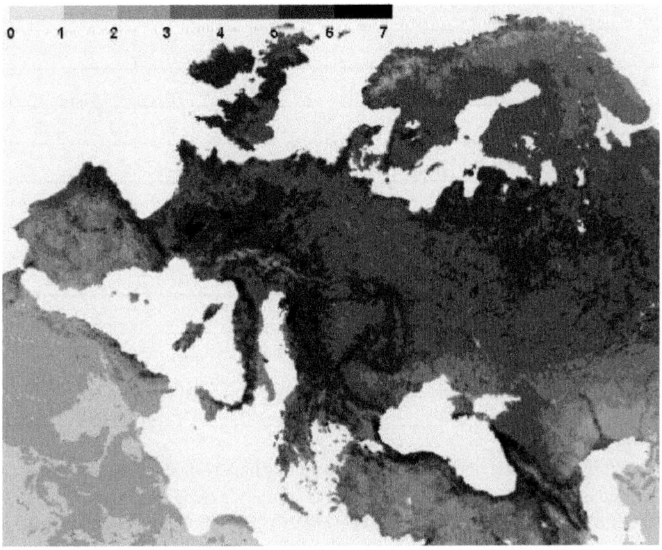

Figure 4.17 Average growing season (May–September) NDVI [NDVI*10] (1982–1999) (EU Project POSITIVE EVK2-CT-1999-00012, Menzel, 2002a).

Scheifinge *et al.*, 2003; Menzel *et al.*, 2003) and increased growing degree days (above certain thresholds, e.g. 0 or 5°C) in the mid and high northern latitudes.

The exact linkage of remote-sensed information to phenological 'ground truth' needs further methodological improvements. However, in mid and higher latitudes of the Northern Hemisphere, satellite-derived estimates of plant seasonality have shown similar recent trends in the length of the growing season and the productivity of the vegetation as shown by direct phenological observations. A strong correlation between 2 week MVC NDVI, GPP (gross primary productivity) at Euroflu sites and traditional phenological recording has been shown by Menzel (2002a) for a beech forest in France (Figure 4.18).

Using spatial average vegetation measures for the start of the growing season, it is also possible to demonstrate their relationship to temperature. Tucker *et al.* (2001) found that the earlier start of the growing season and the increase in growing season NDVI was associated with increase in surface air temperatures. Lucht *et al.* (2002) systematically analysed high northern latitude greening trends over the past two decades, especially with its downward spike after the Mount Pinatubo volcano eruption in 1991. The observed trends towards earlier spring bud burst and increased maximum NDVI were mainly attributed by a biogeochemical vegetation model to changes in temperature. A very smart approach was recently introduced by Jolly *et al.* (2005), where a combined growing season index, constructed from day length, vapour pressure defici and sub-optimal (minimum) temperatures showed good agreement ($r > 0.8$) with NDVI for nine widely dispersed ecosystems. White

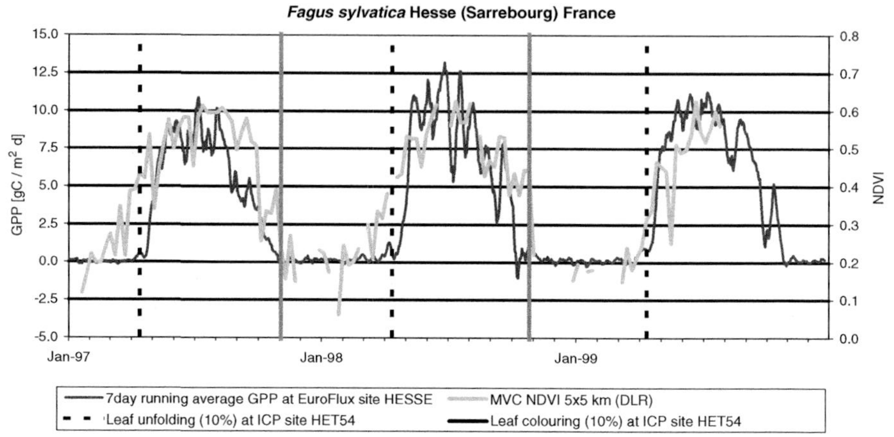

Figure 4.18 Phenological ground observations in a beech (*Fagus sylvatica*) stand at the ICP Level II site (HET54) in France and corresponding GPP (7 days running averages) at the Euroflu site Hesse as well as MVC NDVI of surrounding 5 × 5 km pixels (MVC, NOAA AVHRR; processed by DLR) (Menzel, 2002a).

et al. (2002) used NDVI data to demonstrate that in the eastern United States, the urban heat island effect was associated with a growing season expansion of almost 8 days.

4.5 Possible consequences

The most apparent shifts in phenological phases observed during the last two to three decades in the mid and higher latitudes of the Northern Hemisphere have been an earlier start of various spring phases, such as fl wering or leaf unfolding, and a lengthening of the growing season, mostly due to the earlier start of spring. This lengthening of the active season of vegetation may have several different consequences. An earlier start of fl wering of pollen allergenic plants implies an earlier start of the pollen season, which affects the health of all those suffering from pollinosis (‘hay fever’). As allergic reactions often relate to several plant species, their whole ‘pollen season’ might be lengthened.

If there are favourable conditions during the longer growing season, particularly sufficien rainfall, the total productivity of forest stands and agricultural crops may be altered and possibly improved. The increased forest productivity already observed in Europe may be partly due to the longer growing season, together with direct effects of higher temperatures, higher CO_2 concentration and the consequent better water-use efficien y and modifie forest management practices. Increased nitrogen deposition is, however, probably the most important factor (Spiecker, 1999).

Phenological changes in different taxonomic groups may not have major ecological consequences if they are synchronous with each other and with related climatic processes. For example, as long as the bud burst dates (and thus the critical phase of lowered frost hardiness) and the last spring frost both advance in parallel, the risk of damage by frost will not alter. At the moment, there are indications that due to a smaller advance of spring phenological phases compared to late spring frost dates, the risk of late spring frost damage has not increased in Europe (Menzel *et al.*, 2003; Scheifinge *et al.*, 2003).

The obviously different response between species is a great concern as it implies that we will not proceed through a period of climate warming with unchanged community and species interactions. For example there may be changes in competition between species and in the coincidence of fl wering by potentially hybridising species.

A further question, beyond the scope of this book, concerns the interaction of plant species with vertebrates and particularly invertebrates. If phenology changes differently in species where important synchrony links exist, what of the future of these species? This may affect, for example pollination by insects and seed dispersal as well as all connections via food webs.

References

Aasa, A., Jaagus, J., Ahas, A. & Sepp, M. (2004) The influenc of atmospheric circulation on plant phenological phases in Central and Eastern Europe. *Int. J. Climatol.*, **24**, 1551–1564.

Abu-Asab, M.S., Peterson, P.M., Shetler, S.G. & Orli, S.S. (2001) Earlier plant fl wering in spring as a response to global warming in the Washington, DC, area. *Biodivers. Conserv.*, **10**, 597–612.

Ahas, R., Aasa, A., Menzel, A., Fedotova, V.G. & Scheifinge , H. (2002) Changes in the European spring phenology. *Int. J. Climatol.*, **22** (14), 1727–1738.

Chmielewski, F., Muller, A. & Bruns, E. (2004) Climate changes and trends in phenology of fruit trees and fiel crops in Germany, 1961–2000. *Agric. For. Meteorol.*, **121**, 69–78.

Chmielewski, F. & Rötzer, T. (2001) Response of tree phenology to climate changes across Europe. *Agric. For. Meteorol.*, **108**, 101–112.

Chuine, I., Kramer, K. & Hänninen, H. (2003) Plant development models. In: *Phenology: An Integrative Environmental Science* (ed. M.D. Schwartz), pp. 217–235. Kluwer Academic Publishers, Dordrecht, Boston.

Defila C. & Clot, B. (2001) Phytophenological trends in Switzerland. *Int. J. Biometeorol.*, **45**, 203–207.

Dose, V. & Menzel, A. (2004) Bayesian analysis of climate change impacts in phenology. *Global Change Biol.*, **10**, 259–272.

Dose, V. & Menzel, A. (2006) Bayesian correlation between temperature and blossom onset dates. *Global Change Biol.*, (in press).

Easson, D.L. & Fearnehough, W. (2003) The ability of the Ontario heat unit system to model the growth and development of forage maize sown under plastic mulch. *Grass and Forage Sci.*, **58**, 372–384.

Estrella, N. & Menzel, A. (2006) Responses of leaf colouring of four deciduous tree species to climate and weather in Germany. *Clim. Res.*, (in press).

Federal Biological Research Centre for Agriculture and Forestry (1997) *Growth Stages of Mono- and Dicotyledonous Plants: BBCH Monograph* (ed. U. Meier). Blackwell Wiss.-Verl., Berlin.

Fitter, A.H. & Fitter, R.S.R. (2002) Rapid changes in fl wering time in British plants. *Science*, **296**, 1689–1691.

Hameed, S. & Gong, G. (1993) Variation of spring climate in lower-middle Yangtse River Valley and its relation with solar-cycle length. *Geophys. Res. Lett.*, **21**, 2693–2696.

Hoffmann, H. & Ihne, E. (1882) Phänologischer Aufruf. In: *Beiträge zur Phänologie* (eds E. Ihne & H. Hoffmann), pp. 177–178. J. Ricker'sche Buchhandlung, Giessen, Germany (Republished, 1884).

Hurrell, J.W. (1995) Decadal trends in the North Atlantic Oscillation: regional temperatures and precipitation. *Science*, **269**, 676–679.

Ihne, E. (1883–1941) *Beiträge zur Phänologie/Phänologische Beobachtungen/Phänologische Mitteilungen*, Darmstadt, Selbstverlag des Verfassers.

Jolly, W.M., Nemani, R. & Running, S.W. (2005) A generalized, bioclimatic index to predict foliar phenology in response to climate. *Global Change Biol.*, **11**, 619–632.

Keeling, C.D., Chin, J.F.S. & Whorf, T.P. (1996) Increased activity of northern vegetation inferred from atmospheric CO_2 measurements. *Nature*, **382**, 146–149.

Le Roy Ladurie, E. & Baulant, M. (1980) Grape harvests from the fifteent through the nineteenth Centuries. *J. Interdiscip. Hist.*, **10** (4), 839–849.

Lucht, L., Prentice, I.C., Myneni, R.B., Sitch, S., Friedlingstein, P., Cramer, W., Bousquet, P., Buermann, W. & Smith, B. (2002) Climatic control of the high-latitude vegetation greening trend and Pinatubo effect. *Science*, **296**, 1687–1689.

Matsumoto, K., Ohta, T., Irasawa, M. & Nakamura, T. (2003) Climate change and extension of the *Ginkgo biloba* L. growing season in Japan. *Global Change Biol.*, **9**, 1634–1642.

McCarthy, J.J., Canziani, O.F., Leary, N.A., Dokken, D.J. & White, K.S. (eds) (2001) *Climate Change 2001: Impacts, Adaptation and Vulnerability. Contribution of Working Group II to the Third Assessment Report of the Intergovernmental Panel on Climate Change*, 1000 pp. Cambridge, Cambridge University Press, UK.

Menzel, A. (2000a) Trends in phenological phases in Europe between 1951 and 1996. *Int. J. Biometeorol.*, **44**, 76–81.

Menzel, A. (2000b) *Auswertung phänologischer Beobachtungen an Nutzpflanzen (1951–1998) in Bezug auf Mittelwerte, Extremwerte, Variation und Trends*. Bericht an den Deutschen Wetterdienst, Freising, 30 November 2000.

Menzel, A. (ed.) (2002a) POSITIVE Final Report (February 2000–June 2002) of the EU Project POSITIVE (EVK2-CT-1999-00012), TU Munich.

Menzel, A. (2002b) Phenology: its importance to the global change community. *Clim. Change*, **54**, 379–385.

Menzel, A. (2003a) Phenological anomalies in Germany and their relation to air temperature and NAO. *Clim. Change*, **57**, 243–263.

Menzel, A. (2003b) Phenological data, networks, and research: Europe. In: *Phenology: An Integrative Environmental Science* (ed. M.D. Schwartz), pp. 45–56. Kluwer Academic Publishers, Dordrecht, Boston.

Menzel, A. (2005) A 500 year pheno-climatological view on the 2003 heatwave in Europe assessed by grape harvest dates. *Meteorologische Zeitschrift*, **14**, 75–77.

Menzel, A. & Dose, V. (2005) Analysis of long-term time series of the beginning of fl wering by Bayesian function estimation. *Meteorologische Zeitschrift*, **14** (3), 429–434.

Menzel, A. & Fabian, P. (1999) Growing season extended in Europe. *Nature*, **397**, 659.

Menzel, A., Jakobi, G., Ahas, R., Scheifinge , H. & Estrella, N. (2003) Variations of the climatological growing season (1951–2000) in Germany compared with other countries. *Int. J. Climatol.*, **23**, 793–812.

Menzel, A., Sparks, T.H., Estrella, N. & Eckhardt, S. (2005) 'SSW to NNE' – North Atlantic Oscillation affects the progress of seasons across Europe. *Global Change Biol.*, **11**(6), 909–918.

Murray, M.B., Smith, R.I., Leith, I.D., Fowler, D., Lee, H.S.J., Friend, A.D. & Jarvis, P.G. (1994) Effects of elevated CO_2, nutrition and climatic warming on bud phenology in Sitka Spruce (*Picea sitchensis*) and their impact on the risk of frost damage. *Tree Physiol.*, **14**, 691–706.

Myneni, R.B., Keeling, C.D., Tucker, C.J., Asrar, G. & Nemani, R.R. (1997) Increased plant growth in the northern high latitudes from 1981 to 1991. *Nature*, **186**, 695–702.

Parmesan, C. & Yohe, G. (2003) A globally coherent fingerprin of climate change impacts across natural systems. *Nature*, **421**, 37–42.

Peñuelas, J., Filella, J. & Comas, P. (2002) Changed plant and animal life cycles from 1952 to 2000 in the Mediterranean region. *Global Change Biol.*, **8**, 531–544.

Pfiste , C., Brazdil, R., Glaser, R., Barriendos, M., Camuffo, D., Deutsch, M., Dobrovolny, P., Enzi, S., Guidoboni, E., Kotyza, O., Militzer, S., Racz, L. & Rodrigo, F.S. (1999) Documentary evidence on climate in sixteenth-century Europe. *Clim. Change*, **43**, 55–110.

Reed, B.C., White, M. & Brown, J.F. (2003) Remote sensing phenology. In: *Phenology: An Integrative Environmental Science* (ed. M.D. Schwartz), pp. 365–381. Kluwer Academic Publishers, Dordrecht, Boston.

Robeson, S.M. (2002) Increasing growing-season length in Illinois during the 20th century. *Clim. Change*, **52**, 219–238.

Root, T.L., Price, J.T., Hall, K.R., Schneider, S.H., Rosenzweig, C. & Pounds, J.A. (2003) Fingerprint of global warming on wild animals and plants. *Nature*, **421**, 57–60.

Rötzer, T., Wittenzeller, M., Haeckel, H. & Nekovar, J. (2000) Phenology in central Europe – differences and trends of spring phenophases in urban and rural areas. *Int. J. Biometeorol.*, **44**, 60–66.

Russell, S.C. (1921) Harvest records at Chilgrove, Sussex, 1769–1910. *Q. J. R. Meteorol. Soc.*, **47**, 57–59.

Scheifinge , H., Menzel, A. & Koch, E. (2002) Atmospheric mechanisms governing the spatial and temporal variability of phenological phases across Europe. *Int. J. Climatol.*, **22** (14), 1739–1755.

Scheifinge , H., Menzel, A., Koch, E. & Peter, C. (2003) Trends of spring time frost events and phenological dates in Central Europe. *Theor. Appl. Climatol.*, **74**, 41–51.

Schleip, C. (2005) *Bayesian Analysis of Climate Change Impacts in European Phenology*, 92 pp. Unpublished Master Thesis, Chair of Ecoclimatology, Technical University Munich.

Schnelle, F. & Volkert, E. (1957) Vorschläge zur Errichtung 'Internationaler Phänologischer Gärten' als Stationen eines Grundnetzes für internationale phänologische Beobachtungen. *Meteorologische Rundschau*, **10** (4), 130–133.

Sparks, T.H., Croxton, P.J., Collinson, N. & Grisenthwaite, D.A. (2005) The grass is greener (for longer). *Weather*, **60**, 121–125.

Sparks, T.H., Jeffree, E.P. & Jeffree, C.E. (2001) An examination of the relationship between fl wering times and temperature at the national scale using long-term phenological records from the UK. *Int. J. Biometeorol.*, **44**, 82–87.

Sparks, T.H. & Menzel, A. (2002) Observed changes in the seasons: an overview. *Int. J. Climatol.*, **22**, 1715–1725.

Sparks, T.H. & Tryjanowski, P. (2005) The detection of climate impacts: some methodological considerations. *Int. J. Climatol.*, **25**, 271–277.

Spiecker, H. (1999) Overview of Recent Growth Trends in European Forests. *Water, Air, and Soil Pollut.*, **116**, 33–46.

Tucker, C.J., Slayback, J.E., Pinzon, S.O., Los, S.O., Myneni, R.B. & Taylor, M.G. (2001) Higher northern latitude normalized difference vegetation index and growing season trends from 1982 to 1999. *Int. J. Biometeorol.*, **45**, 184–190.

Walther, G.R., Post, E., Convey, P., Menzel, A., Parmesan, C., Beebee, T.C.J., Fromentin, J.M., Hoegh-Guldberg, O. & Bairlein, F. (2002) Ecological responses to recent climate change. *Nature*, **416**, 389–395.

White, M.A., Nemani, R.R., Thornton, P.E. & Running, S.W. (2002) Satellite evidence of phenological differences between urbanized and rural areas of the eastern United States deciduous broadleaf forest. *Ecosystems*, **5**, 260–273.

Williams, T.A. & Abberton, M.T. (2004) Earlier fl wering between 1962 and 2002 in agricultural varieties of white clover. *Oecologia*, **138**, 122–126.

Willis, J.H. (1944) *Weatherwise*. George Allen and Unwin Ltd, London.

Zhou, L.M., Tucker, C.J., Kaufmann, R.K., Slayback, D., Shabanov, N.V. & Myneni, R.B. (2001) Variations in northern vegetation activity inferred from satellite data of vegetation index during 1981 to 1999. *J. Geophys. Res.* (*Atmospheres*), **106** (D17), 20069–20083.

5 Responses of plant growth and functioning to changes in water supply in a changing climate

William J. Davies

5.1 Introduction: a changing climate and its effects on plant growth and functioning

As rainfall patterns become more unpredictable as climate changes, plants will be subjected to increasing fluctuation in soil moisture availability. These fluctuation are likely to have substantial impacts on plants in natural communities and on crop plants in agriculture (Davies & Gowing, 1999). For example, Silvertown *et al.* (1999) have shown how sensitive plant community composition can be to small changes in soil moisture status. The mechanisms of such changes in composition are likely to be a combination of the responses discussed below. These may be perturbations in plant hydraulics or in plant chemistry, with the driving variable for change being a direct or an indirect result of soil drying or a combination of the two, e.g. reduced soil water availability will reduce water uptake by plants but can also restrict nutrient uptake by roots and transport to the shoots. Changes in N deposition and the resulting nutrient status of ecosystems may also be a direct consequence of environmental change, and other recent work by Gowing and co-workers (Stevens *et al.*, 2004) has shown how changes in N deposition of only 2.5 kg ha^{-1} $year^{-1}$ can result in the addition or removal of a plant species from a 4 m^2 quadrat of an acid grassland community. Other environmental variation as a result of human activities, such as continuing increases in concentrations of ozone in the atmosphere, will also impact significantl on plant water relations and interact with the other important climatic variation highlighted above, but the specifi action of this variable is outside the scope of this review.

Results such as those of Stevens *et al.* (2004) show clearly that reductions in plant growth can be brought about by only very small reductions in water and nutrient availability. Similarly, Boyer (1982) has made an important point that when operating under conditions where irrigation, fertiliser and other management aids are in plentiful supply, US farmers achieve yields that are only around 20% of record yields. This again argues for highly tuned sensitivity of plant growth and development to changes in soil and atmospheric water status.

Excessive precipitation resulting in inundation of soil will reduce the partial pressure of oxygen around the roots of plants, which usually reduces their hydraulic conductivity, thereby reducing water uptake. Therefore, rather counter-intuitively, plant water deficit can result, even when there is plenty of water available in the soil (Jackson *et al.*, 1995). Such changes in plant water status will reduce plant growth, as will any additional flood-induce chemical perturbations, including modification in hormone content of the soil and the plants (e.g. Else & Jackson, 1998) and the accumulation of toxic metabolites.

This brief introduction should be enough to highlight the fact that even subtle changes in the environment are likely to have significan effects on composition and functioning of natural plant communities and on the productivity of agriculture in even the most productive areas of the world. As the climate changes, it is important that we understand the basis of stress-induced changes in plant growth and functioning and if possible intervene, through plant improvement or management programmes, to sustain biodiversity of natural communities where desirable and maintain food production, particularly in some of the most water scarce, populous regions of our planet. This review highlights some of the most sensitive limitations on plant growth and functioning that are imposed by water scarcity. We also focus on the possible exploitation of some of this knowledge to help sustain the production of food under increasingly challenging environmental conditions for farmers.

5.2 Growth of plants in drying soil

5.2.1 Hydraulic regulation of growth

As soil water availability is reduced, water uptake by roots is reduced (see below) and the water potential of the expanding cells will be reduced. Invariably this will limit growth, with the impact on the growth rate of the shoots greater than that on the growth of the root (see, e.g. Sharp *et al.*, 2004). Growth of other plant parts that contribute to crop yield is differentially sensitive to reduced water potential (Westgate & Boyer, 1985) and it may be that reduced sensitivity of growth of some organs to low water potential is explained by solute accumulation in expanding plant parts (Sharp & Davies, 1979). While solute accumulation in roots seems to sustain some growth at low water potential, albeit at a reduced rate, turgor maintenance in shoots does not always sustain growth, and there can even be an inverse relationship between the extent of solute accumulation in plant cells and growth, as carbohydrates accumulate in plant cells as expansion is limited at low water potential. Despite this, the selection of wheat lines for capacity to accumulate solutes has resulted in yield enhancement in water scarce environments (Morgan, 2000). This may not necessarily be a result of continued expansion of vegetative plant parts at low water potential since solute regulation can have other beneficia effects on functioning of plants, such as a delay in the accumulation of potentially damaging concentrations of ABA (abscisic acid) in developing reproductive plant parts.

The beneficia effects of solute regulation on crop yield in certain circumstances, even though turgor maintenance is not sufficien to sustain shoot growth at low water potential, illustrate effectively the complexity of the processes leading to reproductive yielding. Of course, this is something that is well known to plant breeders, and Richards (2004) has recently highlighted the fact that sustained yield in water scarce environments can often be ensured by manipulations of processes that have no direct relationship with drought tolerance or even with plant water relations.

Transfer of solutes between organs to sustain seed yield can be promoted by soil drying, even under circumstances where the effects of reductions in soil water availability are so subtle that no changes in plant water status are obvious. In certain circumstances these changes in allometric relations can even increase seed yield. In a recent paper by Yang *et al.* (2000), high soil nitrogen in the late growth stages of a wheat crop reduced seed yield compared to that of a crop grown with slightly less nitrogen available. This was because high soil N delayed senescence and a high proportion of carbohydrate in the plant was trapped in the stem of the non-senescing plants. Defici irrigation mobilised this reserve from the stems to developing grains such that seed yield of the plants grown with high N was significantl enhanced compared to that of the well-watered, high-N plants. The defici irrigation treatment alone had no impact on seed yield of low-N plants (Yang *et al.*, 2000).

Many environmental stresses will impact on the growth of plant cells via an effect on the hydraulic relations of the cell. These stresses can therefore affect plant growth directly since cell turgor is a motive force for growth, and positive turgors are required to stretch cell walls irreversibly. Changes in cell water relations can also indirectly limit growth by an effect on cell metabolism, which can be altered by changes in the spatial relationship between cell organelles and macromolecules or by changes in the concentration of solutes in the cell (Kaiser, 1987). Stress-induced change in cell wall properties will also affect plant growth rates, and these properties may be altered by the impact of chemical signalling or by a change in the solutes concentrating in the cell wall. Chemical signalling effects are discussed in detail below.

The impact of changing cellular hydraulic relations on growth of cells is commonly visualised via the Lockhart equation. This treatment suggests that growth rate is linearly related to cellular turgor above a threshold value, with the slope of the relationship being a function of cell wall extensibility. Both threshold turgor and cell wall extensibility are define by this model as being under metabolic control (e.g. Pritchard & Tomos, 1993). An alternative model has turgor acting as a switch rather than a proportional controller (e.g. Zhu & Boyer, 1992), with the rate of growth determined by another variable such as the cell wall properties.

It has not proved easy to collect data to support the Lockhart model of growth control, as both growth and cell water relations must be measured in the same population of cells. Even in a single growing organ such as a root tip that might be accessible to water relations assessment, it is clear that cells in different regions of the growing zone are growing at different rates (Figure 5.1) and are differentially sensitive to stresses such as a reduction in cell water potential (see, e.g. Sharp *et al.*, 2004). If this population of cells is treated as responding identically to reduced water potential then an apparent Lockhartian relationship between growth rate and turgor will be observed. This will arise because there is cessation of growth of cells distal to the tip, slowing of growth of cells in another zone only a few millimetres towards the tip and maintenance of growth in the cells closest to the tip (see Spollen

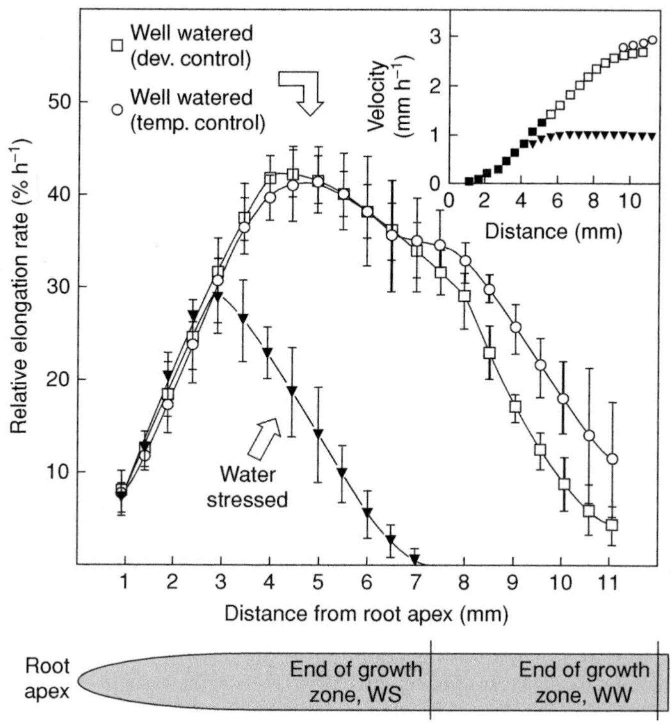

Figure 5.1 Relative elongation rate as a function of distance from the apex of the primary root of maize seedlings growing under well-watered (water potential of −0.03 MPa) or water-stressed (water potential of −1.6 MPa) conditions. Two well-watered controls are shown, a developmental control (roots of the same length as at low water potential, 5 cm) and a temporal control (roots of the same age as at low water potential, 48 h after transplanting). The inset shows longitudinal displacement velocity profile for the same roots. (From Sharp *et al.*, 2004.)

et al., 1993), even though turgor in all zones can be decreased to a uniformly low value across the growing zone at low substrate water potential (Spollen & Sharp, 1991). If we aim to understand the limitation of growth of plants in drying soil then we must understand the mechanistic basis of the growth limitation in different populations of cells such as those described above, and we are making some progress in this regard.

Over a 20-year period, Sharp and co-workers have investigated the control of growth in the primary root of maize seedlings rooted in vermiculite held at different but constant water potentials. The group has identifie preferential deposition of solutes, particularly close to the tip of a growing root, ensuring some turgor maintenance (Voetberg & Sharp, 1991) to drive growth. There is also a signifi cant drought-induced up-regulation of the activity of some putative wall-loosening enzymes (expansins and XET) through the root tip growing zone (Wu *et al.*, 1994, 1996, 2001). The group has shown that low root growth rates in dry substrate can be a result of ethylene accumulation in root tips (LeNoble *et al.*, 2004) but that this can be avoided by ABA accumulation at low water potentials. Most recently, the group has taken a genomics approach to understand the limitation of growth of primary maize roots at low substrate water potential. Importantly, attention is focused on the regions of growth that are define above (Figure 5.1), and perhaps not surprisingly, there are substantial differences in the impact of low water potential on gene expression in the three regions identifie (Sharp *et al.*, 2004). This approach holds out the prospect of providing new insight into the importance of mechanistic responses that are discussed briefl above. In addition, there is some prospect that apparently counter-intuitive responses will be shown to be crucial regulators. One example of this is an apparent role in the root tip for reactive oxygen scavengers (Sharp, personal communication) with ROS (reactive oxygen species) perhaps playing a role in cell wall loosening at moderate to low water potential (Dumville & Fry, 2003) while also damaging membrane integrity as water potential falls further (Sharp *et al.*, 2004).

5.3 Water relations of plants in drying soil

5.3.1 Water movement into and through the plant

Water uptake by roots is extremely sensitive to a reduction in water availability in the soil. This may be mostly a result of partial drying of the root surface and the development of a depletion zone, creating a high root/soil interface resistance for water uptake. This is particularly important when roots are clumped together in compacted soil. Under these circumstances, there may be little to be gained by modifying root membrane properties to increase water uptake in drying soil. This is because the radial resistance to water uptake into roots is in series with the root/soil interface resistance and the resistance to water movement through the bulk soil, which itself increases significantl as the soil dries. For this reason, the maintenance of root growth away from water and nutrient-depleted zones can be an effective way to sustain water uptake in drying soil and as such is an attractive target for those interested in improvement of plants for water scarce environments.

The radial pathway for water movement into roots and the resistances produced by the various components of the pathway have been the source of some controversy in recent years. Steudle and Peterson (1998) describe our current understanding. It is well known that root radial resistance to water uptake is sensitive to the flu of water into the root, with apparent resistance declining as the transpiration flu increases. Steudle and Peterson argue that this may be because of a change in the proportion of total water flu moving though different pathways into the root or because of a change in membrane properties to water fl w – changes consistent with a composite transport model. In some plants, an apoplastic bypass for radial water flu can be important (Freundl *et al.*, 1998), while in others the apoplastic pathway can be effectively blocked due to lignificatio or suberisation. Apoplastic bypass will allow particular chemical species unrestricted movement into the xylem, and we will show below how this may have an important impact on long-distance chemical signalling in droughted plants.

Recent work has suggested that water channels or aquaporins can influenc the radial fl w of water into roots, with the activity of these channels under metabolic control (Maurel & Chrispeels, 2001; Tyerman *et al.*, 2002). Steudle (2000) has suggested that permeation of water through a proteinaceous pore in the membrane can regulate the cell-to-cell pathway as define in his composite transport model to water flu (above). This pathway may dominate water flu when movement is driven largely by osmotic gradients or when the apoplastic pathway becomes blocked, which can occur in response to some soil conditions. Evidence is mounting that a variety of factors will affect aquaporin activity, including pH, pCa and osmotic gradients. Clarkson *et al.* (2000) have shown how various soil conditions such as nutrient status can influenc channel activity and the resulting radial resistance to water movement. In particular, increased nitrogen availability increases the hydraulic conductivity of roots (Clarkson *et al.*, 2000). We show below how similar treatments can have dramatic effects on stomatal sensitivity to soil drying, emphasising again the potential importance of the interaction between soil water and nutrient status on the growth and functioning of plants. Manipulating the nutrient status of soil may provide an effective low-technology possibility for enhancing the drought tolerance of crops in water scarce environments.

The variables that might drive water movement through plants have received considerable attention from researchers in recent years, with some controversy over the motive forces for water movement and whether or not there is sufficien tension in the xylem to account for most water movement, particularly in tall plants (e.g. Zimmermann *et al.*, 1993). The controversy seems to have revolved around the question of whether the micropressure probe can accurately measure the tension in the xylem of the plant without cavitation occurring. Recent technical advances suggest that appropriate tensions to drive water movement do exist even in the tallest plants (Wei *et al.*, 1999) and the results of earlier studies that failed to detect tensions might have been generated because of technical limitations of the early versions of the xylem pressure probe.

A large body of work in the literature has focused upon the impact of the environment and particularly drought stress on the functioning of the hydraulic system of the plant (see e.g. Tyree & Sperry, 1989). Catastrophic xylem embolism as a result of unregulated transpiration and the development of substantial xylem tensions can result in severe, irreversible desiccation of plant shoots. Sperry *et al.* (2002) describe a theory of hydraulic limitation provided by the hydraulic linkages within the plant, and this theory has been successful in predicting the regulation of transpiration in response to variation in moisture status of soils of different types. The theory suggests that many observed stomatal responses to the environment can be shown to be appropriate for the avoidance of hydraulic failure in the plant and for the maximisation of soil water extraction. The theory goes some way to explaining the huge differences in patterns of water use between species. It appears that the pattern of the plant's water potential regulation and, in particular, the thresholds of water potential that stomata appear to regulate are tuned to the soil moisture regime and the hydraulic linkages in the plant. Sperry *et al.* (2002) suggest that the plant's hydraulic equipment is optimised for drawing water from particular hydraulic niches in the soil environment. An interesting question is how the plant achieves a coordination between stomatal regulation (and possibly growth) and the hydraulic capabilities of its xylem architecture. One possibility is a physiological link (e.g. Nardini & Salleo, 2000) and we discuss this below. Clearly, cavitation is a key issue for transport, particularly in tall plants with significan xylem tensions and perhaps particularly in perennial plants where loss of xylem continuity could be responsible for a signifi cant change in community composition in a relatively short time span. We discuss below the relative importance of the control of stomatal behaviour and plant water status by chemical signalling in woody perennials and herbaceous annuals. It seems possible that loss of hydraulic continuity may be less of a controlling influenc on stomatal behaviour of herbaceous plants, but there is little information on this in the literature.

5.3.2 Control of gas exchange by stomata under drought

Stomatal behaviour provides some control over gas exchange by leaves, but the effectiveness with which drought-induced decreases in conductance control transpiration and assimilation is dependent on a number of factors, particularly the coupling between the crop or plant and the environment (Jarvis & McNaughton, 1986). A crop is said to be well coupled when mass and energy exchange between the leaves and the bulk atmosphere is effective so that leaf temperature closely follows air temperature. Under these conditions, stomata will exert good control over crop water loss. For short crops which can be aerodynamically smooth with high boundary layer resistance, coupling is not perfect, and under these conditions stomatal closure can lead to increases in leaf temperature, which drives more transpiration despite the closure of stomata. This means that transpiration will not be well controlled by stomatal closure, and in conditions of low wind speeds that are common, for example in plant canopies, it may be independent of conductance and proportional to incoming radiant energy. It is for this reason that anti-transpirants have not always been shown to be effective when tested in the fiel on short crops (poorly coupled) even though they have affected stomata and controlled water loss when tested in growth chambers where forced air movement over individual plants will often mean that leaves are more effectively coupled to the environment. Through the years, in an attempt to increase water-use efficien y (WUE) of a range of crops, there have been several plant improvement programmes based on selection for reduced stomatal numbers and size. For similar 'coupling' reasons, these programmes have not always been successful.

Analysis by Jones (1992) of interactions between boundary layer resistance, radiation and stomatal conductance in their effects on WUE of single leaves showed that there is an optimal conductance for WUE. Jones (1992) also showed the important impact of night-time respiration and cuticular conductance on WUE with optimal conductance increasing as cuticular conductance and dark respiration increase. Very recent work by Masle *et al.* (2005) where the basis of natural variation in WUE of *Arabidopsis* was investigated shows the impact of a single gene (ERECTA – a putative leucine-rich repeat receptor-like kinase) on WUE. High efficiencie of water use appear to be linked to greater stomatal frequencies plus increases in leaf thickness and mesophyll structure. This work raises the exciting prospect of breeding programmes that might increase assimilation capacity per unit of water used even under non-stressed conditions.

There is one spectacularly successful example in the literature where wheat plants in Australia selected via carbon isotope discrimination for higher WUEs have outyielded commonly used commercial varieties in water scarce environments (see, e.g. Condon *et al.*, 2004). In that programme, resulting lines were tested for yield in a variety of environments. The variety Drysdale was released for southern New South Wales in 2002 and Rees for the northern Australian cropping region in 2003. Yield trials have shown a yield advantage of between 2 and 15% for lines with low-carbon isotope discrimination (high WUE) at yield levels from 5 to 1 t ha^{-1} (Rebetzke *et al.*, 2002) when compared with high discrimination sister lines. The highest yield advantages were found only in the most drought-prone environments. Trials in southern New South Wales demonstrated 23% yield increases for Drysdale compared with Diamondbird, the current recommended variety for this region. There are several mechanisms underlying the results obtained from this successful breeding programme but presumably part of the selection for high WUE lines results from stomatal characteristics and part from modifie photosynthesis, since carbon isotope discrimination can occur firstl during the diffusion of CO_2 from the air into the sub-stomatal cavities and secondly during the biochemical fixatio of CO_2. The yield results suggest that despite some of the predicted effects of poor coupling between wheat crops and the environment there can still be advantages to WUE selection which results from both differences in assimilation capacity and stomatal conductance.

The CO_2 assimilation rate of plants under drought can be substantially restricted by stomatal closure, at least until the relative water content (RWC) of the shoot is significantl reduced, with assimilation capacity unaffected if water is again made available before plant water content has declined too far. In some species or situations, however, assimilation capacity can be more directly sensitive to relatively small changes in RWC, and the resulting carbon gain (and WUE) can be restricted both by stomatal effects and by these more direct effects. Lawlor and Cornic (2002) have discussed the basis of these kinds of non-stomatal limitations and have even gone so far as to describe two types of relationships between CO_2 assimilation rate and RWC with the two types of plants differing in the sensitivity of the direct response of assimilation capacity to developing water deficit It may be, however, that there is more of a continuum of photosynthetic sensitivity, with the kind of selection for WUE that we have described above exploiting interspecifi variation in the robustness of the plant's photosynthetic apparatus. There is now renewed interest in the much-discussed possibility that stomatal behaviour under drought may be controlled by some signalling between photosynthetic capacity and the guard cells. This is interesting in the context of understanding the regulation of WUE but also opens possibilities for exploitation of plant signalling in agriculture, which we discuss below.

5.4 Water relation targets for plant improvement in water scarce environments

We have highlighted above the sensitivity of growth, development and yield of crop plants to a reduction in water availability in the soil (e.g. Boyer, 1982). Boyer and co-workers have shown that even a few days of drought stress at a critical period during the production of yield components can result in complete crop failure (e.g. Boyle *et al.*, 1991). Much reduction of 'yield' in water scarce environments occurs while there is still a lot of water in the soil and well before plants show conventional stress symptoms such as a reduction in shoot water potential (e.g. Richards, 1993). This is because plants can sense and respond to changes in water availability and then regulate growth and functioning. A good example of this is the closure of stomata to avoid shoot dehydration stress, rather than a reduction in conductance in response to reduction in shoot water potential. To sustain yielding as soil dries, which will be necessary as the climate changes and rainfall patterns become more unpredictable, we must initially address these regulatory and developmental processes, rather than focusing on processes that contribute to desiccation resistance as such.

Passioura and co-workers have developed a breeding programme for wheat in Australia, based around the argument that breeding for a narrow xylem vessel in the seminal roots of wheat should increase the resistance to water flu and force plants to use water more slowly in the subsoil (Passioura, 1972). In cereals, seminal roots develop before nodal roots and grow deeper into the subsoil. Because crops in dry land environments can rely largely on subsoil water and this water must pass through the single xylem vessel in each seminal root, then the hydraulics of these roots are crucial to determining water use patterns. If plants use the subsoil water too rapidly during the development of the vegetative plant, then too little will remain for the crucial period of development when grain is filling However, use of subsoil water will be reduced if there is a large hydraulic resistance in the seminal roots.

These ideas at firs appear to be counter-intuitive in their suggestion that making water transport more difficul may somehow enhance yield. However, much detailed physiological work highlights the importance of having some water available for the crucial period of reproductive development in cereals. In Australia, a breeding programme reduced the xylem vessel diameter of two commercial wheat varieties from 65μm to less than 55μm (Richards & Passioura, 1981a,b). In fiel trials narrow vessel selections yielded 8% more than the unselected controls in the driest environments, while yield differences in the wetter environments were largely not significan (Richards & Passioura, 1989) (Figure 5.2). This was probably because

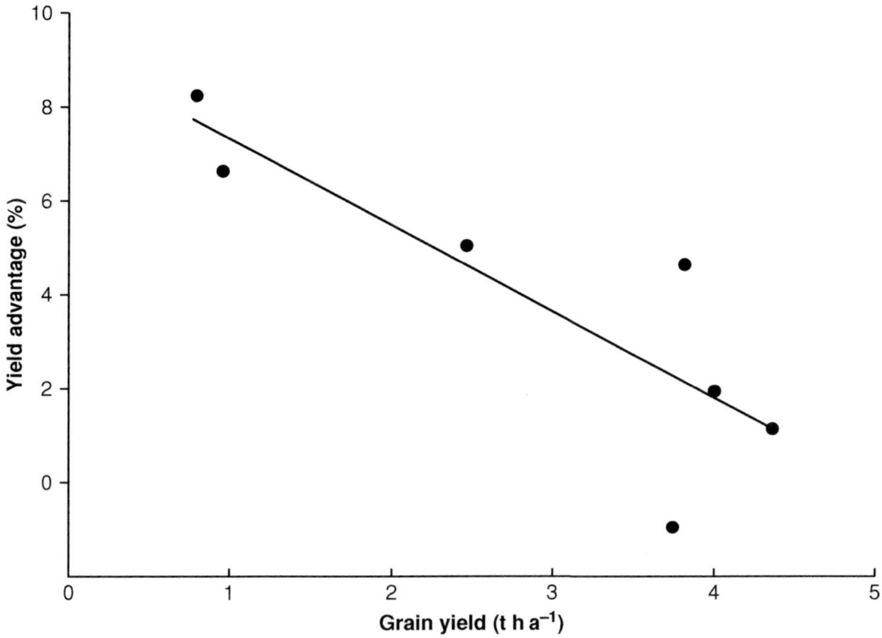

Figure 5.2 Yield advantage of wheat lines selected for narrow xylem vessels. Values in each environment are the yield differences between lines selected for narrow xylem vessels and unselected controls averaged over two genetic backgrounds (cv. Kite and Cook). (From Richards, 2004.)

there was no growth penalty resulting from narrow vessels in plants in wet soil, as the nodal root system, which is well developed in the topsoil, can supply the crop with water under these conditions.

Although plant biologists have given an enormous amount of attention to plant desiccation resistance, arguably these processes are largely irrelevant for crop yielding. If plant cells desiccate, crop yielding will be negligible and even if yield is doubled by plant manipulation, then it is still negligible! One exception to this situation is the combination of responses that allow a perennial crop plant to stay alive under desiccating conditions. This capacity to 'live to figh another day' can be highly advantageous for yield in succeeding growth seasons. The capacity to survive is largely irrelevant in an annual crop plant where a stress-induced delay in development can result in a complete loss of yield (e.g. if the crop is growing in a relatively short frost-free season).

The breeding work at CIMMYT to ensure that drought impacts on the anthesis to silking period in maize do not limit grain production (Bolaños & Edmeades, 1996) is an excellent example of a plant breeding programme for water scarce environments, which has little to do with drought tolerance in the sense in which the word is commonly used in modern plant biology.

A plant improvement programme focused on reducing the sensitivity of the plant's environmental sensing mechanisms or its regulatory response to the stress could act to stabilise vegetative crop yield between years and enhance yield per unit cropping area. It may also be possible to modify management practice to take profi from regulation of plant development, which may, for example, increase WUE. For example, we can apply reduced amounts of water to some crops and exploit the plant's stress-sensing capacity to reduce unnecessary vegetative growth while allowing maintenance of fruit production with a reduced supply of water (see examples for vines etc.; Davies *et al.*, 2002). Such manipulations will be necessary if we are to maintain food production while reducing the amount of water used for irrigation. Currently, 70% of the world's water is used in agriculture. If substantial water savings in agriculture can be achieved without substantial yield penalty, then the use of this water elsewhere can bring substantial benefit to the environment and to society.

We argue here that by focusing our attention on understanding and potentially manipulating the processes that contribute to the regulation of crop growth and water use when there is plenty of water in the soil or when soil moisture deficit are relatively mild, we found that there are prospects of maintaining yield while using substantially reduced quantities of water in agriculture, a highly desirable combination. In the next section, we place emphasis on the gains that can be achieved by an understanding and exploitation of the long-distance chemical signalling processes in plants.

5.5 Control of stomata, water use and growth of plants in drying soil: hydraulic and chemical signalling

5.5.1 Interactions between different environmental factors

Inevitably, most work on the regulation of plant growth and functioning in water scarce environments has focused upon the capacity of the plant to respond to changes in individual components in the edaphic or the aerial environment. While the assumption is that the successful plant will be able to optimise its behaviour with respect to *both* the above ground and the below ground environment, there is comparatively little work on the impact of interacting stresses on plant functioning in droughted conditions. This is even true for well-studied model systems like guard cells, although we are beginning to make some progress in understanding the impact of different environmental factors at different points in the signal transduction chains within single cells (e.g. Hetherington & Brownlee, 2004). For example, the interactive effects of water defici and changing CO_2 concentration on guard cell functioning may be explained by the interaction between the effects of ABA and CO_2 on stomata, caused by the role of intracellular calcium in signalling (e.g. Webb & Hetherington, 1999).

One excellent example of the interactions of different agents in the control of cell growth that is relevant for plant's sensing of environmental change is the markedly

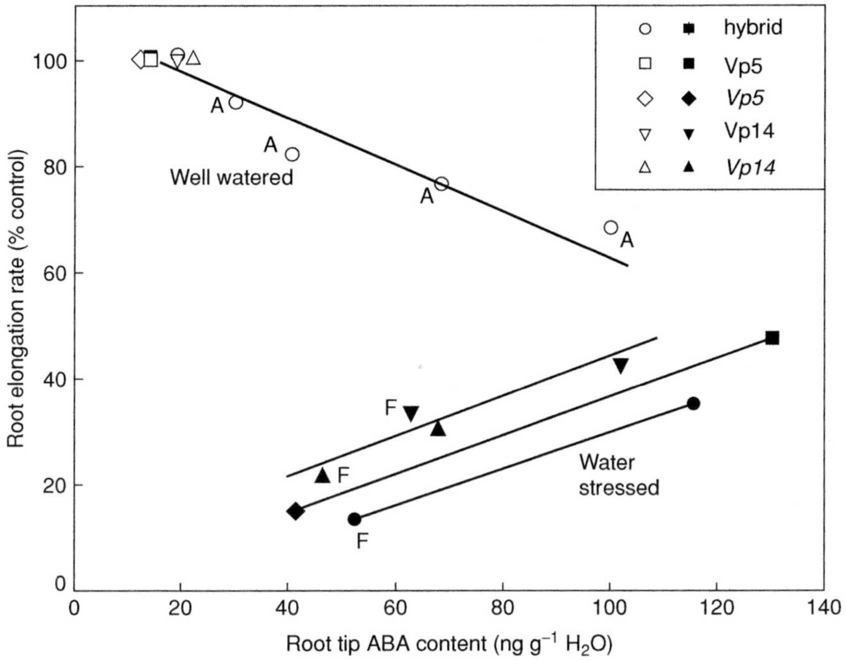

Figure 5.3 Primary root elongation rate as a function of root tip (apical 10 mm) ABA content for various maize genotypes growing under well-watered (water potential of −0.03 MPa; open symbols) or water-stressed (water potential of −1.6 MPa; closed symbols) conditions. At high water potential, the root ABA content of hybrid (cv. FR27 × FRMo17) seedlings was raised above the normal level by adding various concentrations of ABA (A) to the vermiculite. At low water potential, the root ABA content was decreased below the normal level by treatment with fluridon (F) or by using the *vp5* or *vp14* mutants. Data are plotted as a percentage of the rate for the same genotype at high water potential. Elongation rates of the mutants under well-watered conditions were similar to their respective wild types. (From Sharp *et al.*, 2004.)

different impact that the stress hormone ABA has on plant growth at different tissue water potentials. Sharp and co-workers have shown that when water is in plentiful supply, ABA can act as an inhibitor of plant growth, while at low tissue water potentials, ABA accumulation can act to promote plant growth, albeit from a low rate (Figure 5.3; Sharp *et al.*, 2004). This may be because ABA counteracts the growth inhibitory effects of accumulating ethylene at low water potentials (see above). Whatever the explanation, this result will be important in understanding the control of plant functioning by chemical signals, as one effect of ABA signalling as soil dries can be to close stomata and to sustain shoot turgors at values comparable to those of well-watered plants (Figure 5.4; Sobeih *et al.*, 2004). Therefore, we might expect that at mild soil water deficit when shoot turgor is high, ABA may act to restrict growth of shoots (see, e.g. Bacon *et al.*, 1998) and thereby help the plant restrict water use and husband water resources. As soil moisture deficit increase and

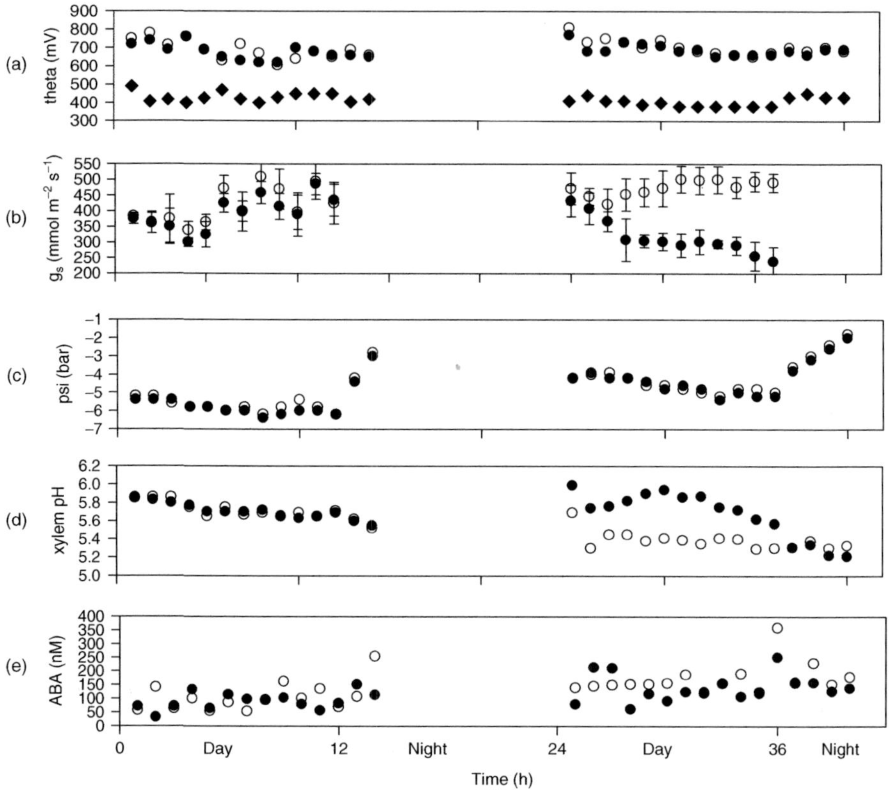

Figure 5.4 Effects of partial root drying on functioning of tomato leaves. (a) moisture content of the upper 6 cm of potting compost from pots watered daily on both sides of the split-pot (○), and from the watered (•) and drying (♦) sides of plants watered daily on one side of the split-pot (b). Stomatal conductance, (c) leaf water potential, (d) xylem sap pH and (e) xylem ABA concentration of fully expanded leaves at node 9. Points are from individual wild-type (cv. Ailsa Craig) plants watered daily on one (•) or both (○) sides of the split-pot (b–e). In (b) points are means ± S.E. of fi e leaflet per leaf. Dark shading on the time axis indicates the night period. (From Sobeih *et al.*, 2004.)

despite signals to cause reductions in stomatal conductance, the plant loses control of shoot turgor and ABA may then act to sustain at least some root growth to help maintain the supply of at least some water.

5.5.2 Measuring the water availability in the soil: long-distance chemical signalling

Our current understanding is that soil drying around roots will not only reduce water uptake and transport to shoots, but will also result in the generation of a range of chemical signals that can have local effects (such as modifie root growth and functioning – see above) and can also have systemic effects throughout the plant. These can include changes in shoot growth and functioning, changes in plant morphology and fl wering and fruiting. Some workers have argued that plants have evolved the capacity to 'measure' changes in the water availability in the soil and to communicate this information to the shoots, where the message triggers regulation of gas exchange and growth. Such a communication system may be interpreted as means of avoiding catastrophic hydraulic breakdown (see above) or as means of husbanding the use of soil water to increase the chances of the plant completing its life cycle before the water resource is exhausted (e.g. Jones, 1976). Both possibilities require that the plant has the capacity to integrate the impact of these edaphic changes with changes in its aerial environment. We return below to examine the possible mechanistic basis of this proposal.

Our assumption is that as soil dries in the rhizosphere, a range of plant responses and changes in the soil will contribute to root-to-shoot signalling and provide the shoot with information on resource availability. Most simply, reduced water uptake can result in reduced root cell turgor with a direct impact on the synthesis, compartmentation and transport of plant hormones to the xylem stream and onto the shoot (e.g. Hartung *et al.*, 1999). Hormonal signals that have received most attention in this regard are ABA and ACC (1-aminocyclopropane-carboxylic acid), which are synthesised in increased quantities in roots as root turgor falls, and cytokinins, the supply of which from roots is generally reduced at lower root water contents (Bano *et al.*, 1993, 1994). There are several forms of cytokinin transported through plants, and it is important to quantify these in any investigation of chemical control of plant functioning under drought. The same is also true for conjugated forms of ABA that are important transport forms, easily converted to free hormone in the shoot (Sauter *et al.*, 2002).

Hartung's research group has emphasised the impact of drought on the recirculation from roots and that of ABA arriving in the phloem from the shoots and have stressed that much ABA arriving in the transpiration stream may actually be shoot-sourced (Peuke *et al.*, 1994). It is also possible that drought-induced changes in soil strength will also contribute directly to modifie hormone transport to the shoots (e.g. Hartung *et al.*, 1994). The long-distance hormone signalling pathway can also be influence by hormones originating in the soil, perhaps as a result of microbiological activity in the rhizosphere (Hartung *et al.*, 1996).

Other than hormones, a whole range of chemical species can act as signals to the shoot, with Wilkinson and co-workers emphasising the importance of xylem and apoplastic pH as regulators of both stomatal behaviour (Wilkinson & Davies, 1997; Wilkinson *et al.*, 1998) and growth (Bacon *et al.*, 1998). While apoplastic pH can have a direct physiological impact on both guard cell functioning and cell expansion, it will also impact significantl on the effect of ABA on both processes. This is because ABA is a weak acid (pK_a 4.8) and is distributed within the apoplast and the symplast of tissues according to the anion-trap concept and the Henderson–Hasselbalch equation. Alkalisation of xylem sap is a common response to soil drying in some plants (see, e.g. for tomato data; Wilkinson *et al.*, 1998), and supplying detached leaves with neutral or alkaline buffers (pH $\geq$7) via the transpiration stream can restrict transpiration (Wilkinson & Davies, 1997; Wilkinson *et al.*, 1998). These buffers can apparently increase apoplastic pH, which will result in higher apoplastic ABA concentrations. pH-induced increases in apoplastic ABA concentration will ultimately close the stomata (Wilkinson & Davies, 1997), and it is possible that increased xylem sap pH could elicit ABA-dependent stomatal closure without the need for increased xylem ABA delivery. In other words there will always be enough ABA to close stomata, even in the well-watered plant, but the degree of conductance reduction is dependent on apoplastic pH. Increased xylem sap pH can also correlate with drought-induced leaf growth inhibition in barley, and feeding leaves alkaline buffers via the xylem inhibits leaf growth (Bacon *et al.*, 1998).

5.5.3 The integrated response to the environment

One way of interpreting the interaction between ABA and pH on stomatal behaviour and growth is to argue that a reduction in plant water status (often resulting in alkalinisation of xylem sap) enhances the sensitivity of both growth and stomatal behaviour to the ABA signal (see, e.g. Tardieu *et al.*, 1992). This will mean that early in the day when leaf-to-air vapour pressure difference (VPD) is low and transpiration rates are restricted, apoplastic pHs will be low and even though the soil may be comparatively dry and the root ABA signal relatively intense, stomata may open to high conductances. As the day progresses, increased VPD and transpiration will reduce water potential and apoplastic pH, generating a stomatal response to an ABA signal that is relatively constant throughout the day (Tardieu *et al.*, 1993). This is effectively a description of the mechanistic basis for optimal stomatal behaviour, which can maximise WUE for the prevailing environmental conditions (Cowan & Farquhar, 1977).

Wilkinson and Davies (2002) have shown that xylem and presumably apoplastic pH can be sensitive to both edaphic and climatic variation, and therefore environmental modulation of this variable provides the plant with one way of integrating the impact of a variety of environmental agents and modifying growth and development. Soil conditions can influenc xylem pH through a variety of factors in addition to the impact of soil drying. Perhaps most significan is the nitrogen status of the soil. Plants that are well fertilised with nitrate tend to show higher xylem and apoplastic pH (e.g. Mengel *et al.*, 1994; Muhling & Lauchli, 2001). Soil water defici generally reduces nitrate uptake and transport to shoots in the xylem (e.g. Shaner & Boyer, 1976), an observation that at firs sight is not consistent with a drought-induced increase in xylem pH. One explanation for a marked drought-induced increase in the pH of the xylem sap and the apoplast in some plants (but not in all species – see below; also see Wilkinson, 2004) may be the proposed shift in nitrate reduction from shoots to roots that may occur in some plants as soil dries and drought develops (Lips, 1997). This may result in more malate and related compounds being loaded into the xylem, which can alkalinise the xylem sap, even if nitrate contents of sap are substantially reduced (Gollan *et al.*, 1992). Whatever the explanation for the changes in pH that we see, the clear interaction between N availability and soil drying in the regulation of stomatal behaviour and growth suggests an important impact of both changing rainfall patterns and increasing N deposition on plant fitness

Lastly, Wilkinson (2004) highlights the different impact of drought on apoplastic pH between, for example, woody and herbaceous plants and argues that slower growing, often woody, perennial species that assimilate most of their nitrate in the root may never transport a significan amount of N as nitrate within the xylem. This may mean that soil water deficit (or inundation) are unlikely to change xylem sap pH of these species and that hydraulic control may dominate. There is some evidence that xylem sap pH of woody species may remain unchanged or even acidify as soil dries (e.g. Thomas & Eamus, 2002).

5.6 Conclusions: a strategy for plant improvement and management to exploit the plant's drought response capacity

We have suggested above that it may be possible to use defici irrigation to exploit the plant's long-distance signalling networks to enhance WUE in agriculture and to increase reproductive crop quality, in part by restricting vegetative crop development and the commitment of resources to this end (Yang *et al.*, 2001; Davies *et al.*, 2002). As soil dries, shoot water status can be sustained by signalling-induced restrictions in stomatal aperture (e.g. Mingo *et al.*, 2003; Sobeih *et al.*, 2004) (Figure 5.4). If as an alternative approach for different circumstances where we want to sustain vegetative growth we can develop genotypes that do not produce chemical leaf growth inhibitors as soil dries or have leaf growth processes that are insensitive to these signals, then we can perhaps also *sustain* biomass accumulation and yield of vegetative plant parts when water supply for agriculture is restricted. This strategy is dependent on identifying the different chemical signals that limit both stomatal conductance and leaf expansion during drought – if indeed there are different regulators of the two processes. While decreased plant water use (caused by the limitation to both stomatal conductance and leaf expansion) can allow the plant to husband immediately available water resources, another strategy might be for the roots to explore deeper parts of the soil profil (Reid & Renquist, 1997). Manipulation of this variable may provide extra water supply to growing shoots and allow maintenance of shoot growth processes at low bulk soil water status.

In many plants, drought increases root and xylem concentrations of the ethylene precursor ACC (Gomez-Cadenas *et al.*, 1996). Although the delivery of ACC from the root system can account for shoot ethylene evolution (Else & Jackson, 1998) and may thus limit leaf growth under drought, the relationship between xylem ACC concentration and leaf growth of plants exposed to drying soil has not been defined We have recently shown that in tomato both xylem ACC and ABA concentrations increased in response to partial rootzone drying (PRD), prior to any decrease in shoot water status (Sobeih *et al.*, 2004). It is therefore appropriate to assay the interaction between these two hormones on leaf expansion using well-hydrated plants. Feeding ABA and ACC simultaneously via the xylem to detached shoots inhibits leaf growth additively (I.C. Dodd, unpublished results 2005), suggesting an important role for ethylene in the inhibition of leaf growth in drying soil, when shoot water status is maintained. In contrast, in plants at low water potential, ABA accumulation is necessary to minimise high rates of ethylene synthesis and ethylene-mediated root growth inhibition (LeNoble *et al.*, 2004).

Under drought the plant hormone ethylene can be involved in both the suppression of root growth during soil drying (see above) and the suppression of leaf growth via long-distance chemical signalling, again emphasising a key role for this hormone in the regulation of plant production in water scarce environments. Our recent work has shown that ethylene evolution of wild-type (WT) tomato plants increased as soil dried but could be suppressed using transgenic ($ACO1_{AS}$) plants containing an antisense gene for one isoenzyme of ACC oxidase. Most importantly, $ACO1_{AS}$ plants also showed no inhibition of leaf growth when exposed to PRD, even though both $ACO1_{AS}$ and WT plants showed similar changes in other putative chemical inhibitors of leaf expansion (xylem sap pH and ABA concentration). It seems likely that the enhanced ethylene evolution under PRD is responsible for leaf growth inhibition of WT plants. $ACO1_{AS}$ plants showed no leaf growth inhibition over a range of soil water contents, which significantl restricted growth of WT plants (Figure 5.5), but it is important to note that this lack of drought sensitivity was only apparent when leaf turgor was maintained by ABA/pH signalling, reducing stomatal conductance in response to PRD.

Transgenic approaches to enhance drought tolerance may be effective but are not always socially acceptable. It may be important, therefore, that certain bacteria occurring on the root surface contain high levels of the enzyme ACC deaminase that will degrade the ethylene precursor ACC. Since a dynamic equilibrium of ACC concentration exists between root, rhizosphere and bacterium, bacterial uptake of rhizospheric ACC (for use as a carbon and nitrogen source) may decrease root ACC concentration and root ethylene evolution and may potentially increase root growth (Glick *et al.*, 1998). Our recent experiments (A. Belimov, unpublished results 2005) with the plant growth-promoting bacterium *Variovorax* showed that pea plants grown with the bacterium added to the soil showed a promotion of root biomass, leaf area and total biomass relative to uninoculated plants in drying soil, suggesting that these effects were mediated by modifying plant ethylene status.

Sustaining leaf growth in water scarce environments has proved to be a particularly intractable problem for plant breeders. We suggest here a model system for plant production in an increasingly water scarce world. It involves a combination of genetic (suppression of ethylene signalling) and agronomical manipulations (promotion of ABA signalling by defici irrigation) to exploit basic plant physiology and plant developmental responses (in this case, soil to root to shoot signalling pathways). The aim is to sustain some turgor as the soil dries via partial stomatal closure and to sustain leaf growth and biomass production via this maintenance

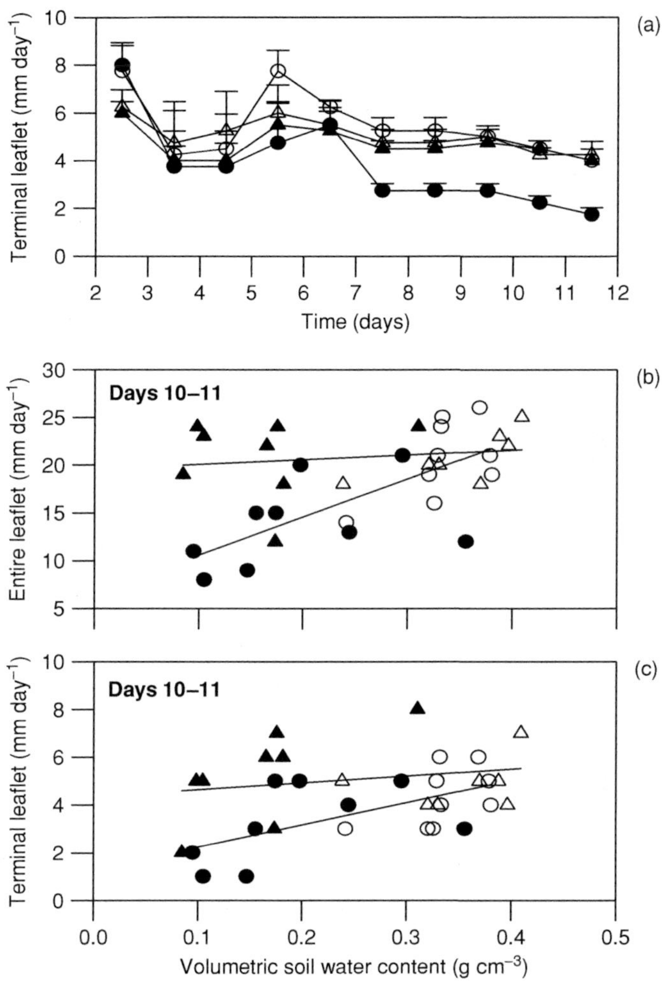

Figure 5.5 Leaf growth responses of wild-type (cv. Ailsa Craig) and transgenic ($ACO1_{AS}$) tomato plants in response to partial rootzone drying. (a) Terminal leafle elongation rates of leaves at node 12 from wild-type (•, ○) or transgenic (▲, △) plants watered daily on one (•, ▲) or both (○, △) sides of the split-pot. Data are means ± S.E. of—seven to nine replicates. (b) Entire leaf and (c) terminal leafle elongation rate (days 10–11) plotted against the pre-watering volumetric water content of the upper 6 cm of soil on day 11. Linear regressions were fitte to each genotype. (From Sobeih *et al.*, 2004.)

of the driving force for growth and the over-riding of any direct chemical inhibition of leaf expansion. This approach is based on an understanding and an exploitation of the plant's fundamental drought avoidance responses. Other, more straightforward uses of novel management techniques based on exploitation of long-distance signalling in the plant, may be used to change the plant's allometric relationships to obtain more harvestable 'crop per drop'.

References

Bacon, M.A., Wilkinson, S. & Davies, W.J. (1998) pH-regulated leaf cell expansion in droughted plants is abscisic acid dependent. *Plant Physiol.*, **118**, 1507–1515.

Bano, A., Dörffling K., Bettin, D. & Hahn, H. (1993) Abscisic acid and cytokinins as possible root-to-shoot signals in xylem sap of rice plants in drying soil. *Aust. J. Plant Physiol.*, **20**, 109–115.

Bano, A., Hansen, H., Dörffling K. & Hahn, H. (1994) Changes in the content of free and conjugated abscisic acid, phaseic acid and cytokinins in the xylem sap of drought-stressed sunfl wer plants. *Phytochemistry*, **37**, 345–347.

Bolaños, J. & Edmeades, G.O. (1996) The importance of the anthesis-silking interval in breeding for drought tolerance in tropical maize. *Field Crops Res.*, **48**, 65–80.

Boyer, J.S. (1982) Plant productivity and environment. *Science*, **218**, 443–448.

Boyle, M.G., Boyer, J.S. & Morgan, P.W. (1991) Stem infusion of liquid culture medium prevents reproductive failure of maize at low water potentials. *Crop Sci.*, **31**, 1246–1252.

Clarkson, D.T., Carvajal, M., Henzler, T., Waterhouse, R.N., Smyth, A.M., Cooke, D.T. & Steudle, E. (2000) Root hydraulic conductance: diurnal aquaporin expression and the effects of nutrient stress. *J. Exp. Bot.*, **51**, 61–70.

Condon, A.G., Richards, R.A., Rebetzke, G.J. & Farquhar, G.D. (2004) Breeding for high water use efficien y. *J. Exp. Bot.*, **55**, 2447–2460.

Cowan, I.R. & Farquhar, G.D. (1977) Stomatal function in relation to leaf metabolism and environment. *Symp. Soc. Exp. Biol.*, **31**, 471–505.

Davies, W.J. & Gowing, D.J.G. (1999) Plant responses to small perturbations in soil water status. In: *Physiological Plant Ecology* (eds M.C. Press, J.D. Scholes and M.G. Barker), pp. 67–90. Blackwell Publishing, Oxford.

Davies, W.J., Wilkinson, S. & Loveys, B. (2002) Stomatal control by chemical signalling and the exploitation of this mechanism to increase water use efficien y in agriculture. *New Phytol.*, **153**, 449–460.

Dumville, J.C. & Fry, S.C. (2003) Solubilisation of tomato fruit pectins by ascorbate: a possible non-enzymic mechanism of fruit softening. *Planta*, **217**, 951–961.

Else, M.A. & Jackson, M.B. (1998) Transport of 1-aminocyclopropane-1-carboxylic acid (ACC) in the transpiration stream of tomato (*Lycopersicon esculentum*) in relation to foliar ethylene production and petiole epinasty. *Aust. J. Plant Physiol.*, **25**, 453–458.

Freundl, E., Steudle, E. & Hartung, W. (1998) Water uptake by roots of maize and sunfl wer affects the radial transport of abscisic acid and the ABA concentration in the xylem. *Planta*, **209**, 8–19.

Glick, B.R., Penrose, D.M. & Li, J.P. (1998) A model for the lowering of plant ethylene concentrations by plant growth promoting bacteria. *J. Theor. Biol.*, **190**, 63–68.

Gollan, T., Schurr, U., & Schulze, E.-D. (1992) Stomatal response to soil drying in relation to changes in xylem sap composition of Helianthus annuus. 1. The concentration of cations, anions and amino acids in, and pH of, the xylem sap. *Plant Cell Environ.*, **15**, 551–559.

Gomez-Cadenas, A., Tadeo, F.R., Talon, M. & Primo-Millo, E. (1996) Leaf abscission induced by ethylene in water-stressed intact seedlings of cleopatra mandarin requires previous abscisic acid accumulation in roots. *Plant Physiol.*, **112**, 401–408.

Hartung, W., Peuke, A.D. & Davies, W.J. (1999) Abscisic acid – a hormonal long distance stress signal in plants under drought and salt stress. In: *Handbook of Crop Stress* (ed. M. Pessarakali) 2nd edn, pp. 731–747. Marcel Dekker, New York.

Hartung, W., Sauter, A., Turner, N.C., Fillery, I. & Heilmeier, H. (1996) Abscisic acid in soils: what is its function and which factors and mechanisms influenc its concentration? *Plant Soil*, **184**, 105–110.

Hartung, W., Zhang, J. & Davies, W.J. (1994) Does abscisic acid play a stress physiological role in maize plants growing in heavily compacted soil? *J. Exp. Bot.*, **45**, 221–226.

Hetherington, A.M. & Brownlee, C. (2004) The generation of Ca^{2+} signals in plants. *Ann. Rev. Plant Biol.*, **55**, 401–427.

Jackson, M.B., Davies, W.J. & Else, M.A. (1995) Pressure-fl w relationships, xylem solutes and hydraulic conductivity in roots of floode tomato plants. *Ann. Bot.*, **77**, 17–24.

Jarvis, P.G. & McNaughton, K.G. (1986) Stomatal control of transpiration: scaling up from leaf to region. *Adv. Ecol. Res.*, **15**, 1–49.

Jones, H.G. (1976) Crop characteristics and the ratio between assimilation and transpiration. *J. Appl. Ecol.*, **13**, 605–622.

Jones, H.G. (1985) Physiological mechanisms involved in the control of leaf water status: implications for the estimation of tree water status. *Acta Hort.*, **171**, 291–296.

Jones, H.G. (1992) *Plants and Microclimate*. Cambridge University Press, p. 428.

Kaiser, W.M. (1987) Effect of water defici on photosynthetic capacity. *Physiol. Plant.*, **71**, 142–149.

Lawlor, D.W. & Cornic, G. (2002) Photosynthetic carbon metabolism and associated metabolism in relation to water deficit in higher plants. *Plant Cell Environ.*, **25**, 275–294.

LeNoble, M.E., Spollen, W.G. & Sharp, R.E. (2004) Maintenance of shoot growth by ABA: genetic assessment of the role of ethylene suppression. *J. Exp. Bot.*, **55**, 237–245.

Lips, S.H. (1997) The role of organic nitrogen ions in plant adaptation processes. *Russ. J. Plant Physiol.*, **44**, 421–431.

Masle, J., Gilmore, S.R. & Farquhar, G.D. (2005) The *ERECTA* gene regulates plant transpiration effi ciency in *Arabidopsis*. *Nature*, **436**, 866–870.

Maurel, C. & Chrispeels, M.J. (2001) A molecular entry into plant water relations. *Plant Physiol.*, **125**, 135–138.

Mengel, K., Planker, R. & Hoffmann, B. (1994) Relationship between leaf apoplast pH and iron chlorosis of sunfl wer. *J. Plant Nutr.*, **17**, 1053–1065.

Mingo, D.M., Bacon, M.A. & Davies, W.J. (2003) Non-hydraulic regulation of fruit growth in tomato plants growing in drying soil. *J. Exp. Bot.*, **54**, 1205–1212.

Morgan, J.M. (2000) Increases in grain yield of whet by breeding for an osmoregulation gene: relationship to water supply and evaporative demand. *Aust. J. Agric. Res.*, **51**, 971–978.

Muhling, K.H. & Lauchli, A. (2001) Influenc of chemical form and concentration of nitrogen on apoplastic pH of leaves. *J. Plant Nutr.*, **24**, 399–411.

Nardini, A. & Salleo, S. (2000) Limitations of stomatal conductance by hydraulic traits: sensing or preventing xylem cavitation. *Trees*, **15**, 14–24.

Passioura, J.B. (1972) The effect of root geometry on the yield of wheat growing on stored water. *Aust. J. Agric. Res.*, **23**, 745–752.

Peuke, A.D., Jeschke, W.D. & Hartung, W. (1994) The uptake and fl w of C, N and ions between roots and shoots in *Ricinus communis* L. III. Long-distance transport of abscisic acid depending on nitrogen nutrition and salt stress. *J. Exp. Bot.*, **45**, 741–747.

Pritchard, J. & Tomos, A.D. (1993) Correlating biophysical and biochemical control of root cell expansion. In: *Water Deficits: Plant Responses from Cell to Community* (eds J.A.C. Smith & H. Griffiths) pp. 53–72. BIOS Scientifi Publishers, Oxford.

Rebetzke, G.J., Condon, A.G., Richards, R.A. & Farquhar, G.D. (2002) Selection for reduced carbon isotope discrimination increases aerial biomass and grain yield of rainfed bread wheat. *Crop Sci.*, **42**, 739–745.

Reid, J.B. & Renquist, A.R. (1997) Enhanced root production as a feed forward response in response to soil water defici in field-gr wn tomatoes. *Aust. J. Plant Physiol.*, **24**, 685–692.

Richards, R.A. (1993) Breeding crops with improved stress resistance. In: *Plant Responses to Cellular Dehydration During Environmental Stress* (eds T.J. Close & E.A. Bray), pp. 211–223. ASPP, Washington.

Richards, R.A. (2004) Physiological traits used in the breeding of new cultivars for water-scarce environments. In: *Proceedings of the Fourth International Crop Science Congress*, Brisbane, Australia.

Richards, R.A. & Passioura, J.B. (1981a) Seminal root morphology and water use of wheat. I. Environmental effects. *Crop Sci.*, **21**, 249–252.

Richards, R.A. & Passioura, J.B. (1981b) Seminal root morphology and water use of wheat. II. Genetic variation. *Crop Sci.*, **21**, 253–255.

Richards, R.A. & Passioura, J.B. (1989) A breeding program to reduce the diameter of the major xylem vessel in the seminal roots of wheat and its effect on grain yield in rain-fed environments. *Aust. J. Agric. Res.*, **40**, 943–950.

Sauter, A., Dietz, K.-J. & Hartung, W. (2002) The possible stress physiological role of abscisic acid conjugates in root to shoot signalling. *Plant Cell Environ.*, **25**, 223–228.

Shaner, D.L. & Boyer, J.S. (1976) Nitrate reductase activity in maize leaves. II. Regulation of nitrate flu at low water potential. *Plant Physiol.*, **58**, 555–559.

Sharp, R.E. & Davies, W.J. (1979) Solute regulation and growth by roots and shoots of water-stressed maize plants. *Planta*, **147**, 43–49.

Sharp, R.E., Poroyko, V., Hejlek, L.G., Spollen, W.G., Springer, G.K., Bohnert, H.J. & Nguyen, H. (2004) Root growth maintenance during water deficits physiology to functional genomics. *J. Exp. Bot.*, **55**, 2343–2352.

Silvertown, J., Dodd, M.E., Dowing, D.J.G. & Mountford, J.O. (1999) Hydrologically define niches reveal a basis for species richness in plant communities. *Nature*, **400**, 61–63.

Sobeih, W., Dodd, I.C., Bacon, M.A., Grierson, D.C. & Davies, W.J. (2004) Long-distance signals regulating stomatal conductance and leaf growth in tomato (*Lycopersicon esculentum*) plants subjected to partial rootzone drying. *J. Exp. Bot.*, **55**, 2353–2364.

Sperry, J.S., Hacke, U.G., Oren, R. & Comstock, J.P. (2002) Water deficit and hydraulic limits to leaf water supply. *Plant Cell Environ.*, **25**, 251–263.

Spollen, W.G. & Sharp, R.E. (1991) Spatial distribution of turgor and root growth at low water potentials. *Plant Physiol.*, **96**, 438–443.

Spollen, W.G., Sharp, R.E., Saab, I.N. & Wu, Y. (1993) Regulation of cell expansion in roots and shoots at low water potentials. In: *Water Deficits: Plant Responses from Cell to Community* (eds J.A.C. Smith & H. Griffiths) pp. 37–52. BIOS Scientifi Publishers, Oxford.

Steudle, E. (2000) Water uptake by roots: effects of water deficit *J. Exp. Bot.*, **51**, 1531–1542.

Steudle, E. & Peterson, C.A. (1998) How does water get through roots? *J. Exp. Bot.*, **49**, 775–788.

Stevens, C.J., Dise, N.B., Mountford, J.O. & Gowing, D.J.G. (2004) Impact of nitrogen deposition on the species richness of grasslands. *Science*, **303**, 1876–1879.

Tardieu, F. & Davies, W.J. (1992) Stomatal response to ABA is a function of current plant water status. *Plant Physiol.*, **98**, 540–545.

Tardieu, F. & Davies, W.J. (1993) Root-shoot communication and whole-plant regulation of water flux In: *Water Deficits: Plant Responses from Cell to Community* (eds J.A.C. Smith & H. Griffiths) pp. 147–162. BIOS Scientifi Publishers, Oxford.

Thomas, D.S. & Eamus, D. (2002) Seasonal patterns of xylem sap pH, xylem ABA, leaf water potential and stomatal conductance of six evergreen and deciduous Australian savanna tree species. *Aust. J. Bot.*, **50**, 229–236.

Tyerman, S. D., Niemetz, C.M. & Bramley, H. (2002) Plant aquaporins: multifunctional water and solute channels with expanding roles. *Plant Cell Environ.*, **25**, 173–194.

Tyree, M.T. & Sperry, J.S. (1989) Vulnerability of xylem to cavitation and embolism. *Ann. Rev. Plant Physiol. Mol. Biol.*, **40**, 19–38.

Voetberg, G.S. & Sharp, R.E. (1991) Growth of the maize primary root at low water potentials. III. Role of increased proline deposition in osmotic adjustment. *Plant Physiol.*, **96**, 1125–1130.

Webb, A.A.R. & Hetherington, A.M. (1999) Convergence of the abscisic acid, CO_2 and extracellular calcium signal transduction pathways in stomatal guard cells. *Plant Physiol.*, **114**, 1557–1560.

Wei, C.F., Tyree, M.T. & Steudle, E. (1999) Direct measurement of xylem pressure in leaves of intact maize plants. A test of the cohesion-tension theory taking hydraulic architecture into consideration. *Plant Physiol.*, **121**, 1191–1205.

Westgate, M.E. & Boyer, J.S. (1985) Osmotic adjustment and the inhibition of leaf, root, stem and silk growth at low water potentials in maize. *Planta*, **164**, 540–549.

Wilkinson, S. (2004) Water use efficien y and chemical signalling. In: *Water Use Efficiency in Plant Biology* (ed. M.A. Bacon), pp. 75–112. Blackwell Press, Oxford.

Wilkinson, S., Corlett, J.E., Oger, L. & Davies, W.J. (1998) Effects of xylem pH on transpiration from wild-type and *flacca* tomato leaves: a vital role for abscisic acid in preventing excessive water loss even from well-watered plants. *Plant Physiol.*, **117**, 703–709.

Wilkinson, S. & Davies, W.J. (1997) Xylem sap pH increase: a drought signal received at the apoplastic face of the guard cell which involves the suppression of saturable ABA uptake by the epidermal symplast. *Plant Physiol.*, **113**, 559–573.

Wilkinson, S. & Davies, W.J. (2002) ABA-based chemical signalling: the co-ordination of responses to stress in plants. *Plant Cell Environ.*, **25**, 195–210.

Wu, Y., Sharp, R.E., Durachko, D.M. & Cosgrove, D.J. (1996) Growth maintenance of the maize primary root at low water potentials involves increases in cell wall extension properties, expansin activity and wall susceptibility to expansins. *Plant Physiol.*, **111**, 765–772.

Wu, Y., Spollen, W.G., Sharp, R.E., Hetherington, P.R. & Fry, S.C. (1994) Root growth maintenance at low water potentials: increased activity of xyloglucan endotransglycosylase and its possible regulation by abscisic acid. *Plant Physiol.*, **106**, 607–615.

Wu, Y., Thorne, E.T., Sharp, R.E. & Cosgrove, D.J. (2001) Modificatio of expansin transcript levels in the maize primary root at low water potentials. *Plant Physiol.*, **126**, 1471–1479.

Yang, J., Zhang, J., Huang, Z., Zhu, Q. & Wang, L. (2001) Remobilisation of carbon reserves is improved by controlled soil drying during grain fillin of wheat. *Crop Sci.*, **40**, 1645–1655.

Zhu, G.L. & Boyer, J.S. (1992) Enlargement in Chara: studies with a turgor clamp. Growth rate is not determined by turgor. *Plant Physiol.*, **100**, 2071–2080.

Zimmermann, U., Benkert, R., Schneider, H., Rygol, J., Zhu, J.J. & Zimmermann, G. (1993) Xylem pressure and transport in higher plants and tall trees. In: *Water Deficits: Plant Responses from Cell to Community* (eds J.A.C. Smith & H. Griffiths) pp. 87–108. BIOS Scientifi Publishers, Oxford.

6 Water availability and productivity

João S. Pereira, Maria-Manuela Chaves, Maria-Conceição Caldeira and Alexandre V. Correia

6.1 Introduction

Plant life and primary productivity depend on water availability. On Earth, nearly 20% of the global land surface is too dry to be cultivated. The quest for water and devising ways to use it efficientl for crop production has shaped civilisations around the world. When shortages in precipitation, often coupled to high evaporative demand, reduce moisture availability for an extended period in a way that will affect negatively the normal life in a region, a drought is said to occur. Drought, however, is not easy to defin or to quantify objectively. In ecological terms, a drought will interfere negatively with ecosystem processes (productivity, biogeochemical cycles) or structure, whereas in agriculture a drought is said to occur when soil water is not enough to meet the needs of the local crops.

Temporary drought, as a climatic anomaly, must be distinguished from the normal occurrence of seasonal low precipitation, which is a permanent feature of some climates. For example, aridity refers to low moisture regions, such as those where the mean annual precipitation is less than half the value of potential evapotranspiration. In semi-arid regions the interannual variability in water availability is larger than in humid regions and adequate rainfall may not occur every year (Ellis, 1994; Loik *et al.*, 2004). In arid lands, the precipitation may come in well-separated events or 'pulses'. The timing of rainfall, the extent of the dry season and the regime of rain pulses determine resource availability and shape ecosystem structure and function (Schwinning *et al.*, 2004).

Droughts have affected human societies since the earliest times and had enormous impacts throughout history. For example, the invasion of Europe by the barbarian tribes from central Asia at the time of the fall of the Roman Empire may have been driven by the drying of pastures (Lamb, 1995). Later, by the beginning of the seventeenth century, the initial difficultie of British settlement in North America may have resulted from coincidental extreme droughts (Stahle *et al.*, 1998).

Today, water is considered again a major issue in international politics. This is the case of highly populated countries, e.g. Egypt, Iraq or Syria, that use large amounts of water resources in irrigation but depend on river infl w from neighbouring countries (Araus, 2004). With the growth of the human population and improvement in the standard of living, especially in developing countries, the need for water is expected to increase. Moreover, large areas of the globe will suffer the intensificatio of water deficit (IPCC, 2001; see also Chapter 1). At present approximately 7% of the world's population lives in areas where water is scarce but this may rise to 67% of the world's population by 2050 (Wallace, 2000). The very dry areas of the globe have more than doubled since the 1970s (Dai *et al.*, 2004). On the other hand, with climate change, plants will be subjected to an increased variability of water availability as the frequency and intensity of extreme droughts may increase (Gutschick & BassiriRad, 2003).

In Portugal, for example, there has been a greater variability in the frequency and intensity of rainfall and a consistent increase in drought frequency in the last 25 years, resulting from warming and a significan reduction of precipitation in late winter and early spring (Miranda *et al.*, 2006). In the Iberian Peninsula, almost all simulations with general circulation models suggest a future reduction in precipitation during spring and summer, i.e., an increase in the length of the dry season (Miranda *et al.*, 2006).

In this chapter we will assess how plant productivity is determined in water-limited environments in the context of climate change scenarios. We will consider the impact of droughts on natural vegetation as well as in agriculture and forestry and the importance of spatial and temporal variability in water supply. Finally, we will discuss wildfires as they are major environmental forces, closely linked to drought, that determine the structure and function of many ecosystems (Bond *et al.*, 2005).

6.2 Water deficits and primary productivity

6.2.1 Net primary productivity

Net primary productivity (NPP) may be quantifie as a linear function of the photosynthetically active radiation absorbed by the canopy (APAR):

$$NPP = \varepsilon \times APAR$$

where ε the radiation conversion efficien y into biomass. The value of APAR depends on incident short-wave solar radiation, leaf area index (LAI) and the canopy structure, which affects the light extinction coefficien (k). The slope of the relationship between plant productivity and APAR, i.e. ε, varies with plant type and environmental conditions (Russell *et al.*, 1989).

Water deficit affect NPP in two ways: (1) reducing APAR (mainly as a result of changes in LAI) and (2) reducing radiation conversion efficien y (ε) through changes in net photosynthesis and whole-plant carbon loss through respiration. Important interactions with nutrients, temperature and atmospheric CO_2 are expected. Reductions in APAR (associated with lower LAI) due to medium-term water deficit result mainly from restricted growth, leaf senescence and branch losses. Leaf expansion is one of the plant processes most sensitive to drought. With water deficits not only are fewer leaves produced but also they are smaller and thicker, thus leading to low LAI (Chaves *et al.*, 2003). On the other hand, as water potentials decrease, leaf shedding will increase through regulated senescence (Munné-Bosch & Alegre,

2004) and/or branch and petiole xylem cavitation (Tyree & Sperry, 1989; Rood *et al.*, 2000; Davis *et al.*, 2002; Vilagrosa *et al.*, 2003).

Differences in ε may result from differences in plant respiration. In general, long-term exposure to water deficit leads to a decline in plant respiration, as a result of decreased metabolism associated with lower photosynthesis, export of assimilates and growth. For example, this is what happens during the dry summer in Mediterranean ecosystems (Rambal *et al.*, 2004). Differences observed among species in the response of respiration to drought, which are often reported in the literature, are apparently due to different growth sensitivity to drought (Lambers *et al.*, 1998). There are also differences between mitochondrial respiration in the light that depends mostly on the amount of primary products directly derived from photosynthesis, and respiration in the dark that also depends on end products of metabolism (Haupt-Herting *et al.*, 2001). On the other hand, the accumulation of osmolytes (e.g. sorbitol) under drought, implying less availability of sugars, may result in a further decrease of respiration, in particular in the alternative path, as was observed in wheat roots in drying soil (Lambers *et al.*, 1998). Studies by Ghashghaie *et al.* (2001) in *Helianthus annuus* and *Nicotiana sylvestris* indicated a progressive decline in respiration with dehydration (from around 2 μmol m^{-2} s^{-1} to less than 0.5 μmol m^{-2} s^{-1}, accompanying the decline in relative water content from 95 to 60%). Although respiration rates decrease under water deficits plant carbon balance may be negatively affected when the ratio of respiring biomass increases relative to assimilatory surface, because shoot growth is more sensitive to water stress than root growth (see also Chapter 5).

The value of ε changes seasonally. For example, we calculated the monthly average $\acute{\varepsilon}$, in terms of gross primary productivity (GPP) as GPP/APAR, from eddy-covariance data in an eucalypt plantation. As GPP = (NPP + R), with R standing for total plant respiration, $\acute{\varepsilon}$ should mimic ε even though not parallel, as the responses of GPP and R to temperature differ. The variation in $\acute{\varepsilon}$ ranged from approximately 4 in winter to near 1 g MJ^{-1}PAR in the summer (Mateus, J., Pita, G. & Rodrigues, A., 2005, personal communication; Figure 6.1). The high monthly $\acute{\varepsilon}$ in winter resulted from moderate temperatures, abundant water and a large number of overcast days. Diffuse light from overcast skies is photosynthetically more effective than direct light and can account for increases in daily $\acute{\varepsilon}$ up to 42% (Rosati & Dejong, 2003). The decline in $\acute{\varepsilon}$ through the season is probably the result of increasing vapour pressure deficit and light saturation at high PAR (Ruimy *et al.*, 1995) as the number of clear-sky and dry days increase from winter to summer. In summer, severe plant water deficit lead to declining carbon assimilation rates (Pereira *et al.*, 1986) and even lower $\acute{\varepsilon}$.

In crops, in addition to the decline in NPP, yield may be further decreased as a result of the negative effect of water deficit on harvest indices, i.e., the ratio between harvestable biomass and NPP. For example, Earl and Davis (2003) showed that water stress reduced substantially the fina grain yield in maize, but the reduction in APAR contributed much less to the yield loss than the decreases in ε and in harvest indices.

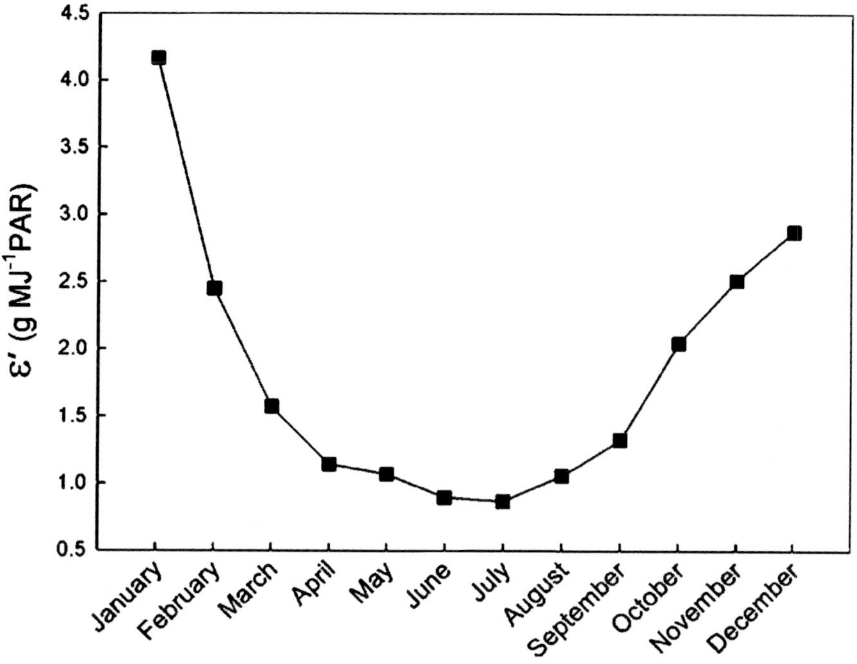

Figure 6.1 Monthly averages of $\acute{\varepsilon}$ as GPP $= \acute{\varepsilon} \times$ APAR, measured with the eddy covariance method, in a *Eucalyptus globulus* plantation in Herdade de Espirra, central Portugal – Lat. 38° 38′ N, Long. 8° 36′W; mean annual temperature, 16°C; mean annual precipitation, 709 mm; stem age, 9 years; leaf area index, 3 (Mateus, J., Pita, G. & Rodrigues, A., 2005, personal communication).

6.2.2 *Water-use efficiency*

The quantificatio of the dependence of plant productivity on water resources may be viewed as the slope of the relationship of net primary production and the amount of water actually lost by transpiration (T) over the year as

$$\text{NPP} = \text{WUE}_t \times \text{water supply} \times \text{proportion of water used by plants},$$

where the season-long water-use efficien y (WUE_t) or transpiration efficien y is the ratio of biomass produced to the corresponding plant transpiration [in g (dry matter) kg^{-1} H_2O or mmol C mol^{-1} H_2O] (Jones, 2004b). Water supply is precipitation plus irrigation, if appropriate, or precipitation during the growing season plus water in the soil at the moment of sowing for annual crops.

Short-term variability in transpiration efficien y is dominated by physiological (stomatal conductance and photosynthesis) and meteorological variables (vapour pressure defici of the air, wind). The transpiration efficien y tends to increase under moderate water stress, as a result of greater stomatal restriction on transpiration and a relatively less sensitive response of the photosynthetic apparatus. On the contrary, high vapour pressure defici in the atmosphere causes a decline in WUE_t because transpiration increases without concomitant change in photosynthesis (Jones, 2004b). This sets an upper limit for WUE_t in any given climate. Reduced transpiration under high irradiance raises the risk of leaf temperature increasing above the optimum for metabolic activity or at least above the threshold that leads to irreversible leaf tissue oxidative stress. Additionally, water-use efficien y (WUE) may decrease under severe water stress, or when water deficit combine with high temperature and high light, due to inhibition of photosynthesis (Chaves *et al.*, 2004; Jones, 2004b). This is also apparent at the whole canopy level, as for example, under the Mediterranean summer drought, where WUE_t decreased with severe water deficit accompanied by a strong decline in carbon assimilation (Reichstein *et al.*, 2002).

At the scale of ecosystems we can integrate both hydrological and physiological components and ecosystem level WUE (WUE_e; Gregory, 2004) is define as:

$$WUE_e = NPP/(E + T + R + D)$$

where E is the direct evaporation from plant and soil surfaces, T is transpiration, R is the liquid water run-off and D is drainage below the rooting zone. Since in hydrological analysis it is common to separate liquid from vapour flu es, the use of water for biomass production has been historically considered as the ratio of NPP to evapotranspiration ($T + E$) (Rosenzweig, 1968; Lieth & Whittaker, 1975). While T represents the amount of water required for primary production, the other terms of the water balance are virtually non-productive. The proportion of water transpired in relation to evapotranspiration [$T/(T + E)$] is a measure of water-supply efficien y (Rockström, 2003).

Reflectin roughly the impact of physiological controls, WUE_e (or rain use efficien y) tends to be maximum under limiting water supply (Huxman *et al.*, 2004), as suggested by Figure 6.2. The great variability in the data is mainly because of species differences and plant metabolism (e.g. C3/C4), differences in nutrition and soil properties and rainfall seasonality. The trend line shown for forests indicates that with high water supply the non-productive flu es of water become more important. This trend was also shown in a eucalyptus plantation where irrigation and fertilisation treatments were applied (Table 6.1). The treatments were irrigation to satisfy the evapotranspiration demand in summer (I), irrigation as in I plus fertilisers added according to plant needs (IL), no irrigation but with fertilisers added (F) and control plots (C) (Madeira *et al.*, 2002). WUE_e decreased substantially (80%) in well-watered as compared to rainfed plots in the normal rainfall year as shown in Table 6.1 (precipitation close to the average, 607 mm). In a previous wet year (precipitation 1200 mm) the differences between well-watered and rainfed plots were negligible (unpublished results), but WUE_e was approximately 12% greater in the fertilised plots than in the non-fertilised plots, both rainfed and irrigated.

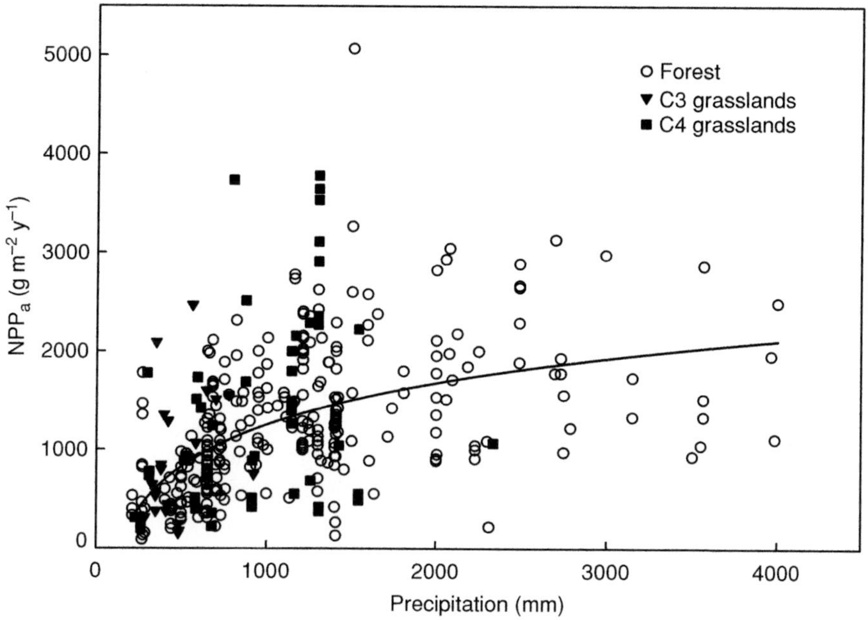

Figure 6.2 Total NPP (g m^{-2} year^{-1}) versus precipitation (mm) across world biomes. The trend line was drawn for forest data only (original data from Olson *et al.* (2001)).

6.3 Variability in water resources and plant productivity

6.3.1 Temporal variability in water resources

Some biomes are characterised by the strong seasonality of water availability. For plant productivity it is not indifferent if the water comes continuously in a regular fashion, or if it comes in widely separated instalments (Harper *et al.*, 2005). In tropical savannas, grasslands and regions with Mediterranean climate, there are several months without rain, occasionally interrupted by sporadic rainfall events. In

Table 6.1 NPP, annual water supply (precipitation + irrigation) and WUE_e, i.e., the quotient of biomass production to water supply in a eucalypt plantation in Furadouro, central Portugal, 6 years after planting* (adapted from Madeira *et al.*, 2002)

Water supply (mm)	Treatments	NPP (aboveground) (kg m^{-2} year^{-1})	WUE_e (g mm^{-1})
613	C	2.08	3.39
613	F	2.39	3.89
1532	I	2.90	1.89
1532	IL	3.25	2.12

*See text for details.

these ecosystems, NPP is often more closely related with the length of wet or dry seasons than with annual rainfall per se (House & Hall, 2001).

The timing of the rainy seasons is also important. Ecosystems with winter rain (Mediterranean) and summer rain (monsoonal) differ in NPP and community composition, for example, in the distribution of plants with C4 and C3 photosynthesis metabolism. As C4 plants are favoured by drought and high temperatures during the growing season, the mixture of C3 and C4 species can be achieved in one of two ways: a temporal separation, with C3 grasses active in winter–spring and C4 grasses active in summer, or by growth-form separation as in the monsoonal system with C4 grasses and C3 woody vegetation (Ehleringer & Cerling, 2001). The Mediterranean type of ecosystems, which have an active winter–spring C3 herbaceous component, do not have a native group of C4 plants because the summer is too dry, even though C4 crops (such as maize) thrive there when irrigated.

C4 crops have an intrinsic transpiration efficien y that is roughly twice that of C3 crops, due to lower stomatal conductance and higher photosynthetic capacity. In rainfed crops, however, actual transpiration efficien y under the usual climatic conditions for the different photosynthetic types is rather conservative. This is because WUE_t is also determined by the prevailing vapour pressure deficit and so for temperate zone C3 crops a less efficien photosynthetic pathway is compensated for by a more humid atmosphere (Rockström, 2003).

In many arid and semi-arid environments, rainfall pulses are a major feature of the climate and the ecosystem goes through repeated cycles of drying and rewetting (Schwinning *et al.*, 2004). During wet periods plant production may occur and reserves are stored for the continuation of ecosystem functioning between rain events (Reynolds *et al.*, 2004). However, some plant groups (e.g. trees) may obtain resources from different depths in the soil (Walter, 1973), behaving in partial independence from specifi rainfall events. Plant responses may be (1) increase in LAI due to germination of annuals and sprouting of perennials, (2) beginning of photosynthesis in perennials as plant water status improves and (3) mineralisation of soil organic matter and improvement of nutrient availability. But not all rainfall events trigger the same responses. The rain thresholds will vary with plant functional group and response type. For example, the amount of water delivered by a given 'rainfall pulse' may not be enough to allow the increase in grass LAI, but permit the mineralisation of soil organic matter. The biological meaning of rainfall pulses will be different for each component of the ecosystem (Reynolds *et al.*, 2004).

Plant responses also depend upon the 'memory' of the system, i.e., the time between rain events will modify the response, and there is often a decoupling between resource availability and their use. For example, as soils dry during the prolonged rainless season, their biological activity declines. When soils are subsequently rewetted by small rain events, there is a sudden 'burst' of decomposition, nutrient mineralisation and CO_2 release – the Birch effect (Cui & Caldwell, 1997; Austin *et al.*, 2004; Jarvis *et al.*, in press) – but not plant activity (Pereira *et al.*, 2003). At this time the herbaceous plants may not be there to utilise the released nutrients, and the deep-rooted perennials cannot use current rainfall until water is enough to reach deeper soil horizons. In these circumstances, the loss of carbon and nitrogen from the soil is inevitable (Pereira *et al.*, 2003; Schwinning & Sala, 2004; Jarvis *et al.*, in press), and so summer rains often have no effect on plant growth. Climate changes towards greater aridity may decrease water and nutrient availability due to enhanced temporal heterogeneity and increased asynchrony of water availability and the growing season (Austin *et al.*, 2004). Rain falling when plant cover is scarce leads to a decrease in the proportion of water that is used by the plants $[T/(T + E)]$ and lower WUE_e.

Severe droughts may have long-lasting effects on ecosystems. For example, during the severe drought of 1994 in Spain there was high mortality of *Quercus ilex* trees and other woody species (Peñuelas *et al.*, 2001). Similar results have been reported for other regions as shown by tree-ring analyses, which allow a precise dating of tree deaths over decades. Episodes of massive tree mortality occurred in northern Patagonia and coincided with exceptionally dry springs and summers during the years 1910s, 1942–1943 and the 1950s (Villalba & Veblen, 1998). Different species may exhibit different sensitivities to drought. Those species that normally reach subsoil water, as *Q. ilex* ssp. *rotundifolia* (David *et al.*, 2004), showed less variability in wood-ring patterns with climate than species that depend more on the use of current precipitation, e.g., *Pinus halepensis* (Ferrio *et al.*, 2003).

In many cases there is not a simple short-term relationship between tree death and annual rainfall. Jenkins and Pallardy (1995) studied the effects of drought on growth and death of trees of the red oak group in Missouri Ozark Mountains and found that trees that were dead at the time of sampling had in all cases been severely affected by drought in the past. Likewise, ring variation could be used to predict the likelihood of tree death following a severe drought in *Pinus edulis* in arid northern Arizona (Ogle *et al.*, 2000). In northeastern Spain Lloret *et al.* (2004) found that the response of *Q. ilex* to the 1994 drought was influence by the effects of a drought 10 years earlier: plants that resprouted weakly after the previous drought were more likely to die in response to the recent event than the more vigorous plants. How vigorously a given plant recovers from stress will influenc its hierarchy in the community and chances of survival. The resilience of ecosystems subjected to recurrent extreme droughts may be seriously affected by the loss of vigour and increasing difficult of regeneration of surviving trees (Lloret *et al.*, 2004).

6.3.2 Variability in space

Spatial variability in water resources may have a large effect on the landscape. In addition to micro-environment patterns the spatial variability in water can be affected by the plants themselves. Plant foliage intercepts rain before it reaches the soil, leading to evaporative losses and to the rearrangement of water input into the soil; roots and litter enhance water infiltratio and reduce run-off, whereas roots may promote redistribution of moisture (Ryel *et al.*, 2004). Work in evergreen oak Mediterranean savannas showed that more water was stored and was available in soils underneath tree crowns than in the open. This may result from better soil properties (e.g. more organic matter; Joffre & Rambal, 1993) and increased rain capture by canopy interception and throughfall (David *et al.*, 2005) as well as from hydraulic redistribution through roots (Ludwig *et al.*, 2003).

An increasing number of studies have reiterated the crucial role of deep rooting for plant survival during the drought season (but see Section 6.3.3). In tropical and temperate zone savannas, the long dry seasons tend to select either for deep-rooting woody perennials that may use subsoil water (Schenk & Jackson, 2005) and/or for herbaceous plants that are strict drought avoiders with their life cycle tuned to the duration of the period with enough soil moisture (Walter, 1973). Although soil water may be exhausted up to the grass/shrub rooting depth during the dry season, enough water is usually available for woody plant transpiration, except in extremely dry sites or after severe droughts.

Deep rooting (>1 m) is more likely to occur in sandy soils, as opposed to clayey or loamy soils (Schenk & Jackson, 2002a) and depends on plant type, increasing from annuals to trees (Schenk & Jackson, 2002b). In extreme arid environments, rooting depth is limited by the small infiltratio depth that results from low-rainfall events on very dry soils (Schenk & Jackson, 2002b).

6.3.3 *In situ water redistribution – hydraulic redistribution*

Root architecture and distribution in the soil is of utmost importance as it determines plant access to water (Ryel *et al.*, 2004). However, roots have also the role of water redistribution. The passive movement of water through roots from wetter, deeper soil layers into drier, shallower layers along a gradient of water potential (Caldwell *et al.*, 1998; Horton & Hart, 1998) is known as *hydraulic lift*. A similar concept was developed to include the downward (Schulze *et al.*, 1998) or even lateral transport of water by roots. Together they are called *hydraulic redistribution* (Burgess *et al.*, 1998). These processes typically occur when stomatal aperture is minimal (e.g. at night), otherwise the atmospheric draw on water for transpiration is stronger than that provided by the water potential gradients in the soil. Hydraulic redistribution seems to be more effective in plants with dimorphic root distributions (e.g. shallow lateral and deep tap roots) and where soil water infiltratio is limited as in more fine-t xtured soils (Ryel *et al.*, 2004).

Hydraulic redistribution has been proposed as a mechanism that can buffer plants against water deficit during seasonal drought (Richards & Caldwell, 1987; Ryel, 2004). The downward water transport increases infiltration reducing run-off losses and may help plants to use water in a more conservative way and may facilitate root growth through dry soil layers (Schulze *et al.*, 1998), as well as allowing nutrient uptake from deep soil horizons (McCulley *et al.*, 2004). In hydraulic lift, water absorbed by deep roots is redistributed back to shallow roots, enabling them to survive and absorb water and nutrients even when the soil is dry and to take advantage of precipitation pulses (Seyfried *et al.*, 2005). The quantity of water redistributed by the upward movement of water may amount to 14–33% of plant daily transpiration (Richards & Caldwell, 1987). The movement of water to the shallower soil layers, where most of the soil nutrients and microbes are, can improve plant water and nutrient status (Caldwell *et al.*, 1998), as well as provide benefit to mycorrhizal mutualists (Querejeta *et al.*, 2003) and neighbouring plants (Dawson, 1993) (but see Ludwig *et al.* (2004); see also Section 6.4).

6.4 Plant communities facing drought

Adaptation to semi-arid environments, namely in the Mediterranean, may be used as a paradigm for the range of plant traits adaptive to water scarcity. In Figure 6.3a, plants of group I have drought-avoiding behaviour without photosynthetic active parts during dry periods but survive in a resistant form. These are a majority in the flor of most semi-arid and arid environments (e.g. annuals, chamaephytes). Another extreme is plants of group II, which are water spenders without tolerance of dehydration, exploiting specifi habitats that permit access to water during most of the year. The other groups in Figure 6.3 consist of 'drought persistent' (i.e. perennial plants that maintain some photosynthesis during the dry periods) according to Noy-Meir (1973). Some of these are true xerophytes, but others may be very vulnerable to climate change such as the lauroid schlerophyllous (group V), which are relicts from the Tertiary, such as *Arbutus* and *Myrtus*, that may be eradicated if rainfall becomes more irregular than in the present period (Figure 6.3b). Groups III and IV succeed either by avoiding dehydration through stomatal closure (group III) or by some dehydration avoidance (e.g. deep rooting) and a variable degree of tolerance to dehydration (Valladares *et al.*, 2004b).

6.4.1 Species interactions with limiting water resources

Species coexistence in a situation of limiting water resources implies either avoiding interactions (niche segregation) or allowing some interaction (niche overlap). For example, the coexistence of different functional types regarding water resources enables plant communities to occupy a larger amount of physical space, exploring more resources (McConnaughay & Bazzaz, 1992). The exploitation of spatially and/or temporally distinct water resources by plants allows the coexistence of different species and life forms in environments where water is scarce (Noy-Meir, 1973; Reynolds *et al.*, 2004).

Heterogeneity in hydrological conditions across topographic gradients may result in niche differentiation as has been observed in many plant communities (e.g. Dawson, 1990). Even in the absence of any obvious topographic variation, species segregation along a niche gradient of soil drying has been shown to occur (Silvertown *et al.*, 1999). In water-limited environments successful competitors have root systems that are able to rapidly proliferate in resource-rich volumes of soil, depleting the resources before competing plants do (Passioura, 1982; Kroon *et al.*, 2003). For example, Eissenstat and Caldwell (1988) showed that *Agropyron desertorum*, an invader bunchgrass of the Great Basin in the United States, exhibited root growth

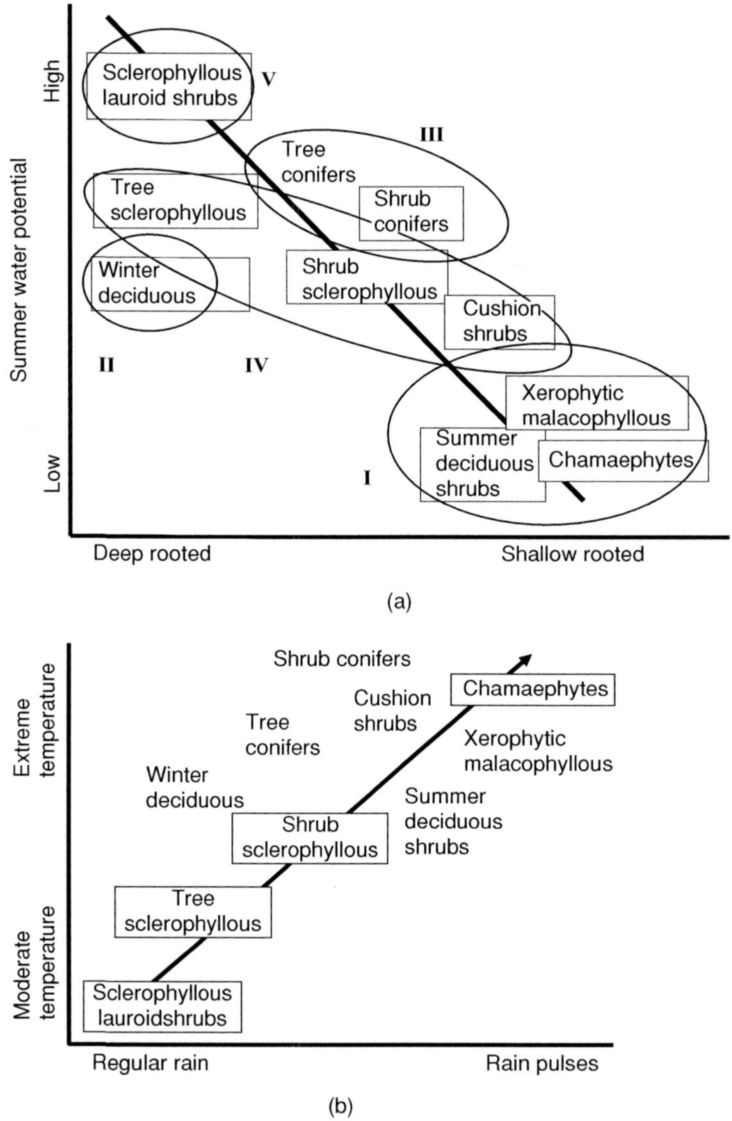

Figure 6.3 (a) Distribution of the main functional groups of Mediterranean woody plants in relation to their strategies regarding water use (see text) and (b) the same functional groups according to their positioning in face of climate conditions and presumed tendencies with climate change. The arrow in (b) indicates the trend of climate change according to most scenarios (adapted from Valladares *et al.* (2004b)).

earlier in the season than *Pseudoroegneria spicata* (native bunchgrass), resulting in more rapid water extraction.

Competition is a relatively frequent plant–plant interaction in semi-arid and arid plant communities (Fowler, 1986). However, as water availability fluctuate temporally and spatially, it can be postulated that the intensity of competition also fluctuates For example, trees and shrubs in semi-arid and arid systems can specialise in using deeper stores of water for drought survival, but they usually have an extensive and fairly dense horizontal root system in the sub-superficia layers (10–30 cm), augmented in wet periods by deciduous rootlets. There is strong competition for water in this layer between direct evaporation, ephemerals and shrubs (and shrub seedlings) and between different species within each plant group (Noy-Meir, 1973; LeRoux *et al.*, 1995). Nevertheless, the stratificatio of soil moisture and root systems tends to minimise competition for water and enables coexistence (Lin *et al.*, 1996). These contrasting results may arise from differences in the seasonality of precipitation, with stratificatio being most effective in environments with most precipitation falling when low-potential evapotranspiration or plant inactivity allows a surplus of water to infiltrat for later use by deep-rooted plants (Sankaran *et al.*, 2004; but see Section 6.3). As mentioned above, the downward redistribution of water (hydraulic redistribution) can be a mechanism for deep-rooted plants to store water below the reach of shallower rooting plants. Competition for water can be avoided by the asynchrony of biological activity, e.g., different phenologies or different growth responses to temperature (Reynolds *et al.*, 2000; Filella & Peñuelas, 2003), as is the case of trees and herbaceous plants in Mediterranean ecosystems.

Positive interactions, or facilitation, occur when one plant species enhances the survival, growth or fitnes of another (Callaway, 1995). Neighbouring plant species may compete with one another for resources but they may also provide benefit for neighbours such as more available moisture, shade, higher nutrient levels and shared resources via mycorrhizae. Under water-stress conditions the shade provided by 'nurse plants' significantl increases seedling survival because of improved water relations. Hydraulic lifted water by deep-rooted plants can facilitate water use by shallow-rooted plants (Dawson, 1993), including tree seedlings (Brooks *et al.*, 2002). But this is not always the case because competition by roots of the dominant plants may eradicate the advantages (Ludwig *et al.*, 2003).

It is unlikely that species coexistence is determined by one mechanism alone. Net effects of one species on another may be the sum of both positive (niche differentiation, facilitation) and negative (competition) effects (Callaway, 1995). For example, nurse plants may facilitate the germination of seedlings of herbaceous plants by reducing soil temperature and increasing water content (Holzapfel & Mahall, 1999), but negatively affect seedling growth by reducing light. The combination of shade and drought may be especially meaningful for plant recruitment during ecological succession in environments with a long dry season. In such cases (dry-shade) seedling mortality may be higher in the shade rather than in the open, as shade not only decreases photosynthetic assimilation but also reduces the assimilate allocation to the roots more than the allocation to the shoots, rendering the plants less capable of avoiding dehydration (Valladares *et al.*, 2004a). The balance between negative and positive effects can also vary as productivity and resource availability increases (Pugnaire *et al.*, 1996). This is supported by Briones *et al.* (1998), who found that competition between three dominant perennial desert species could be absent or reduced in low-precipitation years and be high in years with abundant precipitation.

Productivity of water-limited communities can be affected by species richness and identity. For example, water use and productivity of a community dominated by drought-avoiding species (group I) can be totally different from another dominated by water-spenders species (group II). In an extreme example with species substitution, Farley *et al.* (2005) showed that the afforestation of grasslands and shrublands could reduce the annual run-off on average by 44 and 31%, respectively. Run-off reduction can mirror higher community water use and productivity, although other factors can be involved (e.g. increased canopy interception losses).

Species- and functional-group rich communities can be more productive than poorer ones due to complementarity in resource use or positive interactions (but see Huston, 1997). For example, in a Mediterranean grassland, species-rich communities were more productive and used more available water than poorer ones (Caldeira *et al.*, 2001). Also, asynchronous responses of different species to drought may lead to more stable primary productivity in diverse ecosystems than in less diverse communities (Yachi & Loreau, 1999). Several empirical studies showed that the temporal variability of ecosystems properties, e.g., productivity, decreased with increasing diversity (e.g. Tilman & Downing, 1994; Caldeira *et al.*, 2005).

6.4.2 Vegetation change and drought: is there an arid zone 'treeline'?

In the long-term, the mortality of woody plants may lead to changes in species geographical distribution. For example a simulation with the biogeochemistry–biogeography model BIOME4 (Kaplan *et al.*, 2003) for Portugal, run with climate data from the Hadley Centre HadRM2 regional model, predicted that forest-dominated biomes might decrease from approximately 30 to 17%, whereas shrublands and grasslands might increase from 2 to 24% under a severe climate change scenario with atmospheric CO_2 concentration twice the present (Pereira *et al.*, 2002). Changes would be more pronounced in the drier southern and interior regions where drought might become more severe and species are closer to the boundaries of their climatic distribution ranges. The simultaneous occurrence of severe droughts and wildfire might intensify this process.

Single drought events may also change plant community boundaries. For example, in northern New Mexico, United States, a severe drought during the 1950s shifted the ecotone between ponderosa pine forest and pinyon–juniper woodlands (*Pinus edulis–Juniperus monosperma*) by more than 2 km (Allen & Breshears, 1998). This change was rapid (less than 5 years), and occurred through the mortality of ponderosa pine, while the more drought resistant pinyon–juniper woodland was recruited into the new space. The recent drought of 2003 led to a sudden, regional-scale, tree mortality of the dominant species (*Pinus edulis*) (Breshears *et al.*, 2005). This die-off was more extensive than that in the 1950s possibly because in 2003 it was accompanied by anomalously high air temperatures. The results of Breshears *et al.* (2005) quantify a trigger leading to rapid, drought-induced die-off of overstory woody plants and highlight the potential for such die-off to be more severe and extensive for droughts under warmer conditions.

In arid environments trees can be displaced and substituted by other plant growth-forms such as shrubs, establishing a dry-land treeline (Stevens & Fox, 1991). When rains are infrequent and fail to fully saturate the soil, deep-rooted trees may be at a competitive disadvantage in comparison to shallower rooted functional groups (see Figure 6.3). In hot deserts, deep-rooted plants are largely restricted to habitats with deep-water infiltratio such as washes, wadis or rock clefts (Schenk & Jackson, 2005). For example, in the Taklamakan desert the water-spending desert phreatophytes, such as *Populus euphratica*, have little tolerance of dehydration and their high water demand can only be met by ground water (Gries *et al.*, 2003). Outside these specifi habitats, rooting depth of desert plants is often restricted by shallow infiltratio depths (Schenk & Jackson, 2002a).

Plant hydraulic failure as a result of water stress determines the limit of water deficit that a plant can withstand. It occurs when leaf and xylem water potentials fall below a species specifi xylem cavitation threshold (Jackson *et al.*, 2000) or if soil hydraulic conductance falls to zero due to high rates of plant water extraction or desiccation (Sperry *et al.*, 1998). As water becomes scarcer, leaf water status is maintained above the threshold for xylem runaway cavitation by stomatal control and leaf area adjustments, avoiding loss of hydraulic continuity with soil water (Sperry *et al.*, 2002). Other factors being equal, the hydraulic limits in the soil–leaf continuum depend on the branching structure, overall size of the continuum and root/shoot ratio (Sperry *et al.*, 2002). As trees grow taller, increasing leaf water stress due to gravity and path length resistance may ultimately limit leaf expansion and photosynthesis so that further height growth can increase the risk of xylem cavitation (Koch *et al.*, 2004). The partial dieback of peripheral branches and their attendant foliage may be a last-resort mechanism for whole-plant water conservation to survive drought (Davis *et al.*, 2000). Under severe water deficits trees which have a single stem, may be more vulnerable to hydraulic failure than shrubs, typically with multiple stems. The hydraulic segmentation achieved by the multiple stems system can confin cavitation to the disposable organs that can thus be sacrificed leaving still some viable elements (Rood *et al.*, 2000), functioning as an insurance for long-term survival. On the other hand, repeated dieback of tree canopies with recurrent drought may induce a shrub habit in plants that would otherwise develop into a tree.

6.5 Droughts and wildfires

Fire is a natural component of many ecosystems. Often, it is the fir regime (frequency, intensity and timing) rather than drought that determines primary productivity as well as plant community (Pyne, 1997; Bond *et al.*, 2005). Nevertheless, dry weather enhances the risk of biomass burning. For example, the severe drought of 1994 that damaged large amounts of woody plants in central and southern Spain

(Peñuelas *et al.*, 2001) also resulted in major forest fires which burnt approximately 1.6% of the national forest area.

It is likely that wildfire will become more common in the future worldwide (Bond *et al.*, 2005). The IPCC Third Assessment Report states that the higher the maximum temperatures, the more hot days and heat waves are very likely to occur over nearly all land areas, increasing the risk of forest fire (IPCC, 2001). Pereira *et al.* (2002) simulated the impact of future climate change on the meteorological risk of fir in Portugal. They found a significan increase in fir severity and length of fir season under the future climate, which resulted from a temperature increase and a decrease in precipitation in spring–summer. Likewise, Brown *et al.* (2004) found that prospective drying in the western United States created a future climate scenario with an increase in the number of days of high fir danger.

Vegetation fire are always possible because plant biomass is a good fuel in our oxygen-rich atmosphere. Live biomass, however, does not burn easily because it has a high moisture content. Drought interacts with fires increasing dead branches and leaf shedding. These materials (dead biomass or necromass) represent the fin fuels, which once dehydrated in hot and dry weather, become highly inflammabl and increase the risk of fire Although drought and wildfire share common causes, it cannot be concluded that more or larger fire will occur in more arid regions. For fire to occur and expand, adequate amounts of fin fuel must be present. Wind, topography and human activities (often as the source of ignition) will also play a role (Pyne, 1997). The Iberian Peninsula may serve as a good case study. Fire frequency is highest in the hilly provinces of central and northern Portugal and Galicia (Spain), not in the more arid south (European Commission, 2003; Pereira & Santos, 2003). Wildfire occur where highly productive periods alternate with a hot dry weather, which facilitates ignition. The Mediterranean vegetation '*could... stand as a dictionary definition of a fire-prone environment. Annually, it undergoes a rhythm of winter wetting and summer drying, over which beats a cruder rhythm of drought. Almost always there is fuel in abundance – combustibles that lack only a properly timed spark to burst into flame*' (Pyne, 2005). Likewise, tropical savannas, where a highly productive rainy season alternates with a dry season, are the major contributors for biomass burning globally (Dwyer *et al.*, 2000). In more arid climates, primary productivity is lower, decreasing the amount of fuel and fir incidence (Lloret, 2004).

Extreme events can override the climate tendency. For example, in 2003 Portugal experienced its worst fir season, with a total burnt area of about 5% of the countryside (~4000 km^2; Pereira & Santos, 2003). But 2003 was not a very dry year as the annual precipitation exceeded the 1951–1980 30-year average. The exceptional fir season resulted from a heat wave, i.e., daily temperature maxima rising 5°C above the daily average (period of reference 1961–1990) for at least 6 consecutive days.

Droughts may have dramatic effects in ecosystems where water deficit are uncommon, as happened in the tropical rain forests of Southeast Asia in 1997/1998 where widespread wildfire were triggered by the droughts associated with the El Niño Southern Oscillation (ENSO) phenomenon (Roberts, 2001). Likewise, it was estimated that during the 2001 ENSO period of drought approximately one-third of Amazonian forests became susceptible to fir (Nepstad *et al.*, 2004).

In regions where fir has been present for a long time, such as where a Mediterranean type of climate prevails, the vegetation has evolved under a strong fir influ ence (Lloret, 2004; Pausas *et al.*, 2004; Bond *et al.*, 2005). Plant traits responsible for post-fir persistence operate either at the level of the individual (resprouting) or by stimulating germination from the soil seed bank. Nevertheless, the regeneration depends largely upon environmental conditions before and after the fir as well as the fir regime (Lloret, 2004).

The post-fir persistence plant traits are often associated with differences in drought resistance. Morphological drought-avoiding traits (e.g. higher root/whole-plant biomass, deeper root systems) are more common in resprouters than in non-resprouters (Pausas *et al.*, 2004). Furthermore, fire-induce sprouting does increase drastically the ratio of root to canopy biomass and will promote drought avoidance after fir (Lloret, 2004). On the contrary, woody non-resprouters (e.g., germination stimulated by fire tend to be more drought-tolerant (e.g. higher xylem resistance to cavitation and embolism) and survive on drier sites than do resprouters. It appears that a greater drought resistance may be only coincidental and not causally related.

Fires may induce changes in soil hydraulic properties and nutrient availability, which may exacerbate the impacts of a drought. The effects depend largely on type of biomass burnt and on soil characteristics (type and moisture content), fir characteristics (intensity and duration), as well as on post-fir precipitation (Chandler *et al.*, 1983). In general, low to moderate severity fire may promote a transient increase of pH and available nutrients as well as the enhancement of hydrophobicity, lowering the capability for the soil to soak up water (Certini, 2005). Severe fires however, may have a much stronger impact. They may cause removal of organic matter, the creation of water-repellent layers, which may decrease markedly water infiltratio rates, the deterioration of the soil structure and the increase in bulk density, which will result in further decreases in permeability and in water-holding capacity of the soil (Certini, 2005). One consequence of these changes in soil hydraulics is increased run-off and surface erosion, which, in turn, may induce a decline in nutrient availability, enhanced by volatilisation losses due to heating (Lloret, 2004; Certini, 2005). However, fir may improve nutrient availability, especially in cases where primary productivity is stagnant due to the immobilisation of nutrients in plant biomass or slow-decomposing litter and soil organic matter. In such cases fir may function as a rejuvenation factor at ecosystem level that will stimulate post-fir primary productivity, although this effect may be short-lived (Briggs & Knapp, 1995; Van de Vijver *et al.*, 1999; Santos *et al.*, 2003a).

6.6 Agricultural and forestry perspectives

6.6.1 Agriculture

The world cultivated land is 80% dedicated to rainfed agriculture, with the remaining 20% allocated to irrigation (Rockström, 2003). Nevertheless, irrigated agriculture is a major consumer of water resources and 40% of the food and agriculture commodities are produced in irrigated areas. With the predicted growth in human population and climate change scenarios of increasing water scarcity, especially in the interior of continents and semi-arid regions, achieving a better efficien y of use of water in agriculture has become a major issue for farmers and researchers. Furthermore, land degradation reduces the soil water holding capacity and many irrigation systems waste large amounts of water. For example, more than 50% of the water allocated to irrigation in the southern and eastern Mediterranean may be wasted (Araus, 2004). One of the main aims of the 2000 World Water Council in the Hague was to increase water productivity for food production from rainfed and irrigated agriculture by 30% until 2015 (FAO, 2002). Additionally, increasing plant water use in agriculture is limited because sufficien run-off has to be guaranteed to sustain river ecology and other water uses, especially in drought-prone environments. It was suggested that globally only approximately 17% of the fresh water can be used for agricultural production (Rockström, 2003).

Many practices developed over the history of agriculture aimed at increasing the availability of water (such as irrigation, rainwater harvesting, mulching and contour ploughing) and enhancing the share of crop use in ecosystem water balance (such as ploughing, weeding, adjusting spacing to water availability). Plant selection and breeding for water-limited environments has resulted frequently in greater crop competitiveness with weeds and more thorough use of water resources (Blum, 1984). However, as mentioned above, especially in drought-prone environments, increasing plant water use in agriculture may be limited by other social and ecological needs.

Concerns for a more efficien use of water resources led to the development of new management strategies that bring to the fiel agronomical and plant physiology concepts that may improve crop WUE while maintaining or even improving crop production and quality. New approaches may exploit plant sensing and physiological signalling of mild water deficit that coordinate plant adaptive responses to water shortage, as it is provided by controlled irrigation (Loveys *et al.*, 2004). Attempts to manage crop source/sink balance by fine-tunin agricultural practices are also important (Goodwin & Boland, 2002) as harvest indices are often sensitive to water deficits Plant breeding to develop genotypes with improved water uptake or better WUE without penalising yield is also taking place (see Chapter 5). Plant plasticity under water deficit is large, with some genotypes showing a high potential to deal with periods of water shortage (Centritto *et al.*, 2004; Chaves & Oliveira, 2004).

Among the options to improve productivity in rainfed agriculture, is increasing the ratio between plant transpiration and non-productive evaporation losses through (1) avoiding the early season soil evaporation (or consumption by weeds or fallow) before full emergence of the crop and (2) maintaining high canopy cover throughout the growing season (Rockström, 2003). Increasing WUE_t by increasing yields through improved agricultural management and plant breeding is possible and desirable (Wallace, 2000; Gregory, 2004). However, as discussed above, increasing WUE_t has limitations. Moreover, constitutive high WUE_t is sometimes associated with a low productivity syndrome that may limit the scope for breeding crops for higher transpiration efficien y.

Table 6.2 Improving water economy in rainfed crops*

Strategies	Tools
Optimising canopy development to increase the ratio of crop transpiration/soil evaporation	Agronomic and breeding practices
Reducing water losses by drainage and increasing water capture	Early crop cover, deep root systems (genetic or nutrition)
Improving WUE at the leaf level	Need to overcome pests and diseases and nutrient limitations, breeding
Improving harvest index per unit of water used	Adjusting crop phenology to the environment, specially fl wering time

*Adapted from Passioura (2004).

Climate change may have contrasting impacts on rainfed agriculture depending on geography and technology. While droughts may reduce crop production, warming and elevated atmospheric CO_2 may act positively on production potential. But even in water-limited environments, precipitation may not be the major determinant of crop productivity. In Australia, wheat seldom reaches the yield potential of 20 kg ha^{-1} mm^{-1} of water supply due to a combination of several limiting factors, such as low soil fertility or pests and diseases (Passioura, 2004). To come closer to the yield potential in rainfed crops in a changing climate, adaptation techniques should be adopted in the short-term and in the long-term (Pinto & Brandão, 2002). The former include the adequate choice of cultivars, timely planting, correct densities and harvest dates, as well as proper soil and nutrient management. Based on the Australian experience with dry land wheat, Passioura (2004) lists some practices that can ensure efficien water use (Table 6.2). On the other hand, land degradation may intensify the effects of drought to disaster levels.

The long-term measures will be the search for new genotypes with a better adaptation to heat and drought and increased water- and nutrient-use efficiencies Biotechnology may play a fundamental role in this context, although it must be acknowledged that a significan gestation time is still required before its impact is realised, as far as genetic modifie crops are concerned (InterAcademy Council, 2004). There are, however, major breakthroughs utilising conventional breeding – good examples are the drought tolerant maize and wheat lines developed by CIMMYT through marker-selected breeding. Another example is the New Rice for Africa (NERICA), interspecifi hybrid rice obtained by crossing *Oryza sativa* (Asian rice) with *Oryza glaberrima* (African rice), that gives 35% higher grain yields than the upland African rice varieties, when cultivated with traditional rainfed systems without fertilizer (InterAcademy Council, 2004). In addition to higher yields, the NERICA varieties are richer in protein and they are claimed to be more disease and drought resistant than local varieties of the West African savanna region.

In irrigated agriculture there is a strong need to increase efficien y, avoiding unnecessary water spending while improving product quality (Araus, 2004). These are the objectives of fine-tunin irrigation practices such as defici irrigation, whereby water is supplied below the full plant demand, allowing a mild stress to develop with only small negative effects on yield (FAO, 2002). This strategy may lead to greater economic gains than that by maximising yields. In general, defici irrigation has been more successfully applied to crops less sensitive to water deficit (such as cotton, maize, groundnut, grapevine, peach or pears) than to sensitive crops like potato (Kirda *et al.*, 1999).

Regulated defici irrigation (RDI) is a type of defici irrigation where the amount of water applied is not constant throughout crop development, taking into consideration the needs at each stage. This method is used in high-density orchards to reduce excessive growth and to optimise fruit size and quality (Chalmers, 1986). RDI may also improve the extent of soil water uptake as mild deficit during vegetative growth may have a favourable effect on root growth, improving water acquisition from deeper soil layers, as observed in studies with groundnuts in India (FAO, 2002).

In the partial rootzone drying approach, each side of the root system is irrigated during alternate periods. The plant water status is maintained by the wet part of the root system and stomatal closure is promoted by the dehydrating roots of the other half of the root system (Davies *et al.*, 2000), using less water per plant. This type of defici irrigation will be efficien in canopies where stomatal control over shoot water status through transpiration is important (Kang & Zhang, 2004). This is the case in crops with isohydric behaviour, where stomata do respond to root signalling, most likely through ABA synthesised in the roots and modulated via xylem pH, such as grapevines (Santos *et al.*, 2003b; Loveys *et al.*, 2004; Souza *et al.*, 2005, see Chapter 5).

An efficien monitoring of plant performance is an essential component of the water-saving strategy. Several techniques are available, although most of them are time-consuming and demanding as far as equipment is concerned, such as monitoring soil water or plant water relations (sap fl w meters or leaf water potential). Thermal imaging is emerging as a potential tool to monitor canopy water status. The use of indices such as crop water stress index, calculated from canopy temperatures in relation to references, can give us estimates of stomatal aperture and therefore be used for irrigation scheduling (for a review see Jones, 2004a).

6.6.2 Forestry

As in agriculture, current trends in population growth and improvement of living standards leads to an increase in the global demand for forest products. The consumption of wood-based products and paper increased four times faster than the population during the twentieth century (FAO, 2000). Today these needs are partly covered by cultivated forests, but natural forests will face an increasing pressure for logging. Additionally, deforestation for agriculture and energy is likely to proceed in tropical countries. On the other hand, the forest management paradigm changed during the last decades, emphasising sustainable management and ecosystem services, rather than wood production alone. As a consequence, forests must provide raw materials, preserve biodiversity and provide other ecosystem services such as the mitigation of greenhouse gas emissions through carbon sequestration. Because natural forests are at risk and need to be preserved, there is little doubt that managed forests – both tree plantations and 'renaturalised' forests – will continue to perform these essential roles in society.

In many regions, e.g. central Europe, forest production may have increased with global change due to the effects of increased CO_2 concentration in the atmosphere combined with nitrogen deposition (Bascietto *et al.*, 2004; Kilpeläinen *et al.*, 2005) and a longer growing season due to warming (Myneni *et al.*, 1997). However, severe droughts can offset such gains (Raffalli-Delerce *et al.*, 2004). That is the case of France and Portugal, where assessments of the impacts of climate change in forestry at the regional level have forecasted gains in productivity in the wetter northern regions and losses in the drier southern regions (Loustau *et al.*, 2005; Pereira *et al.*, 2005). In addition to the generalised drought effects on NPP, the change of carbon allocation towards roots will reduce the proportion of NPP available for stem growth, resulting in a greater decline in timber productivity than in NPP.

Changes in climate, e.g. increasing drought severity, will put trees under stress and may influenc the distribution of other organisms, some of them essential for ecosystem function (mycorrhizae) as well as for the preservation of biodiversity. On the other hand, many observations suggest that plants subjected to drought stress may become more susceptible to insect attacks (Mattson & Haack, 1987). For example, plant water stress had a major role in promoting survival and growth of *Phorachantha semipunctata* larvae, an insect pest that attacks *Eucalyptus globulus* outside Australia (Caldeira *et al.*, 2002). The consequent tree mortality may lead to this crop becoming unviable in drought-prone areas.

Maintaining forest productivity with increased aridity may imply diverting to the economically interesting species the largest possible proportion of water supply. This may be achieved using deep-rooting genotypes (if possible), site preparation techniques that can improve water availability (e.g. by removing hardpans that limit rooting depth) and increasing the ratio of transpiration/actual evapotranspiration (T/AET). The main non-productive portion of AET is the evaporation loss of rainfall intercepted by the canopies, which may account for 25–75% of overall evapotranspiration (McNaughton & Jarvis, 1983). Very little has been done to increase T/AET, except manipulating tree density. Yet, as mentioned above, the option of using more water for tree production is constrained by the need to allow enough run-off and drainage to maintain ecological and socioeconomic services such as river f ows and aquifer recharge.

The reduction of stand density (thinning) may decrease the interception losses and increase the amount of water available per remaining tree, enhancing their survival and growth. Thinning, however, may produce changes in the physical environment below the canopy (e.g. increasing light, higher temperature, changes in soil organic matter decomposition rates), which favour the development of understorey vegetation. This will compete with canopy trees, thus offsetting the effects of water reallocation in the stand. Furthermore, in fire-pron environments, the development of understorey vegetation can pose an additional risk, as grasses, shrubs and juvenile trees are more quickly affected by droughts than deep-rooted mature trees, increasing the amount of highly inflammabl biomass (see also Section 6.5). The association of fire with frequent severe droughts and, eventually, with pests and diseases may bring about drastic changes in the environmental settings for forest development, requiring an adaptive approach to forest management.

During the last decades forest management has emphasised sustainability of resource use and ecosystem services. While the current practices are able to cope to some degree with the effects of climate fluctuation and its associated impacts, large gaps still persist in our knowledge of forest ecosystems functioning and their responses to multiple disturbances. Furthermore, given the long timescale of forest growth, the present climate change process may be too rapid for the natural adjustment of forests to the new environments. Improved ecosystem monitoring and research are therefore key steps in management under a rapidly changing climate, and should be incorporated into the management process itself (Dale *et al.*, 2001). The adaptive management approach, which considers learning as a part of the management process, may be essential especially because greater climatic variability and increased frequency of extreme events are expected.

References

Allen, C.D. & Breshears, D.D. (1998) Drought-induced shift of a forest-woodland ecotone: rapid landscape response to climate variation. *Proc. Natl. Acad. Sci. USA*, **95**, 14839–14842.

Araus, J.L. (2004) The problems of sustainable water use in the Mediterranean and research requirements for agriculture. *Ann. Appl. Biol.*, **144**, 259–272.

Austin, A.T., Yahdjian, L., Stark, J.M., Belnap, J., Porporato, A., Norton, U., Ravetta, D.A. & Schaeffer, S.M. (2004) Water pulses and biogeochemical cycles in arid and semiarid ecosystems. *Oecologia*, **141**, 221–235.

Bascietto, M., Cherubini, P. & Scarascia-Mugnozza, G. (2004) Tree rings from a European beech forest chronosequence are useful for detecting growth trends and carbon sequestration. *Can. J. For. Res.*, **34**, 481–492.

Blum, A. (1984) Methods of selection for plant tolerance to environmental stresses. In: *Selection in Mutation Breeding: Proceedings of Consultants Meeting Organized by Joint FAO IAEA Division of Isotope and Radiation Applications of Atomic Energy for Food and Agricultural Development*, Vienna, 21–25 June 1982. International Atomic Energy Agency, Vienna.

Bond, W.J., Woodward, F.I. & Midgley, G.F. (2005) The global distribution of ecosystems in a world without fire *New Phytol.*, **165**, 525–538.

Breshears, D.D., Cobb, N.S., Rich, P.M., Price, K.P., Allen, C.D., Balice, R.G., Romme, W.H., Kastens, J.H., Floyd, M.L., Belnap, J., Anderson, J.J., Myers, O.B., & Meyer, C.W. (2005) Regional vegetation die-off in response to global-change-type drought. *Proc. Natl. Acad. Sci. USA.* **102**, 15144–15148.

Briggs, J.M. & Knapp, A.K. (1995) Interannual variability in primary production in tallgrass prairie: climate, soil, moisture, topographic position, and fir as determinants of aboveground biomass. *Am. J. Bot.*, **82**, 1024–1030.

Briones, O., Montaña, C. & Ezcurra, E. (1998) Competition intensity as a function of resource availability in a semiarid ecosystem. *Oecologia*, **116**, 365–372.

Brooks, J.R., Meinzer, F.C., Coulombe, R. & Gregg, J. (2002) Hydraulic redistribution of soil water during summer drought in two contrasting Pacifi Northwest coniferous forests. *Tree Physiol.*, **22**, 1107–1117.

Brown, T.J., Hall, B.L. & Westerling, A.L. (2004) The Impact of twenty-firs century climate change on wildland fir danger in the western United States: an applications perspective. *Clim. Change*, **62**, 365–388.

Burgess, S.S.O., Adams, M.A., Turner, N.C. & Ong, C.K. (1998) The redistribution of soil water by tree root systems. *Oecologia*, **115**, 306–311.

Caldeira, M.C., Fernandéz, V., Tomé, J. & Pereira, J.S. (2002) Eucalyptus longicorn borer response to tree water stress. *Ann. For. Sci.*, **59**, 99–106.

Caldeira, M.C., Hector, A., Loreau, M. & Pereira, J.S. (2005) Species richness, temporal variability and resistance of biomass production in a Mediterranean grassland. *Oikos*, **110**, 115–123.

Caldeira, M.C., Ryel, R.J., Lawton, J.H. & Pereira, J.S. (2001) Mechanisms of positive biodiversity-production relationships: insights provided by δ^{13}C analysis in experimental Mediterranean grassland plots. *Ecol. Lett.*, **4**, 439–443.

Caldwell, M.M., Dawson, T.E. & Richards, J.H. (1998) Hydraulic lift: consequences of water efflu from the roots of plants. *Oecologia*, **113**, 151–161.

Callaway, R.M. (1995) Positive interactions among plants. *Bot. Rev.*, **61**, 306–349.

Centritto, M., Wahbib, S., Serraj, R. & Chaves, M.M. (2004) Effects of partial rootzone drying (PRD) on adult olive tree (*Olea europaea*) in fiel conditions under arid climate. II. Photosynthetic responses. *Agric. Ecosyst. Environ.*, **106**, 303–311.

Certini, G. (2005) Effects of fir on properties of forest soils: a review. *Oecologia*, **143**, 1–10.

Chalmers, D.J. (1986) Research and progress in cultural systems and management in temperature fruit orchards. *Acta Hort.*, **175**, 215–225.

Chandler, C., Cheney, P., Thomas, P., Trabaud, L. & Williams, D. (1983) *Fire in Forestry: Forest Fire Behavior and Effects*. Wiley, New York.

Chaves, M.M., Maroco, J.P. & Pereira, J.S. (2003) Understanding plant responses to drought: from genes to the whole plant. *Funct. Plant Biol.*, **30**, 239–264.

Chaves, M.M. & Oliveira, M.M. (2004) Mechanisms underlying plant resilience to water deficits prospects for water-saving agriculture. *J. Exp. Bot.*, **55**, 2365–2384.

Chaves, M.M., Pereira, J.S. & Osório, J. (2004) Water use efficien y and photosynthesis. In: *Water Use Efficiency in Plant Biology* (ed. M. Bacon). Blackwell Publishing, London.

Cui, M. & Caldwell, M.M. (1997) A large ephemeral release of nitrogen upon wetting of dry soil and corresponding root responses in the field *Plant Soil*, **191**, 291–299.

Dai, A., Trenberth, K.E. & Qian, T. (2004) A global data set of Palmer drought severity index for 1870–2002: relationship with soil moisture and effects of surface warming. *J. Hydrometeorol.*, **5**, 1117–1130.

Dale, V.H., Joyce, L.A., McNulty, S., Neilson, R.P., Ayres, M.P., Flannigan, M.D., Hanson, P.J., Irland, L.C., Lugo, A.E., Peterson, C.J., Simberloff, D., Swanson, F.J., Stocks, B.J. & Wotton, B.M. (2001) Climate change and forest disturbances. *Bioscience*, **51**, 723–734.

David, T.S., Ferreira, M.I., Cohen, S., Pereira, J.S. & David, J.S. (2004) Constraints on transpiration from an evergreen oak tree in southern Portugal. *Agric. For. Meteorol.*, **122**, 193–205.

David, T.S., Gash, J.H.C., Valente, F.V., Pereira, J.S., Ferreira, M.I. & David, J.S. (2005) Rainfall interception by an isolated evergreen oak tree in a Mediterranean savannah. *Hydrol. Processes.*, (published online December 2005, DOI 10.1002/hyp.6062).

Davies, W.J., Bacon, M.A., Thompson, D.S., Sobeih, W. & Rodriguez, L. (2000) Regulation of leaf and fruit growth in plants growing in drying soil: exploitation of the plants' chemical signalling system and hydraulic architecture to increase the efficien y of water use in agriculture. *J. Exp. Bot.*, **51**, 1617–1626.

Davis, S.D., Ewers, F.W., Sperry, J.S., Portwood, K.A., Crocker, M.C. & Adams, G.C. (2002) Shoot dieback during prolonged drought in Ceanothus (Rhamnaceae) chaparral of California: a possible case of hydraulic failure. *Am. J. Bot.*, **89**, 820–828.

Dawson, T.E. (1990) Spatial and physiological overlap of three co-occurring alpine willows. *Funct. Ecol.*, **4**, 13–25.

Dawson, T.E. (1993) Hydraulic lift and water use by plants: implications for water balance, performance and plant-plant interactions. *Oecologia*, **95**, 565–574.

Dwyer, E., Pinnock, S., Gregoire, J.M. & Pereira, J.M. (2000) Global spatial and temporal distribution of vegetation fir as determined from satellite observations. *Int. J. Remote Sensing*, **21**, 1289–1302.

Earl, H.J. & Davis, R.F. (2003) Effect of drought stress on leaf and whole canopy radiation use efficien y and yield of maize. *Agron. J.*, **95**, 688–696.

Ehleringer, J.R. & Cerling, T.E. (2001) Origins and expansion of C4 photosynthesis. In: *Encyclopedia of Global Environmental Change* (eds H.A. Mooney & J. Canadell). Wiley, London.

Eissenstat, D.M. & Caldwell, M.M. (1988) Competitive ability is linked to rates of water extraction. A fiel study of two aridland tussock grasses. *Oecologia*, **75**, 1–7.

Ellis, J. (1994) Climate variability and complex ecosystem dynamics: implications for pastoral development. In: *Living with Uncertainty: New Directions in Pastoral Development in Africa* (ed. I. Scoones). Intermediate Technology Publication, London.

European Commission (2003) *Forest Fires in Europe: 2002 Fire Campaign*. European Commission, JRC, Ispra, Italy.

FAO (2000) *The Global Outlook for Future Wood Supply from Forest Plantations*. FAO, Rome.

FAO (2002) *Deficit Irrigation Practices*. FAO, Rome.

Farley, K.A., Jobbágy, E.G. & Jackson, R.B. (2005) Effects of afforestation on water yield: a global synthesis with implications for policy. *Global Change Biol.*, **11**, 1565–1576.

Ferrio, J.P., Florit, A., Vega, A., Serrano, L. & Voltas, J. (2003) Δ13C and tree-ring width reflec different drought responses in *Quercus ilex* and *Pinus halepensis*. *Oecologia*, **137**, 512–518.

Filella, I. & Peñuelas, J. (2003) Partitioning of water and nitrogen in co-occurring mediterranean woody shrub species of different evolutionary history. *Oecologia*, **137**, 51–61.

Fowler, N. (1986) The Role of competition in plant communities in arid and semiarid regions. *Ann. Rev. Ecol. Syst.*, **17**, 89–110.

Ghashghaie, J., Duranceau, M., Badeck, F.W., Cornic, G., Adeline, M.T. & Deleens, E. (2001) Delta13c of CO_2 respired in the dark in relation to δ^{13}C of leaf metabolites: comparison between *Nicotiana sylvestris* and *Helianthus annuus* under drought. *Plant Cell Environ.*, **24**, 505–515.

Goodwin, I. & Boland, A.M. (2002) Scheduling irrigation of fruit trees for optimizing water use efficien y, Water Reports, FAO Publication Number 22. FAO, Rome.

Gregory, P.J. (2004) Agronomic approaches to increasing water use efficien y. In: *Water Use Efficiency in Plant Biology* (ed. M. Bacon). Blackwell Publishing, London.

Gries, D., Zeng, F., Foetzki, A., Arndt, S.K., Bruelheide, H., Thomas, F.M., Zhang, X. & Runge, M. (2003) Growth and water relations of *Tamarix ramosissima* and *Populus euphratica* on Taklamakan desert dunes in relation to depth to a permanent water table. *Plant Cell Environ.*, **26**, 725–736.

Gutschick, V.P. & BassiriRad, H. (2003) Extreme events as shaping physiology, ecology, and evolution of plants: toward a unifie definitio and evaluation of their consequences. *New Phytol.*, **160**, 21–42.

Harper, C.W., Blair, J.M., Fay, P.A., Knapp, A.K. & Carlisle, J.D. (2005) Increased rainfall variability and reduced rainfall amount decreases soil CO_2 flu in a grassland ecosystem. *Global Change Biol.*, **11**, 322–334.

Haupt-Herting, S., Klug, K. & Fock, H.P. (2001) A new approach to measure gross CO_2 flu es in leaves. Gross CO_2 assimilation, photorespiration, and mitochondrial respiration in the light in tomato under drought stress. *Plant Physiol.*, **126**, 388–396.

Holzapfel, C. & Mahall, B.E. (1999) Bidirectional facilitation and interference between shrubs and annuals in the Mojave Desert. *Ecology*, **80**, 1747–1761.

Horton, J.L. & Hart, S.C. (1998) Hydraulic lift: a potentially important ecosystem process. *Trends Ecol. Evol.*, **13**, 232–235.

House, J.I. & Hall, D.O. (2001) Productivity of tropical savannas and grasslands. In: *Terrestrial Global Productivity* (eds J. Roy, B. Saugier & H.A. Mooney). Academic Press, San Diego, CA.

Huston, M.A. (1997) Hidden treatments in ecological experiments: re-evaluating the ecosystem function of biodiversity. *Oecologia*, **110,** 449–460.

Huxman, T.E., Smith, M.D., Fay, P.A., Knapp, A.K., Shaw, M.R., Loik, M.E., Smith, S.D., Tissue, D.T., Zak, J.C., Weltzin, J.F., Pockman, W.T., Sala, O.E., Haddad, B.M., Harte, J., Koch, G.W., Schwinning, S., Small, E.E. & Williams, D.G. (2004) Convergence across biomes to a common rain-use efficien y. *Nature*, **429**, 651–654.

InterAcademy Council (2004) *Realizing the Promise and Potential of African Agriculture. Science and Technology Strategies for Improving Agricultural Productivity and Food Security in Africa*. Inter-Academy Council, Amsterdam, The Netherlands.

IPCC (2001) *Climate Change 2001: Synthesis Report*. A contribution of working groups I, II and III to the Third Assessment Report of the Intergovernmental Panel on Climate Change. Cambridge University Press, Cambridge, UK.

Jackson, R.B., Sperry, J.S. & Dawson, T.E. (2000) Root water uptake and transport: using physiological processes in global predictions. *Trends Plant Sci.*, **5**, 482–488.

Jarvis, P.G., Rey, A., Petsikos, C., Wingate, L., Rayment, M., Pereira, J.S., Banza, J., David, J.S., Miglietta, F., Borgetti, M., Manca, G. & Valentini, R. (in press) Drying and wetting of soils stimulates decomposition and carbon dioxide emission: the 'Birch Effect'. *Tree Physiol.*

Jenkins, M.A. & Pallardy, S.G. (1995) The influenc of drought on red oak group species growth and mortality in the Missouri Ozarks. *Can. J. For. Res.*, **25**, 1119–1127.

Joffre, R. & Rambal, S. (1993) How tree cover influence the water balance of Mediterranean rangelands. *Ecology*, **74**, 570–582.

Jones, H.G. (2004a) Application of thermal imaging and infrared sensing in plant physiology and ecophysiology. *Adv. Bot. Res.*, **41**, 107–163.

Jones, H.G. (2004b) What is water use efficien y? In: *Water Use Efficiency in Plant Biology* (ed. M. Bacon). Blackwell Publishing, London.

Kang, S. & Zhang, J. (2004) Controlled alternate partial root-zone irrigation: its physiological consequences and impact on water use efficien y. *J. Exp. Bot.*, **55**, 2437–2446.

Kaplan, J.O., Bigelow, N.H., Prentice, I.C., Harrison, S.P., Bartlein, P.J., Christensen, T.R., Cramer, W., Matveyeva, N.V., McGuire, A.D., Murray, D.F., Razzhivin, V.Y., Smith, B., Walker, D.A., Anderson, P.M., Andreev, A.A., Brubaker, L.B., Edwards, M.E. & Lozhkin, A.V. (2003) Climate change and Arctic ecosystems. 2. Modeling, paleodata-model comparisons, and future projections. *J. Geophys. Res.* (*Atmospheres*), **108**, D 19, 817, 12-1-12-17.

Kilpeläinen, A., Peltola, H., Ryyppö, A. & Kellomäki, S. (2005) Scots pine responses to elevated temperature and carbon dioxide concentration: growth and wood properties. *Tree Physiol.*, **25**, 75–83.

Kirda, C., Moutonnet, P., Hera, C. & Nielsen, D.R. (eds) (1999) *Crop Yield Response to Deficit Irrigation*. Kluwer Academic Publishers, Dordrecht, The Netherlands.

Koch, G.W., Sillett, S.C., Jennings, G.M. & Davis, S.D. (2004) The limits to tree height. *Nature*, **428**, 851–854.

Kroon, H.D., Mommer, L. & Nishiwaki, A. (2003) Root Competition: towards a mechanistic understanding. In: *Root Ecology* (eds H.D. Kroon & E.J.W. Visser). Springer-Verlag, Heidelberg.

Lamb, H.H. (1995) *Climate History and the Modern World*. Routledge, London.

Lambers, H., Chapin, F.S. & Pons, T.L. (1998) *Plant Physiological Ecology*. Springer-Verlag, New York.

LeRoux, X., Bariac, T. & Mariotti, A. (1995) Spatial partitioning of the soil water resource between grass and shrub components in a West African humid savanna. *Oecologia*, **104**, 147–155.

Lieth, H. & Whittaker, R.H. (1975) *Primary Productivity of the Biosphere*. Springer-Verlag, New York.

Lin, G.H., Phillips, S.L. & Ehleringer, J.R. (1996) Monsoonal precipitation responses of shrubs in a cold desert community on the Colorado Plateau. *Oecologia*, **106**, 8–17.

Lloret, F. (2004) Régimen de incendios y regeneración. In: *Ecologia del bosque mediterráneo en un mundo cambiante* (ed. F. Valladares). Ministério de Medio Ambiente, Madrid.

Lloret, F., Siscart, D. & Dalmases, C. (2004) Canopy recovery after drought dieback in holm-oak Mediterranean forests of Catalonia (NE Spain). *Global Change Biol.*, **10**, 2092–2099.

Loik, M.E., Breshears, D.D., Lauenroth, W.K. & Belnap, J. (2004) A multi-scale perspective of water pulses in dryland ecosystems: climatology and ecohydrology of the western USA. *Oecologia*, **141**, 269–281.

Loustau, D., Bosc, A., Colin, A., Ogée, J., Davi, H., François, C., Dufrêne, E., Déqué, M., Cloppet, E., Arrouays, D., Bas, C.L., Saby, N., Pignard, G., Hamza, N., Granier, A., Bréda, N., Ciais, P., Viovy, N. & Delage, F. (2005) Modeling climate change effects on the potential production of French plains forests at the sub-regional level. *Tree Physiol.*, **25**, 813–823.

Loveys, B.R., Stoll, M. & Davies, W.J. (2004) Physiological approaches to enhance water use efficien y: exploiting plant signalling in novel irrigation practice. In: *Water Use Efficiency in Plant Biology* (ed. M. Bacon). Blackwell Publishing, London.

Ludwig, F., Dawson, T.E., de Kroon, H., Berendse, F. & Prins, H.H.T. (2003) Hydraulic lift in *Acacia tortilis* trees on an East African savanna. *Oecologia*, **134**, 293–300.

Ludwig, F., Dawson, T.E., Prins, H.H.T., Berendse, F. & de Kroon, H. (2004) Below-ground competition between trees and grasses may overwhelm the facilitative effects of hydraulic lift. *Ecol. Lett.*, **7**, 623–631.

Madeira, M.V., Fabião, A., Pereira, J.S., Araújo, M.C. & Ribeiro, C. (2002) Changes in carbon stocks in *Eucalyptus globulus* Labill. plantations induced by different water and nutrient availability. *For. Ecol. Manag.*, **171**, 75–85.

Mattson, W.J. & Haack, R.A. (1987) The role of drought stress in provoking outbreaks of phytophagous insects. In: *Insect Outbreaks* (eds P. Barbosa & J.C. Schultz). Academic Press, New York.

McConnaughay, K.D.M. & Bazzaz, F.A. (1992) The occupation and fragmentation of space: consequences of neighboring roots. *Funct. Ecol.*, **6**, 704–710.

McCulley, R.L., Jobbagy, E.G., Pockman, W.T. & Jackson, R.B. (2004) Nutrient uptake as a contributing explanation for deep rooting in arid and semi-arid ecosystems. *Oecologia*, **141**, 620–628.

McNaughton, K.G. & Jarvis, P.G. (1983) Predicting effects of vegetation change on transpiration and evaporation. In: *Water Deficits and Plant Growth* (ed. T.T. Kozlowski). Academic Press, New York.

Miranda, P.M.A., Valente, M.A., Tomé, A.R., Trigo, R., Coelho, M.F.E.S., Aguiar, A. & Azevedo, E.B. (2006) O clima de Portugal nos séculos XX e XXI. In: *Alterações Climáticas em Portugal. Cenários, Impactes e Medidas de Adaptação* (eds F.D. Santos & P. Miranda), pp. 49-113. Gradiva, Lisbon, Portugal.

Munné-Bosch, S. & Alegre, L. (2004) Die and let live: leaf senescence contributes to plant survival under drought stress. *Funct. Plant Biol.*, 203–216.

Myneni, R.B., Keeling, C.D., Tucker, C.J., Asrar, G. & Nemani, R.R. (1997) Increased plant growth in the northern high latitudes from 1981 to 1991. *Nature*, **386**, 698–702.

Nepstad, D., Lefebvre, P., Silva, U.L.-D., Tomasella, J., Schlesinger, P., Solórzano, L., Moutinho, P., Ray, D. & Benito, J.G. (2004) Amazon drought and its implications for forest flammabilit and tree growth: a basin-wide analysis. *Global Change Biol.*, **10**, 704–717.

Noy-Meir, I. (1973) Desert ecosystems: environments and producers. *Ann. Rev. Ecol. Syst.*, **4**, 398–417.

Ogle, K., Whitham, T.G. & Cobb, N.S. (2000) Tree-ring variation in pinyon predicts likelihood of death following severe drought. *Ecology*, **81**, 3237–3243.

Olson, R.J., Scurlock, J.M.O., Prince, S.D., Zheng, D.L. & Johnson, K.R. (eds) (2001) *NPP Multi-Biome: Global Primary Production Data Initiative Products. Data Set.* Oak Ridge National Laboratory Distributed Active Archive Center, Oak Ridge, TN.

Passioura, J.B. (1982) Water in the soil-plant-atmosphere continuum. In: *Encyclopedia of Plant Physiology – Physiological Plant Ecology II* (eds O.L. Lange, P.S. Nobel, C.B. Osmond & H. Ziegler). Springer-Verlag, Berlin.

Passioura, J.B. (2004) Water use efficien y: in the farmer's fields In: *Water Use Efficiency in Plant Biology* (ed. M. Bacon). Blackwell Publishing, London.

Pausas, J.G., Bradstock, R.A., Keith, D.A. & Keeley, J.E. (2004) Plant functional traits in relation to fir in crown-fir ecosystems. *Ecology*, **85**, 1085–1100.

Peñuelas, J., Lloret, F. & Montoya, R. (2001) Severe drought effects on Mediterranean woody flor in Spain. *For. Sci.*, **47**, 214–218.

Pereira, J.M.C. & Santos, M.T.N. (2003) *Fire Risk and Burned Area Mapping in Portugal.* Direcção Geral das Florestas, Lisboa, Portugal.

Pereira, J.S., Correia, A.V. & Correia, A.P. (2005) Florestas e biodiversidade. In: *Alterações climáticas em Portugal* (eds F.D. Santos & P. Miranda). Gradiva, Lisboa, Portugal.

Pereira, J.S., Correia, A.V., Correia, A.P., Branco, M., Bugalho, M., Caldeira, M.C., Souto-Cruz, C., Freitas, H., Oliveira, A.C., Pereira, J.M.C., Reis, R.M. & Vasconcelos, M.J. (2002) Forests and biodiversity. In: *Climate Change in Portugal. Scenarios, Impacts and Adaptation Measures* (eds F.D. Santos, K. Forbes & R. Moita). Gradiva, Lisboa, Portugal.

Pereira, J.S., David, J.S., David, T.S., Caldeira, M.C. & Chaves, M.M. (2003) Carbon and water flu es in Mediterranean-type ecosystems – constraints and adaptations. In: *Progress in Botany* (eds K. Esser, U. Lüttge, W. Beyschlag & J. Murata). Springer-Verlag, Berlin.

Pereira, J.S., Tenhunen, J.D., Lange, O.L., Beyschlag, W., Meyer, A. & David, M.M. (1986) Seasonal and diurnal patterns in leaf gas exchange of *Eucalyptus globulus* trees growing in Portugal. *Can. J. For. Res.*, **16**, 177–184.

Pinto, P.A. & Brandão, A.P. (2002) Agriculture. In: *Climate Change in Portugal. Scenarios, Impacts, and Adaptation Measures* (eds F.D. Santos, K. Forbes & R. Moita). Gradiva, Lisboa, Portugal.

Pugnaire, F., Haase, P. & Puigdefabregas, J. (1996) Facilitation between higher plant species in a semiarid environment. *Ecology*, **77**, 1420–1426.

Pyne, S.J. (1997) *Vestal Fire: An Environmental History, Told Through Fire, of Europe and Europe's Encounter with the World*. University of Washington Press, Seattle.

Pyne, S.J. (2005) Fogo no jardim: compreensão do contexto dos incêndios em Portugal (Fire in the garden: understanding the contexts for fir in Portugal). In: *Incêndios Florestais em Portugal: Caracterização, Impactes e Prevenção* (eds J.M.C. Pereira, F.C. Rego & J.S. Pereira). ISA Press, Lisboa, Portugal.

Querejeta, J.I., Egerton-Warburton, L.M. & Allen, M.F. (2003) Direct nocturnal water transfer from oaks to their mycorrhizal symbionts during severe soil drying. *Oecologia*, **134**, 55–64.

Raffalli-Delerce, G., Masson-Delmotte, V., Dupouey, J.L., Stievenard, M., Breda N. & Moisselin, J.M. (2004) Reconstruction of summer droughts using tree-ring cellulose isotopes: a calibration study with living oaks from Brittany (western France). *Tellus B*, **56**, 160–174.

Rambal, S., Joffre, R., Ourcival, J.M., Cavender-Bares, J. & Rocheteau, A. (2004) The growth respiration component in eddy CO_2 flu from a *Quercus ilex* mediterranean forest. *Global Change Biol.*, **10**, 1460–1469.

Reichstein, M., Tenhunen, J.D., Roupsard, O., Ourcival, J.M., Rambal, S., Miglietta, F., Peressotti, A., Pecchiari, M., Tirone, G. & Valentini, R. (2002) Severe drought effects on ecosystem CO_2 and H_2O flu es at three Mediterranean evergreen sites: revision of current hypotheses? *Global Change Biol.*, **8**, 999–1017.

Reynolds, J.F., Kemp, P.R., Ogle, K. & Fernandez, R.J. (2004) Modifying the 'pulse-reserve' paradigm for deserts of North America: precipitation pulses, soil water, and plant responses. *Oecologia*, **141**, 194–210.

Reynolds, J.F., Kemp, P.R. & Tenhunen, J.D. (2000) Effects of long-term rainfall variability on evapotranspiration and soil water distribution in the Chihuahuan Desert: a modeling analysis. *Plant Ecol.*, **150**, 145–159.

Richards, J.H. & Caldwell, M.M. (1987) Hydraulic lift: substantial nocturnal water transport between soil layers by *Artemisia tridentata* roots. *Oecologia*, **73**, 486–489.

Roberts, S.J. (2001) Tropical fir ecology. *Prog. Phys. Geogr.*, **25**, 286–291.

Rockström, J. (2003) Water for food and nature in drought-prone tropics: vapour shift in rain-fed agriculture. *Philos. Trans. R. Soc. Lond. B Biol. Sci.*, **358**, 1997–2009.

Rood, S.B., Patino, S., Coombs, K. & Tyree, M.T. (2000) Branch sacrifice cavitation-associated drought adaptation of riparian cottonwoods. *Trees: Struct. Funct.*, **14**, 248–257.

Rosati, A. & Dejong, T.M. (2003) Estimating photosynthetic radiation use efficien y using incident light and photosynthesis of individual leaves. *Ann. Bot.*, **91**, 869–877.

Rosenzweig, M.L. (1968) Net primary productivity of terrestrial communities: prediction from climatological data. *Am. Nat.*, **102**, 67–74.

Ruimy, A., Jarvis, P.G., Baldocchi, D.D. & Saugier, B. (1995) CO2 flu es over plant canopies and solar radiation: a review. *Adv. Ecol. Res.*, **26**, 1–68.

Russell, G., Jarvis, P.G. & Monteith, J.L. (1989) Absorption of radiation by canopies and stand growth. In: *Plant Canopies. Their Growth Form and Function* (eds G. Russell, B. Marshall & P.G. Jarvis). Cambridge University Press, Cambridge, UK.

Ryel, R.J. (2004) Hydraulic redistribution. In: *Progress in Botany* (eds K. Esser, U. Lüttge, W. Beyschlag & J. Murata). Springer-Verlag, New York.

Ryel, R.J., Leffle , A.J., Peek, M.S., Ivans, C.Y. & Caldwell, M.M. (2004) Water conservation in *Artemisia tridentata* through redistribution of precipitation. *Oecologia*, **141**, 335–345.

Sankaran, M., Ratnam, J. & Hanan, N.P. (2004) Tree-grass coexistence in savannas revisited – insights from an examination of assumptions and mechanisms invoked in existing models. *Ecol. Lett.*, **7**, 480–490.

Santos, A.J.B., Silva, G.T.D.A., Miranda, H.S., Miranda, A.C. & Lloyd, J. (2003a) Effects of fir on surface carbon, energy and water vapour flu es over campo sujo savanna in central Brazil. *Funct. Ecol.*, **17**, 711–719.

Santos, T.P., Lopes, C.M., Rodrigues, M.L., Souza, C.R., Maroco, J.P., Pereira, J.S., Silva, J.R. & Chaves, M.M. (2003b) Partial rootzone drying: effects on growth and fruit quality of field-gr wn grapevines (*Vitis vinifera*). *Funct. Plant Biol.*, **30**, 663–671.

Schenk, H.J. & Jackson, R.B. (2002a) The global biogeography of roots. *Ecol. Monogr.*, **72**, 311–328.

Schenk, H.J. & Jackson, R.B. (2002b) Rooting depths, lateral root spreads and below-ground/above-ground allometries of plants in water-limited ecosystems. *J. Ecol.*, **90**, 480–494.

Schenk, H.J. & Jackson, R.B. (2005) Mapping the global distribution of deep roots in relation to climate and soil characteristics. *Geoderma*, **126**, 129–140.

Schulze, E.D., Caldwell, M.M., Canadell, J., Mooney, H.A., Jackson, R.B., Parson, D., Scholes, R., Sala, O.E. & Trimborn, P. (1998) Downward flu of water through roots (i.e. inverse hydraulic lift) in dry Kalahari sands. *Oecologia*, **115**, 460–462.

Schwinning, S. & Sala, O.E. (2004) Hierarchy of responses to resource pulses in arid and semi-arid ecosystems. *Oecologia*, **141**, 211–220.

Schwinning, S., Sala, O.E., Loik, M.E. & Ehleringer, J.R. (2004) Thresholds, memory, and seasonality: understanding pulse dynamics in arid/semi-arid ecosystems. *Oecologia*, **141**, 191–193.

Seyfried, M.S., Schwinning, S., Walvoord, M.A., Pockman, W.T., Newman, B.D., Jackson, R.B. & Phillips, E.M. (2005) Ecohydrological control of deep drainage in arid and semiarid regions. *Ecology*, **86**, 277–287.

Silvertown, J., Dodd, M.E., Gowing, D.J.G. & Mountford, J.O. (1999) Hydrologically define niches reveal a basis for species richness in plant communities. *Nature*, **400**, 61–63.

Souza, C.R.D., Maroco, J.P., Santos, T.P., Rodrigues, M.L., Lopes, C., Pereira, J.S. & Chaves, M.M. (2005) Control of stomatal aperture and carbon uptake by defici irrigation in two grapevine cultivars. *Agric. Ecosyst. Environ.*, **106**, 261–274.

Sperry, J.S., Adler, F.R., Campbell, G.S. & Comstock, J.C. (1998) Limitations of plant water use by rhizosphere and xylem conductance: results from a model. *Plant Cell Environ.*, **21**, 347–359.

Sperry, J.S., Hacke, U.G., Oren, R. & Comstock, J.P. (2002) Water deficit and hydraulic limits to leaf water supply. *Plant Cell Environ.*, **25**, 251–263.

Stahle, D.W., Cleaveland, M.K., Blanton, D.B., Therrell, M.D. & Gay, D.A. (1998) The lost colony and Jamestown droughts. *Science*, **280**, 564–567.

Stevens, G.C. & Fox, J.F. (1991) The causes of treeline. *Ann. Rev. Ecol. Syst.*, **22**, 177–191.

Tilman, D. & Downing, J.A. (1994) Biodiversity and stability in grasslands. *Nature*, **367**, 363–365.

Tyree, M.T. & Sperry, J.S. (1989) Vulnerability of xylem to cavitation and embolism. *Ann. Rev. Plant Physiol. Mol. Biol.*, **40**, 19–38.

Valladares, F., Aranda, I. & Sánchez-Gómez, D. (2004a) La luz como factor ecológico y evolutivo para las plantas y su interaccion con el agua. In: *Ecologia del bosque mediterráneo en un mundo cambiante* (ed. F. Valladares). Ministério de Medio Ambiente, Madrid.

Valladares, F., Vilagrosa, A., Penuelas, J., Ogaya, R., Camarero, J.J., Corcuera, L., Sisó, S. & Gil-Pelegrín, E. (2004b) Estrés hídrico: ecofisiolo ía y escalas de la sequía. In: *Ecologia del bosque mediterráneo en un mundo cambiante* (ed. F. Valladares). Ministério de Medio Ambiente, Madrid.

Van de Vijver, C.A.D.M., Poot, P. & Prins, H.H.T. (1999) Causes of increased nutrient concentrations in post-fir regrowth in an East African savanna. *Plant Soil*, **214**, 173–185.

Vilagrosa, A., Bellot, J., Vallejo, V.R. & Gil-Pelegrâin, E. (2003) Cavitation, stomatal conductance, and leaf dieback in seedlings of two co-occurring Mediterranean shrubs during an intense drought. *J. Exp. Bot.*, **54**, 2015–2024.

Villalba, R. & Veblen, T.T. (1998) Influence of large-scale climatic variability on episodic tree mortality in northern Patagonia. *Ecology*, **79**, 2624–2640.

Wallace, J.S. (2000) Increasing agricultural water use efficien y to meet future food production. *Agric. Ecosyst. Environ.*, **82**, 105–119.

Walter, H. (1973) *Vegetation of the Earth*. Springer-Verlag, New York.

Westerling, A.L., Gershunov, A., Brown, T.J., Cayan, D.R. & Dettinger, M.D. (2003) Climate and wildfir in the western United States. *Bull. Am. Meteorol. Soc.*, **84** (5), 595–604.

Yachi, S. & Loreau, M. (1999) Biodiversity and ecosystem productivity in a fluctuatin environment: the insurance hypothesis. *Proc. Natl. Acad. Sci. USA*, **96**, 1463–1468.

7 Effects of temperature and precipitation changes on plant communities

M.D. Morecroft and J.S. Paterson

7.1 Introduction

The response of a plant community to climate change is not simply the sum of the responses of the component species. It will also be determined by interactions between species (animals as well as plants), colonisations and changes in soil processes and microclimate. The importance of these factors is illustrated by the fact that many species can be successfully cultivated outside their natural climatic limits, provided other conditions are suitable and the plant is freed from competition (Figure 7.1).

Species responses to climate change are frequently presented in terms of changes in their distribution patterns, but it is important to remember that changes in distribution are inextricably linked with changes in community composition. The appearance or disappearance of a species in a particular place is de facto a change in community composition; it is also conditional on the outcome of community-scale processes, such as competition. At the large scale, all species, vegetation types and biomes have distributions that can be broadly related to climate, but they do not occur in all places where the climate is suitable. Climate define an envelope within which a species or vegetation type *may* exist, but other factors such as soil, management and successional stage, together with the constraints of dispersal and competition, control whether it is actually present in a particular place. This principle is analogous to that of the fundamental niche, as define by Hutchinson (1957) and contrasts with the realised niche, which is the full set of conditions – biotic as well as abiotic – under which a species really *does* occur. It is therefore important to understand the processes that control community structure and function in order to be able to predict the impacts of climate change.

Rising temperatures at a global scale are consistently predicted in general circulation models and although regional variations may be substantial and important (even potentially including regional cooling), it is possible to identify some general principles about the impact of 'global warming'. As the spatial patterns of climatic variables change, distributions of plant species and communities would be expected to follow them, albeit that the relationship may be complex and considerable time lags may occur. So, in broad terms, a warming of climate would be expected to lead to a shift of species distributions towards higher latitudes and higher altitudes in comparison to their present locations. Palaeoecological studies have demonstrated climate-induced changes in plant distributions of this sort during the Holocene period (e.g. Huntley & Birks, 1983; Birks, 1989; Jackson & Whitehead, 1991). In contrast, changes in precipitation patterns and aspects of climate such as wind speed have

(a)

(b)

Figure 7.1 *Alchemilla alpina* L. (alpine ladies mantle) plants growing at (a) 845 m and (b) 480 m on Great Dun Fell, northern England, United Kingdom, in individual pots, without competition. *A. alpina* only grows naturally at the higher altitude; at 480 m it is absent from the community, despite being able to grow better at the lower altitude. The plants were from a common origin and were initially the same size; they were allowed to grow for 14 months at the different altitudes (Morecroft & Woodward, 1996).

wider regional variation in the nature of change and less certainty in the models (IPCC, 2001). This makes it harder to draw generalised conclusions and fin analogous present day climates for future predictions. However, changes in precipitation patterns, in particular, may be of great ecological significanc where they do occur (Weltzin *et al.*, 2003).

This chapter considers the general principles that determine how plant communities are likely to change with climate change and provides examples from recent research addressing the issue. It cannot, however, be a comprehensive survey. There are very wide regional variations in both the predicted changes in climate and the character of plant communities, and it is not possible to deal comprehensively with all of them. There is also a wide disparity in the degree to which different communities and different regions have been studied. Most of the published research on the topic has addressed temperate, boreal and polar regions. This reflect both the distribution of most of the countries with a strong research base and perceptions of the susceptibility of communities to undesirable changes. It is telling that a search of the ISI Science Citation Index in June 2005 revealed that 31% of papers found using the keywords 'plant community' and 'climate change' were concerned with arctic or alpine communities!

7.2 Methodology

A number of approaches to studying the effects of climate change on plant communities have been adopted; these can be broadly categorised as follows:

1. Direct long-term monitoring
2. Experimental manipulations of climate
3. Inference from spatial patterns
4. Inference from palaeoecological studies
5. Modelling

The advantages and disadvantages of each of these techniques are summarised in Table 7.1 (see also Chapter 3). None by itself gives a complete understanding and in many cases a combination of approaches is necessary; for example models can only be validated by testing their output against observations, and attribution of temporal changes in plant communities to climate change is strengthened where it is supported by experimental testing.

Experimental approaches are important as they offer the opportunity to test causation of change in a rigorous, controlled way, which is simply not possible with purely observational studies, where other factors may vary in parallel with climate and be impossible to separate. They do, however, have to be interpreted with care and should not be regarded as exact simulations of future conditions. It is not possible to simulate landscape-scale processes, such as dispersal, in plot-scale experiments,

Table 7.1 Broad categories of techniques and approaches to the study of climate change impacts on community composition, and some of their main advantages and disadvantages

Technique	Advantages	Disadvantages
Direct long-term monitoring of change	1. Identifie real changes in real communities.	1. Generally requires many years to identify a trend. 2. Attribution of change to climate effects is often problematic.
Experimental manipulations of climate	1. Attribution to experimental treatments is unambiguous, given appropriate controls. 2. Climates outside of the existing range can be simulated.	1. Impossible to accurately simulate all aspects of climate change. 2. Processes operating at larger than plot scale (e.g. dispersal) tend to be excluded from study.
Inference from spatial patterns	1. Large-scale patterns and processes can be investigated. 2. Results are immediately available.	1. Attribution of spatial pattern to climate may not be clear. 2. Present-day climates may provide no suitable analogues for future climates.
Inference from palaeoecological studies	1. Allows very long-term trends to be identified 2. Deals with real changes in real communities. 3. Results are available as soon as processing of samples is completed.	1. Attribution of changes to climate can be difficult 2. Past habitats may differ substantially from present ones, making extrapolation difficult 3. Some species present very little material for study.
Modelling	1. Allows simulation of future climates and other circumstances, without present analogues.	1. Processes are either simplifie or modelling is based on correlations alone. The inherent assumptions may not hold under different circumstances.

and the treatment conditions are not exact simulations of future climates (Dunne *et al.*, 2004). This is particularly true of temperature manipulations, which have been carried out using a wide range of technologies, from passive techniques, such as ‘open-top chambers’, (which create a warming effect by trapping heat close to the ground in an analogous way to a greenhouse – the open top simply makes this less extreme and allows precipitation to reach the community) to those involving soil heating cables and infrared lamps. The passive techniques are cheaper and easier to apply in remote locations, but are inevitably less controlled. Shaver *et al.* (2000) review the strengths and weaknesses of each, in more detail. Manipulations of rainfall are in some ways more straightforward, but there can be problems with drought simulation. In this case some kind of shelter or ‘roof’ is necessary to shield experimental plots from rainfall. These may, however, create their own greenhouse effect, causing increased temperature. For this reason systems have been developed (e.g. Grime *et al.*, 2000; Beier *et al.*, 2004) to ensure that the plot is only covered during rainfall. In forest systems this is less necessary, as the canopy shades the covers, minimising heating effects. To study the responses of real, natural communities, experiments need to be carried out in situ in the field as it is virtually impossible to reproduce a natural community in a controlled environment. Controlled environment experiments can, however, be useful tools with which to study mechanisms of individual plant responses to climate or the interactions between a small number of species; which may in turn advance understanding of community processes.

Potential shifts in species distributions have been modelled by determining the present-day climate envelope and by projecting distributions on the basis of future climate scenarios (e.g. Huntley *et al.*, 1995; Bakkenes *et al.*, 2002; Pearson *et al.*, 2002). The technique is inevitably limited by how closely current distributions correlate with climate and will always be problematic in those species whose distributions are most dependent on other factors such as soil type or management history. Pearson *et al.* (2002) used the SPECIES model to predict European distributions of 32 species; one-third of the distribution data set was not used to develop the model, but to test it. They found generally good performance in predicting these present-day distribution patterns, with Pearson correlation coefficient (r) varying between 0.605 and 0.948 (mean = 0.841). The value of such models has nevertheless been debated (e.g. Pearson & Dawson, 2003; Hampe, 2004; Pearson & Dawson, 2004) as they do not explicitly address the role of biological interactions, the potential for evolving climatic tolerance and limitations on dispersal. To some extent this is a matter of exercising caution in interpretation: at best, these models indicate where a species *may* survive in future, rather than where it *will* occur. Despite the caveats, the climate envelope approach has proved to be a useful tool for visualising the sort of changes in distribution that are likely to occur and for highlighting species that are potentially at risk (Figure 7.2; Harrison *et al.*, 2001). However, to understand and predict the probable, rather than simply the possible, consequences of climate change for plant communities, a greater degree of understanding of community dynamics is clearly necessary. Mechanistic models, incorporating plant physiological processes, have made important contributions to understanding the large-scale distribution of biomes and the role of vegetation in the global carbon balance (e.g. Cramer *et al.*, 2001; Cox *et al.*, 2004). This approach has, however, limited applicability at the level of changes in species composition of particular communities, because of the impracticality of specifi parameterisation for more than a very few species so only more generic data are used for large-scale models.

7.3 Mechanisms of change in plant communities

7.3.1 Direct effects of climate

The most straightforward case of community composition changing in response to climate is where a species is simply unable to survive the new physical conditions and dies out. The converse situation – where a species that was formerly unable to survive

Figure 7.2 Example of present and future modelled distribution under climate change scenarios using a climate envelope approach: the distribution of *Rubus chamaemorus* (cloudberry) in the British Isles) under present and simulated future climates. Figure supplied by Dr Pam Berry (Oxford University Environmental Change Institute) from the RegIS2 Project.

in a particular environment, was enabled to do so – is more complex, in that dispersal is necessary in order for the species to reach the potential new habitat. In practice, as noted in the Introduction, many distributions are not directly define by the physical limits of survival of a species, but rather by climate mediated by competition (Figure 7.1; see also, e.g. Woodward & Pigott, 1975). There are, however, some examples of direct climatic effects. A classic interpretation of the northern limit of *Tilia cordata*, small-leaved lime, in Great Britain is that it is unable to set seed in cooler climates, which is a result of slow growth of the pollen tube (Pigott & Huntley, 1978, 1980, 1981). There is palaeoecological evidence that this northern limit has shifted over the course of time in ways that are consistent with responses to temperature. This particular example provides a reminder that it is important to take into account the whole life cycle of an organism in assessing its climate sensitivity. The survival of mature plants is no guarantee that they are able to reproduce. A more recent example of a system in which direct effects of climate may be more important than interspecifi interactions is the Alaskan tussock tundra, studied experimentally by Hobbie *et al.* (1999). They both manipulated temperature and removed individual species and concluded that the direct effects of climate had a greater impact on species than the removal of other members of the community.

7.3.2 Interspecific differences in growth responses to climate

In many, perhaps most cases, impacts of climate change on community composition are likely to be caused by differential effects on the growth of different species, rather than the direct elimination of species by climatic factors. It is therefore important to understand the reasons underlying interspecifi differences in plant responses to climate. Growth rates are intrinsically different in different species, and it is useful to be able to identify broad categories of plants. One approach to classifying 'functional types' is that of Grime (1979), who identifie contrasting plant 'strategies'. Objections to the theory that underpins this classificatio have been voiced (Tilman, 1988; Grace, 1991; Grubb, 1992) but it provides a useful framework with which to consider patterns of growth in this context. Plants that are adapted to grow in environmental conditions that tend to limit growth, such as low nutrient supply, low temperatures or low water supply, are identifie as *stress tolerators*. They have intrinsically low growth rates but are typically long-lived and capable of surviving adverse conditions. They can be distinguished from *competitors*, which have the capacity for high growth rates in nutrient-rich, high light conditions, giving a competitive advantage over slower growing forms. A third major group are the *ruderals*, which are typical of disturbed habitats and exploit short-lived niches; they have high reproductive rates, short generation times and high dispersal capacity (ruderals correspond to 'r-selected species' in the r–k system of McArthur and Wilson (1967)). These three 'strategies' – stress tolerators, competitors and ruderals – are extremes of a continuum with many intermediates. Vegetation types associated with cold climates tend to be dominated by stress tolerant species. With rising temperatures, faster growing competitor species may gain an advantage and displace the slow-growing species, for example, by growing taller and reducing the light available to them. The rate of nutrient cycling may also increase with higher temperatures, again favouring species that can respond quickly to increasing nutrient supplies and increase their growth rate and maximum size. In situations where climate change may cause water shortage, on the other hand, fast-growing species are more likely to decrease in frequency and the advantage would shift towards slower growing, more drought tolerant species. Where extreme events – such as droughts, flood and high winds – disrupt the continuity of vegetation cover, creating gaps, the advantage may shift towards the ruderal species with their rapid reproductive rates (see Section 7.3.6 for an example of this). Within these broad categories there are many more specifi differences between species and adaptations to particular conditions that will modify the outcome of competition. For example, where water supply diminishes, a deeper rooting species will tend to gain competitive advantage, similarly a species whose phenological development is more advanced by temperature increase will extend its growing season and annual productivity compared to one that is comparatively insensitive.

7.3.3 Competition and facilitation

Experimental evidence has been accumulating in recent years that warming can induce a change in community composition through changing the outcome of competition, rather than through direct impacts on the plant physiology. For example, Cornelissen *et al.* (2001) used a combination of experimental and transect studies to show that macrolichens in the Arctic are likely to be to be out-competed as a result of increasing vascular plant growth, caused by rising temperatures and nutrient enrichment. Kudo and Suzuki (2003) showed an acceleration of the impacts of competition amongst alpine shrubs in Japan when temperatures were raised by 1.5–2.3°C over the growing season. The two dominant evergreen species in the canopy, *Ledum palustre* and *Empetrum nigrum* increased vegetative growth and height whereas the sub-dominant *Vaccinium vitis-idaea* did not respond and became further suppressed. Heegaard and Vandvik (2004) demonstrated that snow bed species are excluded from more exposed locations by competition, rather than by unsuitability of microclimate: a reduction in snow lie as a consequence of climate change is likely to increase the competitive pressure on these species.

Competition is not the only process acting when species interact. There is clear evidence that in some circumstances, particularly in extreme environments, the opposite process – facilitation – can occur between species, for example, by ameliorating microclimate conditions. There is evidence that in cold mountain environments, facilitation becomes progressively more important as temperatures drop with altitude (Callaway *et al.*, 2002; Kikvidze *et al.*, 2005). This can be explained if plant growth is limited principally by non-resource aspects of the extreme environment, such as low temperature and high wind speeds, which can be ameliorated by proximity to neighbours. In contrast in less extreme climates, growth may be more limited by the availability of resources that species may compete for, such as nutrients or water supply. Climate change would therefore be expected to tip the balance towards competition playing a larger role than facilitation in marginal situations, and there is some evidence for this (Klanderud & Totland, 2005). It is important to bear in mind, however, that facilitation and competition are not mutually exclusive and can both occur at the same time – for example, species competing for nutrients may at the same time benefi from microclimate amelioration (Dormann & Brooker, 2002). Most work on facilitation in recent years has come from studies of low-temperature communities, but similar processes may operate in dry habitats (see also Chapter 6). Lloret *et al.* (2004) have recently shown that the positive effect of canopy cover on seedling establishment of the dominant shrub *Globularia alypum* in a dry Mediterranean community is increased in drought conditions.

7.3.4 Changing water availability and interactions between climate variables

Warming is likely to be a feature of climate change in most parts of the world, but changes in precipitation patterns are more complex, with substantial regional and seasonal variations likely (IPCC, 2001; see also Chapter 1). Even where precipitation patterns do not change, a rise in temperature alone will inevitably have an impact on the water balance of vegetation: evapotranspiration rates rise and there may be effects on the duration of snow cover during winter. A long-running warming experiment in the Rocky Mountains (United States) has shown a shift away from herbaceous species, towards the shrub *Artemisia tridentata* (sagebrush) (Harte & Shaw, 1995; Harte, 2001; Perfors *et al.*, 2003; Saavedra *et al.*, 2003). The treatment, (overhead infrared heating) warms the top 150 mm of soil by approximately 1.5°, but the main cause of the vegetation change appears to be earlier snow melt in the spring, which extends the growing season by approximately 20 days. This reduces the water availability for herbaceous species, such as *Delphinium nuttallianum*, during the later spring and reduces their capacity for reproduction. *A. tridentata* is more resistant to desiccation and is able to increase growth in response to the longer growing season.

Where precipitation patterns do change, the impacts may be more significan than those of temperature, but they are not necessarily straightforward. In particular, the seasonal distribution and variability can be more important than the total amount (Weltzin *et al.*, 2003). Great Britain makes an interesting case study. Current models of climate change (Hulme *et al.*, 2002) indicate that in addition to a general warming trend, precipitation will tend to increase overall. However, in much of the country, especially in the southeast, there is likely to be a shift towards less rainfall in summer and more in winter, together with more variability. This implies more regular summer droughts, but also more floodin and waterlogging in winter. Experiments in grassland communities (Sternberg *et al.*, 1999; Grime *et al.*, 2000; Morecroft *et al.*, 2004) have shown that regular summer droughts can bring about changes in the composition of plant communities, at least as important as those that are caused by warming. Deep-rooted species tend to increase at the expense of shallower rooting species, consistent with a greater capacity to extract water from deeper depths. Many grass species are shallow-rooted and die back during drought; this creates gaps in the community. This in turn can allow ruderal species, with their high rates of reproduction and growth to establish following a drought. Thus, a more frequent incidence of droughts would be expected to increase the proportion of ruderals within some communities. There are, however, at least two complications to what is apparently a straightforward shift in community composition. Firstly old grasslands, dominated by slow-growing, 'stress tolerant' species may be highly resistant to drought even in regions where droughts have historically been uncommon (Grime *et al.*, 2000). The species that dominate these grasslands have survived many fluctuation of climate over centuries and are only likely to be displaced after repeated exposure to changed conditions. In contrast more recent, disturbed grasslands are much more susceptible in the short-term; they may, however, show a greater capacity to revert to their former status if conditions allow (this property is often termed *resilience* – in contrast to *resistance*, where change does not readily occur in the firs place). Wetter winters may also compensate for the effect of drier summers. For example, Morecroft *et al.* (2004) showed that some of the effects of a consistent experimental summer drought treatment (no rainfall during July and August) on species composition of a mid-successional ex-arable grassland ($\sim$ 10-year-old at the start of the experiment) may have been mitigated by a period of unusually wet winters. In the early stages of the experiment, a generally dry period in the mid-1990s, short-lived species with ruderal characteristics increased. Subsequently they declined during a period of extremely wet autumns and winters, and the hypothesis is that gaps in the sward closed more quickly in the wet conditions, preventing establishment of the ruderal species.

7.3.5 Interactions between climate and nutrient cycling

Nutrient relations and soil properties are major factors controlling plant communities, together with climate. Nutrient-poor and nutrient-rich sites have different sets of species associated with them, and the addition of nutrients may change community composition. There are numerous examples of this, from agronomic research, including the nineteenth century Park Grass Experiment at Rothamsted (United Kingdom), to studies of the impacts of atmospheric nitrogen deposition (Bobbink *et al.*, 1998; Cunha *et al.*, 2002). However, nutrient cycling is not independent of climate, and a number of experimental assessments of the impacts of climate change on nutrient cycling processes have been made in recent years. Typically an increase in temperature increases the speed of decomposition and release of nutrients through the process of mineralisation, although there is wide variation between different soils and habitats. A meta-analysis has been published by Rustad *et al.* (2001), and Emmett *et al.* (2004) have investigated nutrient changes within a multi-site study of heathlands across Europe. The water content of soils, which is affected by precipitation, evapotranspiration and drainage (which is in turn largely controlled by topography and the permeability of the underlying rock), is a major factor controlling decomposition, with very dry and very wet soils having low nutrient availability. This reflect on the one hand, the requirement of invertebrate and microbial decomposer communities for water and on the other, their sensitivity to anaerobic conditions. In circumstances where climate change results in either water-logging or extreme drying of soils, nutrient availability to plants will fall, at least whilst those conditions persist. The long-term impacts may, however, be different to short-term effects. For example, high levels of nitrogen mineralisation were found in the months following each drought treatment application in the long-term climate change simulation experiment on an ex-arable grassland on calcareous soil at Wytham, United Kingdom (Jamieson *et al.*, 1998). Overall, it may well be that changes in soil water will prove to have more effect on nutrient cycling than rising temperature (Emmett *et al.*, 2004). Whatever the cause, where a change in nutrient supply occurs it is likely to cause changes to species composition, especially where the nutrient in question is limiting growth. Dormann *et al.* (2004) have demonstrated a critical role of competition for nutrients in determining the impacts of climate change on a High Arctic community. Nutrient availability increased with a warming treatment, and the dwarf shrub *Salix polaris* responded more positively to this nutrient supply than the woodrush, *Luzula confusa.*

Nitrogen has historically been seen as the nutrient that most frequently limits production in semi-natural situations. However, nitrogen availability has increased in many semi-natural communities because of the effects of atmospheric pollutants, especially ammonia and nitrogen dioxide. This deposition is itself causing change in the communities (Bobbink *et al.*, 1998; Krupa, 2003). Atmospheric deposition may make nitrogen-limited systems more sensitive to the effects of warming by allowing growth responses to temperature to take place, and hence, potentially, changes in the balance of competition.

7.3.6 Role of extreme events

Occasional extreme climatic conditions, such as droughts, high temperatures, exceptional wind speeds and abnormal freezing temperatures may exert a dramatic effect on plant communities. What makes an 'extreme event' extreme for a particular community depends on the rarity of the event and difference from normal conditions – rather than the absolute value of any particular climate variable. So, for example, temperatures of –40°C would have a devastating ecological impact across many temperate regions, but are common in much of the boreal forest biome and species are adapted to them. Extreme events are a separate scientifi issue from extreme environments. Although extreme events are hard to predict, most climate modelling exercises indicate that an increase in their frequency is likely with climate change (Easterling *et al.*, 2000).

Exceptional droughts in areas that do not regularly experience them are amongst the most likely drivers of ecological change under climate change. This is partly because reductions in rainfall are predicted for some regions, but also because, as noted above, warmer temperatures tend to increase evapotranspiration, compounding any

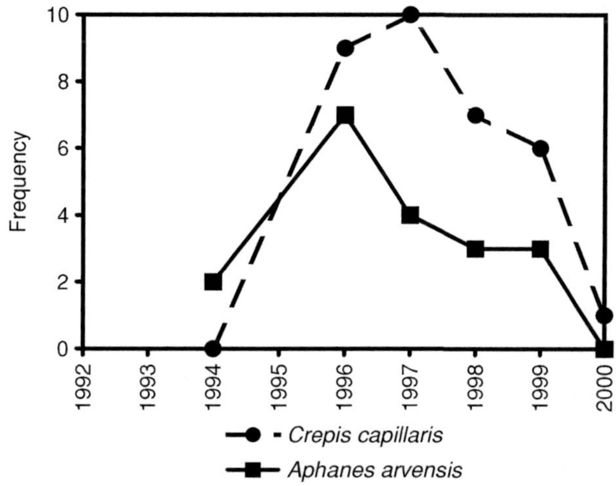

Figure 7.3 Increase in frequency of two ruderal species in mid successional calcareous grassland at Wytham, southern England, following drought in 1995. Frequency indicates number of 400 × 400 mm quadrats in which species occur within a 10 × 10 m plot.

reduction in rainfall. A longer growing season also increases annual vegetation demand for water. Following the drought in 1976, major changes were recorded in a long-running study at Lady Park Wood, a temperate deciduous woodland on the border of Wales and England in the United Kingdom. The death of old beech and young birch trees was particularly important, causing the character of the community to change dramatically in some parts of the site (Peterken & Mountford, 1996). In grassland communities, at the same time, there was a temporary increase in species with ruderal characteristics, which were able to colonise gaps, grow and reproduce rapidly – taking advantage of a window of opportunity (Grime *et al.*, 1994). A similar pattern was seen in another drought in 1995 (Figure 7.3; Morecroft *et al.*, 2002) and is consistent with experimental results (see above); however, it is notable that both 1976 and 1995 were also associated with dry winters. Such patterns of 'outbreak' (patterns of increase followed by decrease) in grasslands can persist for many years. Silvertown *et al.* (2002) presented evidence that a drought in 1929 was the trigger for outbreaks of several grassland species (generally with ruderal characteristics) that continued for up to 50 years, in the long-running Park Grass Experiment at Rothamsted, United Kingdom. There are very few other studies that have run sufficientl long to allow such community dynamics to be recognised. It is a salutary warning that time lags are inherent in ecological systems and plant communities may never truly be in equilibrium.

A secondary effect of increasing temperatures and more frequent drought is inevitably an increasing incidence of fire Fire is an intrinsic feature of many ecosystems from boreal forest to savannah, and many species are adapted to it (Bond &

Keeley, 2005). However, an increased frequency may well cause larger changes in community changes than the direct climatic effects of high temperatures and low rainfall (see also Chapter 6).

7.3.7 Dispersal constraints

With a rise in temperature at any particular location, species adapted to warmer climates would be expected to tend to increase in abundance compared to those adapted to relatively cooler conditions. If this process carries through to its logical conclusion, this would lead to a general shift in distributions to higher altitudes and latitudes (in the hottest regions, species would presumably persist if they could survive, possibly with selection for increased high-temperature tolerance). The changes would be expected firs at range margins where species with contrasting distribution patterns compete with each other, or where new niches become available for colonisation. Where organisms can disperse readily, as is the case in some animal species, such as the well-studied butterflie (Parmesan *et al.*, 1999), there is evidence that this is happening. However, in many plant species, dispersal is intrinsically slow. Evidence from the pollen record shows that many species took thousands of years to recolonise areas after the end of the most recent glaciation (Davis, 1987; Huntley, 1991) and indeed a true equilibrium has never been reached in some species. A good example of this is the beech (*Fagus sylvatica*) tree in Great Britain. Historically this was only found in the southeast part of the country – closest to the continent of Europe, from which colonisation occurred. It has, however, been planted over a much wider part of the country and been shown capable of growing and reproducing: therefore, given sufficien time, it would inevitably have naturally spread further north and west. This presents conservationists with a dilemma as the southeast of the country is likely to become less suitable for the species in future, with increasing summer drought (Broadmeadow, 2002), and so British beech woodland communities may be best conserved in areas in which it is not 'native'. Slow rates of dispersal present conservationists with another dilemma: whether to transplant threatened species to new suitable habitats, which they would not be able to reach quickly enough without human intervention.

Human activities have made important changes to dispersal patterns, which need to be considered alongside the impacts of climate change. On the one hand, people have directly transported species to different parts of the world, and in many situations the introduction of non-native species has created major changes in communities. Climate change may exacerbate this situation as previously unviable introductions may survive and flouris in future. This includes species that have survived for long periods in the warmer local climates associated with larger cities, but have not hitherto expanded into rural areas. On the other hand, the fragmentation of landscapes, particularly where large expanses of monoculture crops separate small patches of semi-natural habitat, may act as a barrier and hinder dispersal of species to new environments. This makes it less likely that species will be able to track changes in their climate envelope by colonising new locations; however, it may also improve the chances of their persisting within existing locations by restricting the ingress of new competitors from warmer regions.

7.3.8 Interactions with animals

In considering how individual plant responses to climate change translate into changes in plant communities, it is important not to neglect the role of the animal community with which plants interact. Animals impact on plants in many ways, but amongst the most important are herbivory, pollination and as agents of dispersal. One of the major issues in understanding the ecological effects of climate change is how differential impacts on interacting species may lead to a disruption of relationships. In the case of pollination, it is important that bud-break of fl wers coincides with a period when the pollinator is active. If the phenologies of plant and animal respond differently to climate change, the synchrony between them may be lost. This is a particularly serious risk as many pollinators are flyin insects, such as bees, with strongly seasonal life cycles. Fitter and Fitter (2002) showed that the fl wering dates of insect-pollinated species in the United Kingdom were more sensitive to interannual variations in temperature than those of wind-pollinated species. This suggests that insect-pollinated plant species have been selected to respond to temperature so as to maximise chances of pollination, but we do not know whether this system will be robust to long-term changes in climate. If the distribution of a pollinator animal species changes before that of the plant does – because of its greater mobility – this may lead to a disruption of the relationship, with no pollinator available to a plant, particularly if the relationship between them is specific Seed dispersal is subject to similar considerations; however, dispersers are more often vertebrates, which are less likely to have a strongly seasonal life cycle than invertebrates. There may therefore be less effect of temperature rises on dispersal than on pollination. Climate affects the biochemical composition and structure of plants in ways that can have a major impact on their nutritional quality to herbivores, which may in turn affect the impact of the herbivores on the plants. Different growth responses in different plant species may therefore modify the outcome of competition because of positive and negative feedbacks through the animal population. Animals also have important indirect effects on the plant community such as their role in decomposition and hence nutrient release, and this also needs to be borne in mind.

7.4 Is community change already happening?

Is there any evidence that communities and distributions are starting to change in response to the (comparatively small) changes in climate that have already been recorded? The last few years have seen a number of important studies investigating whether ecological changes consistent with the impacts of climate change can be detected, including some major reviews and meta-analyses (Walther *et al.*, 2002; Parmesan & Yohe, 2003). Phenological changes, such as earlier leafin are perhaps the clearest documented changes and are frequently cited (see Chapter 4), but they do not necessarily imply anything about change in community composition. There is, however, good evidence of shifts in range boundaries of species under some circumstances. Walther *et al.* (2002) summarised the evidence for latitudinal and altitudinal shifts in distribution, dividing these into nine categories, of which four relate to plants:

1. treelines shifting to higher altitudes (Wardle & Coleman, 1992; Meshinev *et al.*, 2000; Kullman, 2001),
2. shrubs expanding into areas of tundra from which they were formerly absent in Alaska (Sturm *et al.*, 2001),
3. European alpine plants expanding their distributions to higher altitudes (Grabherr *et al.*, 1994),
4. expansions of the distribution of Antarctic plants, including the colonisation of bare ground (Kennedy, 1995).

It is notable that all of these examples are from high-latitude or high-altitude areas, characterised by low temperatures and short growing seasons. In contrast, some of the animal groups – such as the mobile and well-studied butterflie (Parmesan *et al.*, 1999) – have shown major changes in distribution across a wider range of climatic zones, particularly in temperate regions. Plant communities of low-temperature environments are frequently identifie as being at risk from climate change. As they are adapted to low temperatures and generally have relatively slow growth rates and low competitive abilities, one might therefore conclude that change would show up here first because of greater vulnerability. However, slow growth rates, combined with high longevity of species, make for very stable communities in which change tends to happen slowly – because it takes a long time for new competitor species to gain a foothold. It may actually be that other more disturbed, intrinsically variable communities will show change first Low-temperature communities are particularly vulnerable where the climate envelope they occupy is likely to disappear altogether. For example, where alpine plants already occupy only the uppermost region of a mountain, there is no scope for dispersal to higher altitudes. It may simply be that more research of this sort has been carried out in the cool temperate and boreal regions, reflectin a better historical record of species distributions than many warmer regions. It is also true that some of these changes relate to boundaries of major vegetation types, such as rising treelines, which are amongst the easiest to detect and unambiguous to interpret. It may take longer to recognise more subtle changes in relative composition of different species within communities that are not close to obvious range margins. It is also important to realise that there are relatively few monitoring schemes for vegetation composition which have continued for long enough to detect long-term trends.

It is therefore likely that substantially more communities are experiencing climate-induced change than those that have been documented. However, isn't it fair to say that the documented changes to date have not been of the same magnitude as the major changes associated with changing land use, such as deforestation or the draining of wetlands? What is certainly true, however, as Parmesan and Yohe (2003) discuss is that climate change is a long-term trend that cannot be easily reversed or halted in the way in which land use change (potentially) can be. The considerable inertia of many communities also means that the full consequences of climate change would still take many years to work through, even if it were possible to stabilise climate in the next few years (a highly unlikely scenario!). Better monitoring and a better understanding of the processes that are at work are needed if we are to be able to predict future consequences and devise strategies to minimise adverse affects.

Acknowledgements

We are grateful to Dr Pam Berry (Oxford University, Environmental Change Institute) for supplying Figure 7.2 and participants in the UK Environmental Change Network, especially Michèle Taylor (CEH), for their contributions to the work on drought in the United Kingdom reported here. Dr James Morison provided valuable comments and advice. J.S.P. is supported by a research studentship from the UK Forestry Commission.

References

Bakkenes, M., Alkemade, J.R.M., Ihle, F., Leemans, R. & Latour, J.B. (2002) Assessing effects of forecasted climate change on the diversity and distribution of European higher plants for 2050. *Global Change Biol.*, **8**, 390–407.

Beier, C., Emmett, B., Gundersen, P., Tietema, A., Penuelas, J., Estiarte, M., Gordon, C., Gorissen, A., Llorens, L., Roda, F. & Williams, D. (2004) Novel approaches to study climate change effects on terrestrial ecosystems in the field drought and passive nighttime warming. *Ecosystems*, **7**, 583–597.

Birks, H.J.B. (1989) Holocene isochrone maps and patterns of tree-spreading in the British Isles. *J. Biogeogr.*, **16**, 503–540.

Bobbink, R., Hornung, M. & Roelofs, J.G.M. (1998) The effects of air-borne nitrogen pollutants on species diversity in natural and semi-natural European vegetation. *J. Ecol.*, **86**, 717–738.

Bond, W.J. & Keeley, J.E. (2005) Fire as a global 'herbivore': the ecology and evolution of flammabl ecosystems. *Trends Ecol. Evol.*, **20**, 387–394.

Broadmeadow, M. (2002) *Climate Change: Impacts on UK Forests*. Forestry Commission, Edinburgh.

Callaway, R.M., Brooker, R.W., Choler, P., Kikvidze, Z., Lortie, C.J., Michalet, R., Paolini, L., Pugnaire, F.I., Newingham, B., Aschehoug, E.T., Armas, C., Kikodze, D. & Cook, B.J. (2002) Positive interactions among alpine plants increase with stress. *Nature*, **417**, 844–848.

Cornelissen, J.H.C., Callaghan, T.V., Alatalo, J.M., Michelsen, A., Graglia, E., Hartley, A.E., Hik, D.S., Hobbie, S.E., Press, M.C., Robinson, C.H., Henry, G.H.R., Shaver, G.R., Phoenix, G.K., Jones, D.G., Jonasson, S., Chapin, F.S., Molau, U., Neill, C., Lee, J.A., Melillo, J.M., Sveinbjornsson, B. & Aerts, R. (2001) Global change and arctic ecosystems: is lichen decline a function of increases in vascular plant biomass? *J. Ecol.*, **89**, 984–994.

Cox, P.M., Betts, R.A., Collins, M., Harris, P.P., Huntingford, C. & Jones, C.D. (2004) Amazonian forest dieback under climate-carbon cycle projections for the 21st century. *Theor. Appl. Climatol.*, **78**, 137–156.

Cramer, W., Bondeau, A., Woodward, F.I., Prentice, I.C., Betts, R.A., Brovkin, V., Cox, P.M., Fisher, V., Foley, J.A., Friend, A.D., Kucharik, C., Lomas, M.R., Ramankutty, N., Sitch, S., Smith, B., White, A. & Young-Molling, C. (2001) Global response of terrestrial ecosystem structure and function to CO_2 and climate change: results from six dynamic global vegetation models. *Global Change Biol.*, **7**, 357–373.

Cunha, A., Power, S.A., Ashmore, M.R., Green, P.R.S., Haworth, B.J. & Bobbink, R. (2002) Whole ecosystem nitrogen manipulation: an updated review, JNCC Report 331. JNCC, Peterborough, UK.

Davis, M.B. (1987) Invasions of forest communities during the Holocene: beech and hemlock in the Great Lakes region. In: *Colonization, Succession and Stability – 26th Symposium of The British Ecological Society* (eds A.J. Gray, M.J. Crawley & P.J. Edwards), pp. 373–394. Blackwell Science, Oxford.

Dormann, C.F. & Brooker, R.W. (2002) Facilitation and competition in the High Arctic: the importance of the experimental approach. *Acta Oecol.*, **23**, 297–301.

Dormann, C.F., van der Wal, R. & Woodin, S.J. (2004) Neighbour identity modifie effects of elevated temperature on plant performance in the High Arctic. *Global Change Biol.*, **10**, 1587–1598.

Dunne, J.A., Saleska, S.R., Fischer, M.L. & Harte, J. (2004) Integrating experimental and gradient methods in ecological climate change research. *Ecology*, **85**, 904–916.

Easterling, D.R., Meehl, G.A., Parmesan, C., Changnon, S.A., Karl, T.R. & Mearns, L.O. (2000) Climate extremes: observations, modeling, and impacts. *Science*, **289**, 2068–2074.

Emmett, B.A., Beier, C., Estiarte, M., Tietema, A., Kristensen, H.L., Williams, D., Penuelas, J., Schmidt, I. & Sowerby, A. (2004) The response of soil processes to climate change: results from manipulation studies of shrublands across an environmental gradient. *Ecosystems*, **7**, 625–637.

Fitter, A.H. & Fitter, R.S.R (2002) Rapid changes in fl wering time in British plants. *Science*, **296**, 1689–1691.

Grabherr, G., Gottfried, M. & Pauli, H. (1994) Climate effects on mountain plants. *Nature*, **369**, 448–448.

Grace, J.B. (1991) A clarificatio of the debate between Grime and Tilman. *Funct. Ecol.*, **5**, 583–587.

Grime, J.P. (1979) *Plant Strategies and Vegetation Processes*. Wiley, Chichester, UK.

Grime, J.P., Brown, V.K., Thompson, K., Masters, G.J., Hillier, S.H., Clarke, I.P., Askew, A.P., Corker, D. & Kielty, J.P. (2000) The response of two contrasting limestone grasslands to simulated climate change. *Science*, **289** (5480), 762–765.

Grime, J.P., Willis, A.J., Hunt, R. & Dunnett, N.P. (1994) Climate–vegetation relationships in the Bibury road verge experiments. In: *Long-Term Experiments in Agricultural and Social Sciences: Proceedings of a Conference to Celebrate the 150th Anniversary of Rothamsted Experimental Station, Held at Rothamsted, 14–17 July 1993* (eds R.A. Leigh & A.E. Johnston), pp. 271–285. CAB International, Wallingford, UK.

Grubb, P.J. (1992) A positive distrust in simplicity – lessons from plant defenses and from competition among plants and among animals. *J. Ecol.*, **80**, 585–610.

Hampe, A. (2004) Bioclimate envelope models: what they detect and what they hide. *Global Ecol. Biogeogr.*, **13**, 469–476.

Harrison, P.A., Berry, P.M. & Dawson, T.P. (2001) Climate change and nature conservation in Britain and Ireland: modelling natural resource responses to climate change, the MONARCH Project. UK Climate Impacts Programme, Oxford.

Harte, J. (2001) Global warming and terrestrial ecosystems – response. *Bioscience*, **51**, 332–333.

Harte, J. & Shaw, R. (1995) Shifting dominance within a montane vegetation community: results of a climate-warming experiment. *Science*, **267**, 876–880.

Heegaard, E. & Vandvik, V. (2004) Climate change affects the outcome of competitive interactions – an application of principal response curves. *Oecologia*, **139**, 459–466.

Hobbie, S.E., Shevtsova, A. & Chapin, F.S. (1999) Plant responses to species removal and experimental warming in Alaskan tussock tundra. *Oikos*, **84**, 417–434.

Hulme, M., Jenkins, G.J., Lu, X., Turnpenny, J.R., Mitchell, T.D., Jones, R.G., Lowe, J., Murphy, J.M., Hassell, D., Boorman, P., McDonald, R. & Hill, S. (2002) Climate change scenarios for the United Kingdom: the UKCIP02 Scientifi Report. Tyndall Centre for Climate Change Research, School of Environmental Sciences, University of East Anglia, Norwich.

Huntley, B. (1991) How plants respond to climate change: migration rates, individualism and the consequences for plant communities. *Ann. Bot.*, **67** (Suppl. 1), 15–22.

Huntley, B., Berry, P.M., Cramer, W. & McDonald, A.P. (1995) Modelling present and potential future ranges of some European higher plants using climate response surfaces. *J. Biogeogr.*, **22**, 967–1001.

Huntley, B. & Birks, H.J.B. (1983) *An Atlas of Past and Present Pollen Maps for Europe: 0–13 000 Years Ago*. Cambridge University Press, Cambridge, UK.

Hutchinson, G.E. (1957) Concluding remarks. *Cold Spring Harb. Symp. Quant. Biol.*, **22**, 415–457.

IPCC (2001) *Climate Change 2001: The Scientific Basis*. Cambridge University Press, Cambridge, UK.

Jackson, S.T. & Whitehead, D.R. (1991) Holocene vegetation patterns in the Adirondack Mountains. *Ecology*, **72**, 641–653.

Jamieson, N., Barraclough, D., Unkovich, M. & Monaghan, R. (1998) Soil N dynamics in a natural calcareous grassland under a changing climate. *Biol. Fertil. Soils*, **27**, 267–273.

Kennedy, A.D. (1995) Temperature effects of passive greenhouse apparatus in high-latitude climate change experiments. *Funct. Ecol.*, **9**, 340–350.

Kikvidze, Z., Pugnaire, F.I., Brooker, R.W., Choler, P., Lortie, C.J., Michalet, R. & Callaway, R.M. (2005) Linking patterns and processes in alpine plant communities: a global study. *Ecology*, **86**, 1395–1400.

Klanderud, K. & Totland, R. (2005) The relative importance of neighbours and abiotic environmental conditions for population dynamic parameters of two alpine plant species. *J. Ecol.* **93**, 493–501.

Krupa, S.V. (2003) Effects of atmospheric ammonia (NH_3) on terrestrial vegetation: a review. *Environ. Pollut.*, **124**, 179–221.

Kudo, G. & Suzuki, S. (2003) Warming effects on growth, production, and vegetation structure of alpine shrubs: a f ve-year experiment in northern Japan. *Oecologia*, **135**, 280–287.

Kullman, L. (2001) 20th century climate warming and tree-limit rise in the southern Scandes of Sweden. *Ambio*, **30**, 72–80.

Lloret, F., Penuelas, J. & Estiarte, M. (2004) Experimental evidence of reduced diversity of seedlings due to climate modificatio in a Mediterranean-type community. *Global Change Biol.*, **10**, 248–258.

MacArthur, R. & Wilson, E.O. (1967) *The Theory of Island Biogeography*. Princeton University Press, Princeton, NJ.

Meshinev, T., Apostolova, I. & Koleva, E. (2000) Influenc of warming on timberline rising: a case study on *Pinus peuce* Griseb. in Bulgaria. *Phytocoenologia*, **30**, 431–438.

Morecroft, M.D., Bealey, C.E., Howells, O., Rennie, S. & Woiwod, I.P. (2002) Effects of drought on contrasting insect and plant species in the UK in the mid-1990s. *Global Ecol. Biogeogr.*, **11**, 7–22.

Morecroft, M.D., Masters, G.J., Brown, V.K., Clarke, I.P., Taylor, M.E. & Whitehouse, A.T. (2004) Changing precipitation patterns alter plant community dynamics and succession in an ex-arable grassland. *Funct. Ecol.*, **18**, 648–655.

Morecroft, M.D. & Woodward, F.I. (1996) Experiments on the causes of altitudinal differences in the leaf nutrient contents, size andg δ^{13}C of *Alchemilla alpina*. *New Phytol.*, **134**, 471–480.

Parmesan, C., Ryrholm, N., Stefanescu, C., Hill, J.K., Thomas, C.D., Descimon, H., Huntley, B., Kaila, L., Kullberg, J., Tammaru, T., Tennent, W.J., Thomas, J.A. & Warren, M. (1999) Poleward shifts in geographical ranges of butterfl species associated with regional warming. *Nature*, **399**, 579–583.

Parmesan, C. & Yohe, G. (2003) A globally coherent fingerprin of climate change impacts across natural systems. *Nature*, **421**, 37–42.

Pearson, R.G. & Dawson, T.P. (2003) Predicting the impacts of climate change on the distribution of species: are bioclimate envelope models useful? *Global Ecol. Biogeogr.*, **12**, 361–371.

Pearson, R.G. & Dawson, T.P. (2004) Bioclimate envelope models: what they detect and what they hide – response to Hampe. *Global Ecol. Biogeogr.*, **13**, 471–473.

Pearson, R.G., Dawson, T.P., Berry, P.M. & Harrison, P.A. (2002) A spatial evaluation of climate impact on the envelope of species. *Ecol. Model.*, **154**, 289–300.

Perfors, T., Harte, J. & Alter, S.E. (2003) Enhanced growth of sagebrush (*Artemisia tridentata*) in response to manipulated ecosystem warming. *Global Change Biol.*, **9**, 736–742.

Peterken, G.F. & Mountford, E.P. (1996) Effects of drought on beech in Lady Park Wood, an unmanaged mixed deciduous woodland. *Forestry*, **69**, 125–136.

Pigott, C.D. & Huntley, J.P. (1978) Factors controlling distribution of *Tilia cordata* at northern limits of its geographical range. 1. Distribution in northwest England. *New Phytol.*, **81**, 429–441.

Pigott, C.D. & Huntley, J.P. (1980) Factors controlling the distribution of *Tilia cordata* at the northern limits of its geographical range. 2. History in Northwest England. *New Phytol.*, **84**, 145–164.

Pigott, C.D. & Huntley, J.P. (1981) Factors controlling the distribution of *Tilia cordata* at the northern limits of its geographical range. 3. Nature and causes of seed sterility. *New Phytol.*, **87**, 817–839.

Rustad, L.E., Campbell, J.L., Marion, G.M., Norby, R.J., Mitchell, M.J., Hartley, A.E., Cornelissen, J.H.C. & Gurevitch, J. (2001) A meta-analysis of the response of soil respiration, net nitrogen mineralization, and aboveground plant growth to experimental ecosystem warming. *Oecologia*, **126**, 543–562.

Saavedra, F., Inouye, D.W., Price, M.V. & Harte, J. (2003) Changes in fl wering and abundance of *Delphinium nuttallianum* (Ranunculaceae) in response to a subalpine climate warming experiment. *Global Change Biol.*, **9**, 885–894.

Shaver, G.R., Canadell, J., Chapin, F.S., Gurevitch, J., Harte, J., Henry, G., Ineson, P., Jonasson, S., Melillo, J., Pitelka, L. & Rustad, L. (2000) Global warming and terrestrial ecosystems: a conceptual framework for analysis. *Bioscience*, **50**, 871–882.

Silvertown, J., McConway, K.J., Hughes, Z., Biss, P., Macnair, M. & Lutman, P. (2002) Ecological and genetic correlates of long-term population trends in the park grass experiment. *Am. Nat.*, **160**, 409–420.

Sternberg, M., Brown, V.K., Masters, G.J. & Clarke, I.P. (1999) Plant community dynamics in a calcareous grassland under climate change manipulations. *Plant Ecol.*, **143**, 29–37.

Sturm, M, Racine, C, Tape, K. (2001) Climate change – increasing shrub abundance in the Arctic. *Nature*, **411**, 546–547.

Tilman, D. (1988) *Plant Strategies and the Dynamics and Structure of Plant Communities*. Princeton University Press, Princeton, NJ.

Walther, G.R., Post, E., Convey, P., Menzel, A., Parmesan, C., Beebee, T.J.C., Fromentin, J.M., Hoegh-Guldberg, O. & Bairlein, F. (2002) Ecological responses to recent climate change. *Nature*, **416**, 389–395.

Wardle, P. & Coleman, M.C. (1992) Evidence for rising upper limits of 4 native New-Zealand forest trees. *N.Z. J. Bot.*, **30**, 303–314.

Weltzin, J.F., Loik, M.E., Schwinning, S., Williams, D.G., Fay, P.A., Haddad, B.M., Harte, J., Huxman, T.E., Knapp, A.K., Lin, G.H., Pockman, W.T., Shaw, M.R., Small, E.E., Smith, M.D., Smith, S.D., Tissue, D.T. & Zak, J.C. (2003) Assessing the response of terrestrial ecosystems to potential changes in precipitation. *Bioscience*, **53**, 941–952.

Woodward, F.I. & Pigott, C.D. (1975) Climatic control of altitudinal distribution of *Sedum rosea* (L.) Scop and *Sedum telephium* L1. Field observations. *New Phytol.*, **74**, 323–334.

8 Issues in modelling plant ecosystem responses to elevated CO_2: interactions with soil nitrogen

Ying-Ping Wang, Ross McMurtrie, Belinda Medlyn and David Pepper

8.1 Introduction

Models are essential in the study of plant responses to climate change. Most experimental studies are small in spatial scale (individual plants, microcosms, mesocosms) and short in time span (days to years). For example, the largest experimental carbon dioxide concentration ([CO_2]) studies cover less than 1 ha and have been 10 years or less in duration (Hendrey *et al.*, 1999; Ainsworth & Long, 2005). Many science and policy questions that we need to answer, on the other hand, are typically phrased in terms of responses of biomes over several decades: for example, how will crop and forest production be affected in the next 50 years? Will the terrestrial biosphere continue to act as a net carbon sink over the next century? Models are necessary to bridge this gap between experimental and policy time and space scales (e.g. Prentice *et al.*, 2001; Medlyn & McMurtrie, 2005).

8.1.1 Modelling challenges

However, to realistically model plant responses to climate change, we must meet several significan challenges. Firstly, the current rapid increase in atmospheric [CO_2] is shifting ecosystems into a completely new set of environmental conditions, meaning that empirical models, based on existing conditions, are of limited use. Instead, models must be process-based, that is, developed from an understanding of the underlying physiological processes and their responses to changes in [CO_2] and climate. This understanding is gradually advancing, as detailed in other chapters of this volume, but there are still significan gaps.

A second challenge is how to include relevant processes whose timescale is long compared with the duration of experimental studies. The direct effects of [CO_2] on photosynthesis and stomatal conductance feed into a sequence of processes with increasingly long response timescales, such as carbohydrate and nutrient allocation, water balance, nutrient cycling, interspecifi and intraspecifi competition. Processes that respond on timescales of decades are important for many policy questions but are intractable to experimental study, and developing accurate representations of these processes poses a difficul scientifi problem.

A third important challenge is model–data fusion; that is, how to ensure that models are soundly based on experimental data. Until recently, terrestrial models have tended to be fairly loosely based on experimental outcomes. Increasingly, however, scientists are coming to recognise the need to rigorously incorporate experimental data in models, and are using data assimilation techniques to allow a formal exchange of information between models and data (e.g. Braswell *et al.*, 2005; Raupach *et al.*, 2005; Williams *et al.*, 2005).

8.1.2 Chapter aims

In this chapter we focus on modelling issues rather than model outputs. The main aim is not to compare different models of plant responses to climate change, but rather, to identify some of the major obstacles to developing credible models and discuss how these obstacles can be overcome. We illustrate, using the example of nitrogen cycling, how the three challenges described above – developing process-based models, representing processes with long response timescales and model–data fusion – can be met. Firstly, we discuss how nitrogen cycling is represented in ecosystem models. We then review alternative hypotheses of how nitrogen cycling might be affected by increasing [CO_2] and discuss how these hypotheses can be embedded in models. Finally, we apply a model of ecosystem carbon (C) and nitrogen (N) cycling to data from a large-scale elevated [CO_2] experiment and use this example to illustrate how the techniques of model–data fusion can be used to investigate alternative hypotheses.

Nitrogen cycling is used as an example for several reasons. As noted above, nutrient cycling processes generally become important on timescales that are longer than most experiments, but that are highly relevant to human society. Thus, the question of how nutrient cycling might be affected by increasing [CO_2] is very difficul to test experimentally but is key to predicting plant responses on decadal to century timescales (Luo *et al.*, 2004). Unless nitrogen cycling is included explicitly, model results are open to question. For example, Cramer *et al.* (2001) reported the predictions of a net terrestrial C sink over the next 100 years by six dynamic global vegetation models (DGVMs). On average, these models predicted a current net terrestrial C sink of 1.6 Gt C year^{-1} increasing to approximately 4 Gt C year^{-1} by 2050 and then declining to 3.5 Gt C year^{-1} by 2100. However, only two of these models included nitrogen cycling, and the predictions were criticised by Hungate *et al.* (2003) on the grounds that the additional amount (7.7–37.5 Pg N) of N required to sequester that additional amount (350–890 Pg C) of C is signifi cantly greater than their upper estimates (6.1 Pg N) of the N addition over the next 100 years.

Of the plant nutrients, nitrogen is the most widely limiting, and its effects on plant productivity and soil organic matter (SOM) cycling are the best understood. Although phosphorus also limits production widely, particularly in Australia, our ability to quantify its role in ecosystem function is much poorer than for nitrogen, so there are far fewer ecosystem models incorporating phosphorus cycling. However, nitrogen cycling models can be generalised to consider other nutrients; Kirschbaum *et al.* (1998) have demonstrated how this can be done for phosphorus and sulphur with the G'DAY (generic decomposition and yield) ecosystem model.

8.2 Representing nitrogen cycling in ecosystem models

8.2.1 Overview of ecosystem models

A large number of models of plant growth have been used to study responses to climate change. These models cover a wide range of time and space scales (Nightingale *et al.*, 2004), but most of the models can be broadly classifie into three different types: stand-scale models, regional-scale models and dynamic global vegetation models.

Stand-scale models are applied to homogeneous areas of crops or forests – they are parameterised for a single plant type and soil type. Typically, such models include the processes of radiation interception, photosynthesis, exchange of latent and sensible heat, carbon allocation and water and heat fl w in soil; some also include water and nutrient cycling processes. The structure and performance of several such models in simulating forest growth and exchanges were compared by Hanson *et al.* (2004). Regional-scale models generally include the same processes as stand-scale models, but are applied to a region or continent, using some spatially explicit information. Many regional-scale models are generalisations of stand-scale models (e.g. Biome-BGC, Century; Schimel *et al.*, 1997). In such models, the spatially explicit information on vegetation and soil types and look-up tables for parameter values for each vegetation or soil type are input to the model. DGVMs, on the other hand, attempt to predict the spatial distribution of vegetation types and their productivity, by including processes of competition, establishment and mortality. The performance of four regional-scale models and two DGVMs was compared by Gordon *et al.* (2004).

As a general rule, the level of complexity employed in these models is highest in the stand-scale models and lowest in the DGVMs, owing to the increased computing load required for global simulations. In fact, it is common to use detailed stand-scale models to help develop simpler process representations for inclusion in regional models or DGVMs (e.g. Luxmoore *et al.*, 2000). In this chapter we employ a stand-scale model of ecosystem carbon and nitrogen cycling, G'DAY (Comins & McMurtrie, 1993). G'DAY has been widely applied to investigate the effects of atmospheric [CO_2] and temperature on forest and grassland ecosystems (Comins & McMurtrie, 1993; McMurtrie & Comins, 1996; McMurtrie *et al.*, 2000, 2001; Medlyn *et al.*, 2000; Pepper *et al.*, 2005). The representation of nitrogen cycling in G'DAY model is based on the CENTURY model (Parton *et al.*, 1987). The CENTURY approach has been incorporated into many models, including several regional-scale models and DGVMs (e.g. HYBRID and SDGVM; Cramer *et al.*, 2001). Thus, using stand-scale models to conduct detailed studies of interactions between atmospheric [CO_2] and nitrogen cycling helps us develop a general understanding of key processes and parameters that can be incorporated into regional and global models.

8.2.2 Modelling nitrogen cycling

We firs outline the nitrogen cycling process, and then discuss how this process is represented in the G'DAY model. Nitrogen cycling is illustrated diagrammatically in Figure 8.1a. The soil solution contains nitrogen ions in mineral form. Plants and microbes compete for these ions. The nitrogen taken up by plants is distributed among the plant components, foliage, stems and roots. Some nitrogen is retranslocated before the plant parts senesce; the rest is input to the soil via litterfall. The litter is decomposed by soil fauna, with some nitrogen being mineralised and some being sequestered in SOM. SOM is gradually broken down, releasing the sequestered nitrogen. The rates of decomposition of litter and SOM depend on the initial composition of litter, the physical soil environment, particularly soil temperature and moisture, and the size and activity of the decomposer community. External inputs and outputs to the nitrogen cycle include atmospheric deposition, N fixatio and losses to leaching, denitrificatio and volatilisation.

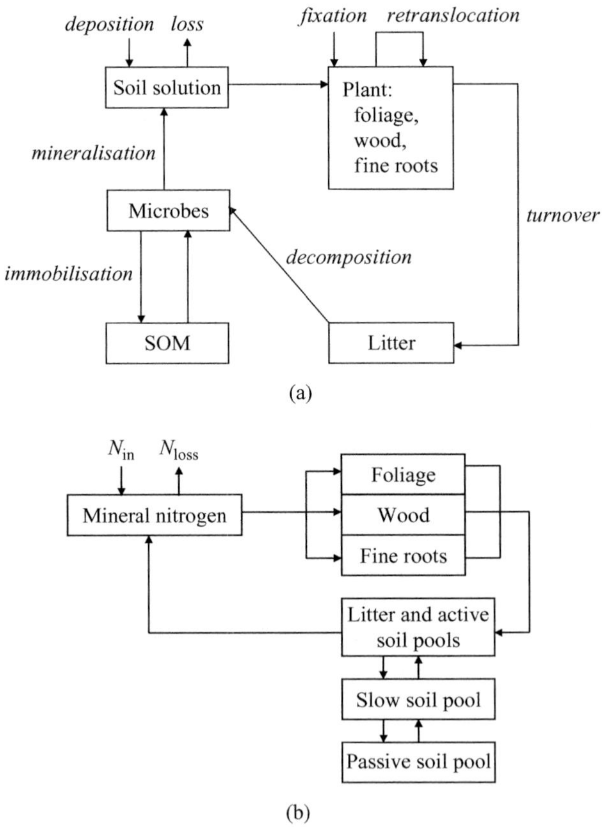

Figure 8.1 (a) Key nitrogen cycling processes (after Aber & Melillo, 2001). (b) Representation of nitrogen cycling in the G'DAY model (after Medlyn *et al.*, 2000), where the litter pool includes above-ground or below-ground metabolic and structural litter carbon.

How these above-mentioned processes are represented in the G'DAY model is shown in Figure 8.1b. The model tracks the C and N contents of 10 pools in total: three plant pools (foliage, stem and roots), four litter pools (metabolic and structural above- and below-ground litter) and three SOM pools with different turnover rates (active, slow and passive). Mineralised nitrogen taken up by plants is allocated to foliage, stem and roots according to the growth rate of each compartment. Stemwood is assumed to have a constant N concentration, while foliage and root N concentrations vary with N uptake, with root N concentration proportional to foliar N concentration. Photosynthetic rate depends on the foliar N concentration and atmospheric [CO_2]. A fxed fraction of N is retranslocated from foliage and roots before senescence. The longevity of foliage, stem and roots is assumed constant. Plant litter is separated into metabolic and structural pools according to its lignin/N ratio. The fl ws of carbon from litter into SOM pools, and among SOM pools, depend on soil temperature, moisture and texture. The N/C ratios of the SOM pools are assumed to increase linearly between prescribed minimum and maximum values as the N concentration of the soil solution increases. Flows of nitrogen in the soil, which depend on the fl ws of carbon and pool N/C ratios, are used to evaluate nitrogen mineralisation or immobilisation. External inputs of N via atmospheric deposition are assumed to be constant, while the losses of N through leaching and volatilisation are proportional to soil inorganic N. The model is thus a fairly abstract representation of the nitrogen cycle, particularly of the processes of litter decomposition and SOM formation. There is no explicit representation of the microbial biomass and its composition.

8.2.3 Major uncertainties

All ecosystem models include some uncertain assumptions. To correctly interpret model output, it is important to identify the uncertain assumptions and to quantify their impact on model predictions. Previous work with the G'DAY model has identifie several important uncertainties in the model, many of which relate to the indirect effects of elevated [CO_2] on nitrogen cycling processes (Kirschbaum *et al.*, 1994; McMurtrie & Comins, 1996; McMurtrie *et al.*, 2000).

At the plant level, growth at elevated [CO_2] may increase demand for nitrogen, which could induce shifts in carbon allocation, with roots being favoured at the expense of above-ground plant parts, in order to increase nitrogen uptake (see Chapter 2). However, patterns of carbon allocation among plant organs under elevated [CO_2] are highly variable among experiments (Curtis & Wang, 1998) and thus constitute a major source of model uncertainty. For example, two large-scale forest free-air CO_2 enrichment (FACE) experiments have shown different responses of allocation to increased [CO_2]. Stem growth in a *Pinus taeda* plantation was consistently increased by growth in elevated [CO_2] (Finzi *et al.*, 2002), but in a plantation of *Liquidambar styraciflua*, additional carbon was allocated to fin roots rather than to stem (Norby *et al.*, 2004).

Another key uncertainty is how changes in soil nitrogen cycling processes will feedback to the [CO_2] response. Three main mechanisms by which soil feedbacks could modify plant responses to rising CO_2 have been proposed: the 'litter quality' feedback, the 'litter quantity' feedback and the 'stimulation of N mineralisation' feedback (Berntson & Bazzaz, 1996; McMurtrie *et al.*, 2000; Medlyn & McMurtrie, 2005). The 'litter quality' feedback hypothesises that reduced N content of live plant tissue leads to reduced N content of plant litter, which retards decomposition and N release from litter, reducing plant N availability, and causing a negative feedback on plant growth (Melillo *et al.*, 1991; Norby *et al.*, 2001). The 'litter quantity' feedback hypothesises that increased litter input to the soil enhances soil C content, tending to increase soil N immobilisation and reduce plant N availability (Diaz *et al.*, 1993). The 'stimulation of N mineralisation' feedback hypothesises that an increased flu of C to the soil, as litter input or root exudates or transfer to mycorrhizae, stimulates microbial activity and thus N mineralisation and N fixatio rates, enhancing plant N availability (Zak *et al.*, 1993). The firs two mechanisms are thought to represent negative feedbacks, while the third results in a positive feedback. It is uncertain which of these mechanisms will predominate in a given ecosystem.

All three soil feedbacks are sensitive to assumptions about the biochemical processes by which soil N is incorporated into SOM. Because these processes (e.g. microbial biomass production, abiotic incorporation, mycorrhizal assimilation) are not well understood (e.g. Aber *et al.*, 1998), many ecosystem models represent N immobilisation in an empirical way. For instance, the G'DAY (McMurtrie *et al.*, 2001) and CENTURY (Parton *et al.*, 1993) models assume that the N/C ratios of newly formed SOM vary between prescribed minimum and maximum values as functions of soil inorganic N content. If soil N/C ratios are assumed to be fi ed, then increased C f ows to the soil at high CO_2 must be accompanied by an increase in N immobilisation, strongly limiting N availability for plants. However, if soil N/C ratios decline, then soil C storage may be increased without a concomitant reduction in N mineralisation. The assumption about how soil N/C ratios change at high [CO_2] thus has important consequences for model output.

8.3 How uncertain assumptions affect model predictions

In this section we quantify the effect of the uncertainties described above on model predictions. We focus on the G'DAY model's predictions of net primary production (NPP), net ecosystem production (NEP), annual N uptake, nitrogen-use efficien y (NUE) and ecosystem carbon storage (ΔC) under alternative assumptions about the impact of elevated [CO_2] on nitrogen cycling processes. We ran simulations of G'DAY with parameters representing seven alternative scenarios for how high [CO_2] affects litter quality, litter quantity, below-ground C allocation, N acquisition and soil N/C ratio. The scenarios are listed in Table 8.1.

The G'DAY model was parameterised for a temperate forest stand dominated by loblolly pine (*P. taeda* L.), planted in 1983 at Duke Forest, NC (35°58′N, 79°05′W), where average annual temperature is 15.4°C and average annual precipitation is 1140 mm. Further site details are in Ellsworth (1999) and Hendry *et al.* (1999). The experiment, which has been running since 1996, includes six FACE plots, three maintained at ambient [CO_2] (365 ppm) and three at elevated [CO_2] (565 ppm). The

Table 8.1 List of scenarios considered in simulations with the G'DAY model of response to step increase of [CO_2] from 365 to 565 ppm at the Duke Forest. f_T is leaf retranslocation factor, f_a is leaf carbon allocation factor, N_{in} is external N input, υ_{ao} and υ_{so} represent the maximal N:C ratios of the newly formed active and slow organic matter in the soil, respectively, α_N is the slope of the J_{max} of leaf and its N/C ratio.

	Scenario	f_T	f_a	N_{in}	υ_{ao}	υ_{so}	α_N
0	Ambient [CO_2] of 365 ppm*	0.3	1	0.3	0.33	0.066	85
1	Increased litter quantity + decreased litter quality (base case)	0.3	1	0.3	0.33	0.066	85
2	1 + higher quality litter	0.1	1	0.3	0.33	0.066	85
3	1 + increased root allocation	0.3	0.85	0.3	0.33	0.066	85
4	1 + increased N input	0.3	1	1.3	0.33	0.066	85
5	1 + decreased N/C ratio of active SOM	0.3	1	0.3	0.25	0.066	85
6	5 + decreased N/C ratio of slow SOM	0.3	1	0.3	0.25	0.05	85
7	2 + 3 + 4 + 6 + decreased α_N	0.1	0.85	1.3	0.25	0.05	63.75

*Scenario 0 represents the control simulation where ambient [CO_2] is maintained at 365 ppm.

Duke Forest site is of particular interest because during the firs 3 years of CO_2-enrichment of the prototype FACE site, there were large CO_2-fertilisation effects on canopy photosynthesis (+40%), NPP (+20–30%) and C storage (Finzi *et al.*, 2002; Schäfer *et al.*, 2003), whereas during the fourth year there was evidence of N limitation. The N limitation was removed, however, in plots that received N fertilizer in the fift year (Oren *et al.*, 2001).

Meteorological data required by G'DAY are daily maximum and minimum air temperatures, total solar radiation and precipitation. For the G'DAY model, mean daily saturation water vapour pressure defici (D) was calculated using a sinusoidal pattern of temperature over a 24-h cycle under the assumption that air is saturated at the daily minimum temperature. Simulations were based on daily meteorological measurements over a 4-year period. Simulations were run over 100 years, which we represent by 25 cycles of the 4-year meteorological data file Simulations were initiated by running G'DAY to quasi-equilibrium (when average NEP is zero) under the baseline climate at Duke Forest, and then (at time $t = 0$) imposing a step increase in [CO_2] from 365 to 565 ppm (Figure 8.2). Because the simulations below have identical initial soil and plant C and N, and average annual NPP and zero average annual NEP, it is possible to directly compare simulations under different scenarios.

For each scenario we evaluated annual NPP, annual N uptake, NUE and the increase in ecosystem C storage during the firs 4 years at high CO_2 (initial), and after 20 and 100 years of CO_2 enrichment. A summary of the numerical results is given in Table 8.2 and model output for several of the scenarios is shown in Figure 8.2.

8.3.1 Scenario 1 (base case): increased litter quantity and decreased litter quality

For our base-case scenario, we assume that no parameter values change under elevated [CO_2]. In this scenario, the only direct impact of elevated [CO_2] is through

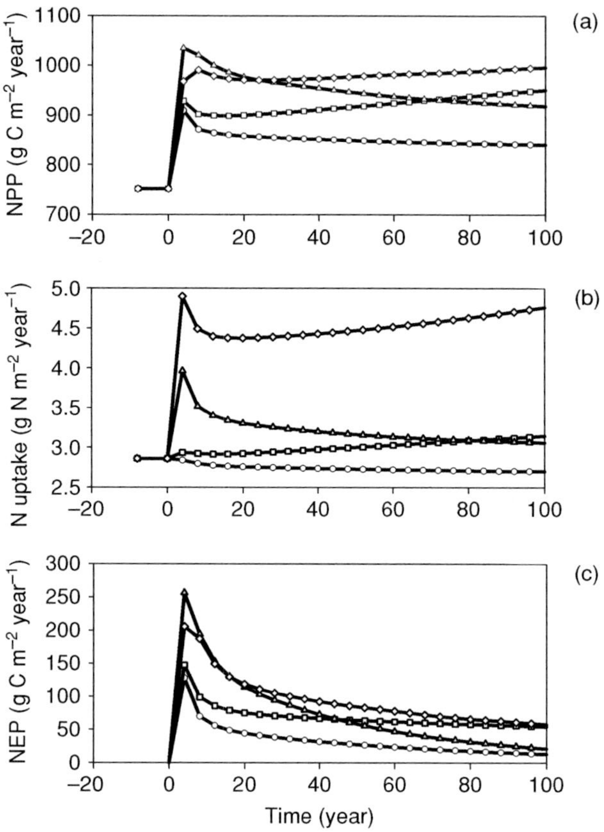

Figure 8.2 Simulated responses of (a) NPP, (b) N uptake and (c) NEP at the Duke Forest to a step increase in [CO_2] from 365 to 565 ppm at time $t = 0$ for Scenarios 1 (circle), 4 (square), 6 (triangle) and 7 (diamond). Each simulation was initiated by running G'DAY to equilibrium at [CO_2] of 365 ppm.

its effect on photosynthesis. Indirect effects include an increase in litter quantity, because NPP increases, and a decrease in litter quality, as the N/C ratios of live foliage and fin roots decrease and the retranslocation percentage is unchanged.

Figure 8.2a shows the simulated CO_2 response as 4-year averages of NPP. Immediately following the atmospheric [CO_2] increase, NPP increases due to CO_2 stimulation of photosynthesis. The 21% increase in NPP in the firs 4 years declines to +14% over the next two decades and to +12% after 100 years. This temporal pattern of a large transient CO_2-fertilisation effect giving way to a smaller nutrient-limited response has been reported previously (e.g. Comins & McMurtrie, 1993; Hudson *et al.*, 1994; McMurtrie & Comins, 1996). This so-called progressive nitrogen limitation has been the subject of much recent debate (e.g. Oren *et al.*, 2001; Luo *et al.*, 2004; Reich *et al.*, 2006). Past work with the G'DAY model has shown that the

Table 8.2 Simulated responses to step increase of [CO_2] from 365 to 565 ppm during the firs 4 years at high [CO_2] (initial), and after 20 and 100 years under 7 scenarios (details given in Table 8.1)

	At time 4 years			At time 20 years				At time 100 years			
Scenario	NPP (g C m^{-2} $year^{-1}$)	N_{up} (g N m^{-2} $year^{-1}$)	NUE (NPP/ N_{up})	NPP (g C m^{-2} $year^{-1}$)	N_{up} (g N m^{-2} $year^{-1}$)	NUE (NPP/ N_{up})	ΔC (g C m^{-2})	NPP (g C m^{-2} $year^{-1}$)	N_{up} (g N m^{-2} $year^{-1}$)	NUE (NPP/ NPP/N_{up})	ΔC (g C m^{-2})
0	752	2.86	263	752	2.86	263	0	752	2.86	263	0
1	910	2.84	320	858	2.76	311	1386	841	2.70	311	3333
2	852	2.96	288	816	2.99	273	722	808	2.95	273	1877
3	901	2.92	309	840	2.89	291	1234	825	2.84	290	2771
4	929	2.94	316	900	2.92	308	1942	951	3.14	303	6909
5	999	3.41	293	865	2.81	308	1857	848	2.75	309	3842
6	1035	3.96	302	977	3.30	303	3395	918	3.06	304	7508
7	968	4.90	197	971	4.37	222	3160	996	4.76	209	9458

CO_2-fertilisation effect varies over time, because responses on different timescales are determined by different ecosystem-level feedbacks and hence by different sets of key model parameters. After a step change in [CO_2], plant and soil pools in the G'DAY model reach equilibrium on various timescales that reflec the time constants of different pools (McMurtrie & Comins, 1996). The decline in NPP on the decadal timescale, as seen in Figure 8.2a, corresponds to the timescale for equilibration of slow SOM (cf. McMurtrie & Comins, 1996). This is associated with a decline in plant N availability (Figure 8.2b) as N is immobilised into the slow SOM pool. Net N immobilisation into slow SOM ceases once this pool equilibrates after approximately 100 years. The model predicts a sharp increase in NEP to levels of +128 g C m^{-2} year^{-1} over the firs 4 years, declining to +44 g C m^{-2} year^{-1} after 20 years and +14 g C m^{-2} year^{-1} after 100 years (Figure 8.2c). The cumulative increment in C storage over 100 years is 3.3 kg C m^{-2}, of which 15% is accumulated in the firs 4 years and 42% in the firs 20 years (Table 8.2).

Thus, the responses of NPP and NEP to an increase in atmospheric CO_2 concentration can vary on different timescales, depending on the turnover rates of different plant and soil carbon pools. In G'DAY and many other models, the SOM turnover rates are functions of soil temperature and moisture, and the same temperature and moisture functions are assumed for all soil pools. The latter assumption has been found to be incorrect (Knorr *et al.*, 2005), so that there is still considerable uncertainty about how to model the effects of temperature and moisture on decomposition (e.g. Kirschbaum, 1995; Kelly *et al.*, 2000).

8.3.2 Scenario 2: Scenario 1 + higher litter N/C ratio

Scenario 2 is used to evaluate the litter quality feedback by changing a single assumption used in Scenario 1: the leaf N retranslocation fraction f_T, which is 0.3 at ambient [CO_2] of 365 ppm, is reduced to 0.1 in [CO_2] 565 ppm. This means that leaf litter N concentration, and hence litter quality are higher under Scenario 2 than Scenario 1, and so soil N mineralisation rates and plant N uptake rates are enhanced compared with Scenario 1 (see Table 8.2). The increase in plant N uptake is countered, however, by the reduced N retranslocation prior to leaf senescence, which tends to decrease the amount of N available per unit C f xed under Scenario 2 compared with Scenario 1. The net effect shown in Table 8.2 is that simulated NPP is lower under Scenario 2 than in Scenario 1.

The difference between Scenarios 2 and 1 reflect the litter quality effect. Table 8.2 indicates that both NPP and C storage are higher for Scenario 1, in which CO_2-enriched litter has lower N/C ratio than in Scenario 2. This result is consistent with a conclusion by McMurtrie *et al.* (2000) that the CO_2 fertilisation effect is larger when litter N/C ratio declines at high [CO_2]. Their explanation for this result was that in a N-limited system, an increase in ecosystem NUE (define as NPP per unit N uptake) is required for an increase in NPP at high [CO_2]. However, NUE is inversely related to litter N/C ratio, and hence can only increase if litter N/C ratio declines. Notice that ecosystem NUE does increase for both Scenarios 1 and 2 at high CO_2, but that the increase is larger for Scenario 1 than for Scenario 2 (Table 8.2). Our conclusion that a decrease in the N/C ratio of CO_2-enriched litter tends to increase the CO_2-stimulation of NPP is noteworthy because it runs counter to the popular hypothesis that declining litter quality at high CO_2 represents a negative feedback on productivity (Norby *et al.*, 2001).

8.3.3 *Scenario 3: Scenario 1 + increased root allocation*

Scenarios 1 and 2 assumed that NPP was partitioned to foliage, wood and fin roots in the ratios $\eta_f:\eta_w:\eta_r = 0.275:0.45:0.275$, respectively. Scenario 3 illustrates the effect of increasing root allocation to 31.625% at the expense of foliage ($\eta_f:\eta_w:\eta_r = 0.23375:0.45:0.31625$). This change is accommodated in the G'DAY model by increasing the carbon allocation factor ($f_a = 2\eta_f/(\eta_f + \eta_r)$). This modest change in leaf/root allocation was chosen because its net effect is to keep simulated leaf area index after 4 years of CO_2-enrichment unchanged from its equilibrium value at ambient [CO_2]. Simulated values of NPP and C storage are slightly smaller under Scenario 3 than under Scenario 1 (see Table 8.2). This effect is due to the reduction in leaf N/C ratio. It should be noted that in the G'DAY model, the root biomass does not affect nitrogen uptake. In reality it is likely that the root biomass would directly impact on soil nitrogen uptake, but this impact is difficul to quantify and is omitted from the model.

8.3.4 *Scenario 4: Scenario 1 + increased N input*

Figure 8.2 illustrates the simulated response to a step increase in external N input, N_{in}, at time zero from the baseline rate of 0.3 g N m^{-2} year^{-1} to the increased rate of 1.3 g N m^{-2} year^{-1}. Changes in simulated plant N uptake following N fertilisation (Figure 8.2b) reflec the extent to which the additional N is immobilised in SOM. The additional N input of 1 g N m^{-2} year^{-1} results in an increase in average N uptake over the firs 4 years of only +0.097 g m^{-2} year^{-1}, relative to N uptake over the corresponding period under Scenario 1, which indicates that approximately 90% of the increased N input is initially immobilised or lost from the system. Over the 100-year simulation the increase in N uptake, relative to Scenario 1, amounts to 28% of the total N addition of 100 g m^{-2}. These increases in N uptake result in large sustained NPP and NEP responses (Figures 8.2a and 8.2c). The increase in simulated C storage over the 100-year period is 6.9 kg m^{-2}, which is more than double the increase achieved without extra N input.

8.3.5 *Scenario 5: Scenario 1 + decreased N/C ratio of new active SOM*

Lower soil N/C ratios will result if, firstl , CO_2-fertilised foliage produces litter with reduced N/C ratio and, secondly, the chemical composition of SOM resembles that of litter substrate (Aber *et al.*, 1990; Baldock *et al.*, 1992). For Scenario 5 the maximum N/C ratio of newly formed active SOM (υ_{ao}) is reduced by 25% at elevated

$[CO_2]$. In this scenario, simulated NPP and N uptake increase dramatically initially because of reduced N immobilisation in active SOM, but the effect on NPP and N uptake is transient (see Table 8.2) because the active pool is a fast turnover pool that equilibrates quickly (Table 8.2).

8.3.6 Scenario 6: Scenario 5 + decreased N/C ratio of new slow SOM

The consequences of reduced N/C ratio of slow as well as active SOM under Scenario 6 are presented in Figure 8.2, where the maximum values of N/C ratio of newly formed active (υ_{ao}) and slow SOM (υ_{so}) are both reduced by 25% at elevated $[CO_2]$. The effect on NPP is similar to that under Scenario 5 over the firs 4 years, but by the twentieth year both NPP and N uptake are much larger than that under both Scenarios 5 and 1. Effects on C storage are shown in Table 8.2. The increase in ecosystem C storage over the 100-year period is more than double that in Scenario 1.

8.3.7 Scenario 7: Scenario 2 + 3 + 4 + 6 + decreased slope of relation between maximum leaf potential photosynthetic electron transport rate and leaf N/C ratio

This scenario was used to evaluate the effects of multiple changes in model parameters on model predictions, and was also used in the application of model–data fusion in Section 8.4. The parameter α_N, which is the slope of the relationship between maximum leaf potential electron transport rate (J_{max}) and leaf N/C ratio, was reduced by 20% to represent the process of photosynthetic acclimation to rising $[CO_2]$ (e.g. see Chapter 2). After a step increase in atmospheric $[CO_2]$, simulated NPP and N uptake increase steeply during the firs 4 years, then decrease for approximately 10 years, after which they increased gradually throughout the next 90 years. These transient responses of NPP and N uptake reflec the initial soil N limitation to plant growth and the time required for the slow SOM pool to equilibrate. Nitrogen uptake and NPP are significantl increased compared to the base case, as a result of higher leaf litter N/C ratio, N deposition and lower immobilisation per unit SOM being formed. The whole system is approaching a new steady state at the end of 100 years, as shown by the steady decrease in NEP after 10 years, with significantl more carbon being sequestered over the 100 years compared with all other scenarios.

Overall, we found that processes operating at a range of timescales can affect the responses of a N-limited forest ecosystem to increased $[CO_2]$. Effects of changes in the N/C ratio of active SOM are short-term (less than a few years), while the effects of changes in the N/C ratio of slow SOM (υ_{so}) are long-term (decade to century). Over the long-term, a 25% decrease in υ_{so} can nearly double the amount of additional C stored (ΔC) by the Duke Forest (cf. Scenarios 6 and 1). Compared with the base case (Scenario 1), increases in litter N/C ratio, root allocation, N/C ratio of active or slow SOM have negative influence on NPP and additional C stored, while an increase in N input has a positive influenc on NPP and ΔC after 100 years. Increases in leaf litter N/C ratio, root allocation and N input or decreases in N/C ratios of active or slow SOM result in higher N uptake and NPP by plants for the N-limited Duke Forest.

It is possible that all fi e parameters considered in Scenarios 1–6 may change in high [CO_2] fiel experiments such as the Duke FACE. If so, a diversity of responses may be observed in measured NPP, NEP and N uptake, as depicted in the range of simulated responses in Figure 8.2. The challenge for modellers is to use measurements of NPP, NEP, N uptake and other variables to infer how model parameters have changed, and hence to identify how model parameters have changed in high [CO_2] experiments. In the next section, we use model–data fusion to address this challenge.

8.4 Model–data fusion techniques

The results in Table 8.2 show that predicted responses of plant ecosystems to increasing [CO_2] are subject to considerable uncertainty, because our understanding of the interactions between plant growth and soil nitrogen availability is still incomplete. We can only reduce this uncertainty by incorporating additional experimental evidence into our models. Sometimes it is possible to directly test model assumptions experimentally. For example, the 'litter quality' hypothesis has been refuted based on many experiments showing that, although litter nitrogen concentration is generally reduced by growth in elevated [CO_2], there is little effect on decomposition rate (Norby *et al.*, 2001). However, other assumptions are more difficul to test experimentally, either because the parameters involved are difficul to measure directly (e.g. slow soil N/C ratio) or because they respond on a timescale longer than most experiments. In these difficul cases, experimental evidence may still be used to inform models by using model–data fusion techniques.

Model–data fusion is a set of quantitative methods that improve model predictions based on observations. Applications of model–data fusion require (a) a model that describes the underlying physical, chemical and biological processes, (b) experimental observations and (c) an optimisation tool. The optimisation tool is used to fin optimal estimates of model parameters or states by minimising the differences between model predictions and experimental observations. Finding the optimal parameters can help us improve predictions or test alternative hypotheses embedded in the models. Model–data fusion can be used in several different ways: to estimate parameter values (Braswell *et al.*, 2005; Williams *et al.*, 2005) or in a sensitivity study that can be used to identify the observations required to estimate model parameters or to test our hypotheses (Wang *et al.*, 2001).

In this section we illustrate the techniques of model–data fusion, using the example of the G'DAY model applied to the Duke FACE experiment. As shown in the previous section, there is considerable uncertainty about long-term model predictions of [CO_2] response at this site, because of several carbon/nitrogen feedbacks operating at timescales varying from weeks to decades. The model outcomes depend on several key parameters, some of which cannot be measured directly. Here we focus on four of the six parameters listed in Table 8.1: the maximum N/C ratios of newly formed active and slow SOM, υ_{ao} and υ_{so}, the carbon allocation factor f_a and the fraction of nitrogen retranslocated from senescent foliage f_T. Observations of changes in these parameters due to increased [CO_2] would indicate which of the hypothesised mechanisms of ecosystem N cycling response to [CO_2] actually occurred. Although most of these parameters are difficul to measure directly, it is possible to estimate them indirectly from other measurements using model–data fusion techniques. Not all measurements are equally useful in estimating these parameters. Therefore, in the exercise that follows, we aim to identify which measurements are required to evaluate these key parameters accurately. This information is useful because it indicates where experimental effort should be concentrated to gain maximum advantage from data.

The potential measurements we considered were divided into three groups: group A – monthly net N mineralisation, group B – yearly carbon and nitrogen pool sizes of foliage, group C – yearly carbon and nitrogen pool sizes of fin roots and active SOM. To identify which of these groups of measurements are required to accurately estimate how the four parameters (f_T, f_a, υ_{ao} and υ_{so}) change at high [CO_2], we carried out what is known as a twin experiment. In this type of experiment, the model is run with a given set of parameter values. Noise is added to the model output to generate a set of hypothetical 'measurements'. The optimisation technique is then applied to these 'measurements' to attempt to recover the original parameter set. The success or otherwise of the optimisation indicates the usefulness of a particular type of measurement in determining parameter values.

We use Scenario 7 in Table 8.1 for this study and assume that changes in N_{in} and α_N can be measured independently. This scenario is chosen because we want to fin out what measurements are required to detect changes in any of the four key N cycling parameters under increased [CO_2]. 'Measurements' were created by adding random errors to the 100-year model output of monthly net mineralisation (A) and sizes of carbon/nitrogen pools in shoots, roots and active SOM. The amplitude of the random measurement error was assumed to be equal to one standard deviation of 100-year output for each of the seven output variables.

We used a commercial optimisation package, PEST (Dogherty, 2001), to estimate the four parameters, υ_{ao}, υ_{so}, f_a, and f_T, from the firs 5, 10, 25, 50 or 100 years of 'measurements'. Confidenc intervals of the estimates of two parameters can be geometrically represented by an ellipse. The centre of the ellipse represents the estimates of two parameters, the projected lengths onto x and y axes represent 1.96 times the standard errors of the optimal estimates (95% confidenc intervals). The estimates of two parameters are positively (negatively) correlated if the slope is positive (negative) (see Draper & Smith, 1981). Figure 8.3 shows the 95% confidenc intervals of the estimates of f_a and f_T, or υ_{ao} and υ_{so} when 10 years of measurements of the various potential measurement groups A, B, C (open regions) or all measurements combined (A + B + C; shaded regions) were used in the optimisation. Notice that the ellipses in Figure 8.3 are centred on the actual parameter values employed for Scenario 7 ($f_T = 0.1, f_a = 0.85$, $\upsilon_{ao} = 0.25$, $\upsilon_{so} = 0.05$), indicating that PEST successfully recovers the original parameter set.

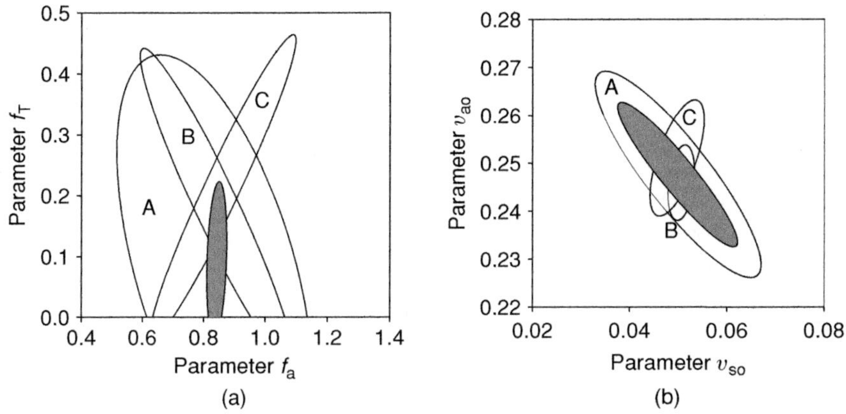

Figure 8.3 95% confidenc interval of parameters (a) f_a and f_T or (b) υ_{ao} and υ_{so} when 10 years' measurements of A, B, C or A + B + C (dark grey region) were used in the optimisation.

The results showed that measurement A cannot be used to provide independent estimates of all four parameters, as the correlation coefficient between any two of the four parameters except υ_{ao} and f_T, υ_{ao} and f_a, υ_{ao} and υ_{so} are significantl positive (>0.47) or negative (<–0.6) (see Table 8.3). Measurement B (annual shoot C and N pools) can be used to provide independent estimates of υ_{ao} and υ_{so}, but not f_a and f_T, because the estimates of the latter two parameters are strongly negatively correlated as the slope of the major axis of the ellipse is negative ($r = -0.88$; see Figure 8.3 and Table 8.3). Figure 8.3a shows that information about parameters f_a and f_T from measurement A is rather repetitive of that from measurement B, as ellipse B is largely located within ellipse A and the major axes of the two ellipses are approximately parallel. Although measurement C alone (annual root and active SOM C and N pools) cannot provide independent estimates of all four parameters, the information it provides is complementary to that provided from measurement A for all four parameters, and from measurement B for f_a and f_T. Therefore measurement A + C will provide nearly as much information about all four parameters as measurement A + B + C. Measurement A + B + C provides better estimates of f_a and f_T (lower correlation between the parameters) than measurement A + C, possibly because the

Table 8.3 Correlation coefficien between estimates of pairs of the four parameters when 10 years of measurements A, B, C or A + B + C were used in the optimisation

	A	B	C	A + B + C
(f_a, f_T)	–0.79	–0.88	0.79	0.03
(f_a, υ_{ao})	–0.14	0.41	0.20	–0.09
(f_a, υ_{so})	0.60	0.09	0.47	0.08
(f_T, υ_{ao})	0.29	–0.12	0.56	0.36
(f_T, υ_{so})	–0.65	0.35	0.86	–0.22
(υ_{ao}, υ_{so})	–0.69	0.24	0.37	–0.77

constraints from measurement A on the estimates of parameters f_{a} and f_{T} are too weak, as suggested by the relatively longer length of both axes of ellipse A than length of the other two ellipses (see Figure 8.3).

Interpretation of the optimisation results as shown in Figure 8.3 is also biologically plausible. Parameter f_{a} affects the fraction of carbon allocated to leaf relative to roots, while parameter f_{T} describes the fraction of leaf nitrogen that is translocated before leaf senescence. Measurements of A alone do not directly quantify the carbon or nitrogen fl ws from leaf to root (f_{a}) or nitrogen translocation within leaves. The net mineralisation rate in the G'DAY model depends on the amount and N/C ratio of litter. An increase in f_{a} reduces the fraction of carbon allocated to leaf relative to root and results in a decrease in leaf litterfall, whereas an increase in f_{T} results in a decrease in the N/C ratio of the leaf litter. As a result, estimates of f_{a} and f_{T} are negatively correlated if only measurement A is used in the optimisation.

Measurements of above-ground (B) or below-ground (C) carbon or nitrogen pool sizes provide better constraints on the estimates of all four parameters than measurement A (see Figure 8.3), because changes in pool sizes with time depend on flu es into and out of each pool, and both parameters f_{a} and f_{T} affect the carbon and nitrogen flu es into the foliage and roots. Increases in f_{a} or f_{T} will result in more carbon or nitrogen available for growth in the above-ground, and less for below-ground. Therefore, measurements of B and C provide complementary constraints on the estimates of f_{a} and f_{T}.

Estimates of some model parameters can be influence by the correlation between other parameters. For example, measurement B provides better constraints on the estimates of the two soil parameters υ_{so} and υ_{ao} than the other two measurements (A or C) (see Figure 8.3b), even though measurement C directly measures the changes in soil carbon and nitrogen in the active SOM. Estimates of f_{a} and f_{T} using measurement C are strongly correlated ($r = 0.94$), and the additional correlation between f_{T} and υ_{so} using measurement C may also contribute to poorer estimates of υ_{ao} and υ_{so} than obtained using measurement B, as suggested by the larger uncertainties in the estimates of both parameters (see Figure 8.3b).

If all measurements (A + B + C) for 10 years were used in the optimisation, the correlation between estimates of υ_{ao} and υ_{so} is still high (−0.77). This correlation generally decreases when longer time series of measurements are used in the optimisation, but is still quite significan even when 100 years' measurements (A + B + C) are used (see Figure 8.4). Therefore, additional measurements would be needed to provide independent estimates of υ_{ao} and υ_{so}, such as measurements of carbon and nitrogen in slow SOM and their changes over decades or more. The implication of this result is that measurements of changes in rapid turnover pools in elevated [CO_2] experiments are not sufficien to identify important parameters in the G'DAY model. The parameter υ_{so} has a large impact on predicted ecosystem carbon storage on the decadal timescale (cf. Scenarios 5 and 6), and cannot be resolved from the measurements of changes in short-term pools of carbon and nitrogen in plants and soil.

Will measurements of A, B or C over a period longer than 10 years provide better constraints on the four parameters? Figure 8.5 shows that yearly measurements of

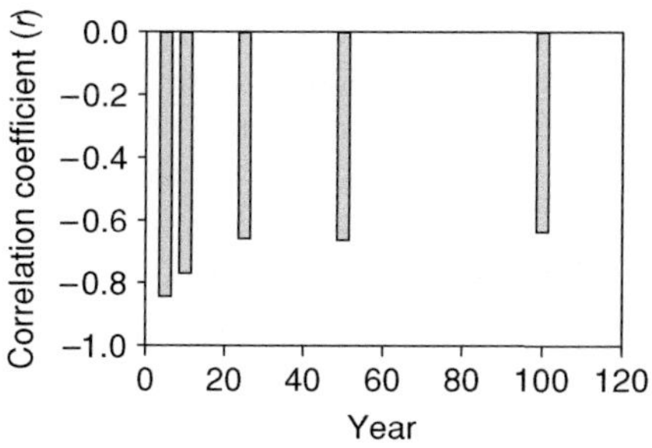

Figure 8.4 Correlation between the optimal estimates of υ_{ao} and υ_{so} when different years of measurements (A + B + C) were used in the optimisations.

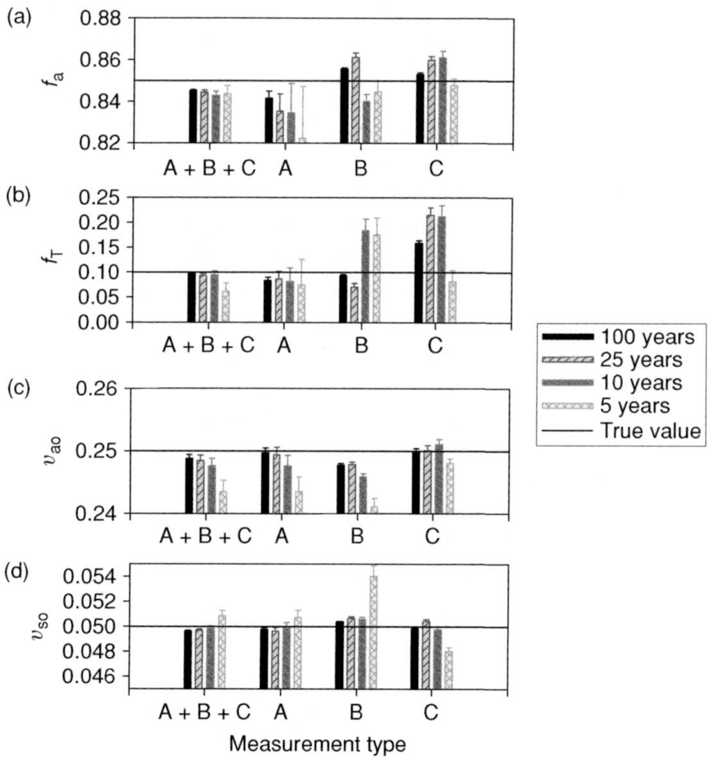

Figure 8.5 Optimal estimates of all four parameters when 5, 10, 25 or 100 years' measurements (A, B, C or A + B + C) were used in the optimisation. The horizontal line on each plot represents the 'true' value of the parameters used in the forward simulations.

B or C over 100 years alone do not provide reliable estimates of all four parameters, this is because measurements B or C cannot provide independent estimates of two or three of the four parameters. Optimal estimates of all four parameters are quite close to their respective 'true' values as used in the forward simulation if 100 years of monthly measurements of A are used in the optimisation. On the other hand, measurements of A + B + C over a period longer than 10 years do not provide significantl more information about the four parameters, as the estimates of four parameters using 10 years' measurements of A + B + C are quite close to their respective 'true' values.

8.5 Discussion

This study highlights the transient nature of the responses of terrestrial ecosystems to increased atmospheric [CO_2] in N-limited environments, and emphasizes that the response on the decadal timescale is related to the turnover rate of 'slow' SOM. Simulations of different scenarios also show a wide range of responses to increasing atmospheric [CO_2] by a terrestrial ecosystem. This uncertainty depends on nitrogen availability and how quickly various pools equilibrate with increased [CO_2]. Since most terrestrial ecosystems in the world are N-limited, predictions of responses to increasing [CO_2] over the next 100 years by models without considering soil nitrogen feedbacks on plant growth are likely to be overestimates.

Then how can we provide more realistic predictions of the terrestrial C sink over the next 100 years? The answer to that question lies in reducing uncertainties in the estimates of model parameters, the model and in observation errors. Observation errors include both instrument and sampling errors, and are a subject outside the scope of this chapter. However, for a given set of observations, we can use model–data fusion techniques to quantify the uncertainties in the estimates of model parameters and model errors and to identify what other observations are required to improve model predictions.

Model–data fusion includes parameter estimation and data assimilation as discussed by Raupach *et al.* (2005). Parameter estimation has been used by terrestrial scientists to fi models to measurements for many decades; data assimilation was firs used in meteorological weather forecast for estimating initial conditions and is a relatively new technique for terrestrial ecosystem modellers. Some applications have shown encouraging results (Braswell *et al.*, 2005; Williams *et al.*, 2005). Our application in this study has identifie that yearly measurements of above- and below-ground C and N pools can provide reliable estimates for three of four key parameters that may vary in response to elevated [CO_2] but that additional measurements are required to provide independent estimates of N/C ratios of new active and slow SOM (υ_{ao} and υ_{so}).

Errors in both measurements and models can degrade the estimates of model parameters and our confidenc in model predictions. In the twin experiment, we considered our model to be perfect. As discussed earlier, estimates of the model parameters depend on the sensitivities of model predictions to those parameters.

Therefore, the estimates of parameters are model specific but they can be used in other models if similar formulations are used to describe the physical or biological processes in a terrestrial ecosystem, such as leaf photosynthesis by the model of Farquhar *et al.* (1980) and the soil biogeochemistry by the CENTURY model (Parton *et al.*, 1987). Any systematic errors in the model can result in biases in the parameter estimates. To overcome this problem, one can develop an error model to account for systematic and random errors in the model predictions, and treat model errors separately from measurement errors (see Wang & McGregor, 2003) or make corrections to model predictions if model errors can be estimated independently (Abramowitz *et al.*, 2005). An even better approach is to identify the causes for systematic model errors, such as incorrect formulation or important processes omitted, and to make necessary modification to the model. The latter approaches often require much greater efforts, and should be one of the important goals in model–data fusion.

Model–data fusion can also be used to verify and reject our hypotheses. It is difficul to validate a model prediction at the decadal or century scale using fiel measurements, but we can test the theory and various hypotheses embedded in the model against results to improve our model and formulate a new set of hypotheses. As many elevated [CO_2] or climate change experiments often consist of measurements at different time and spatial scales, errors of some measurements may be correlated. Model–data fusion provides an efficien way of identifying what information can be extracted from those noisy and correlated measurements and possible deficien y in model structure or formulations that represent our hypothesis and what additional measurements may be required.

Terrestrial ecosystems may respond to increasing CO_2 concentration in the atmosphere by several mechanisms, and different mechanisms will be important in different ecosystems, depending on the dominant vegetation type or depending on whether growth is nutrient- or water limited. This chapter has focused on three feedback mechanisms that can affect the long-term CO_2 response of nitrogen-limited forests. It would be quite difficul to determine which of these mechanisms are operating from fiel measurements alone. However, we have shown how interactions between measurements and modelling studies through model–data fusion can help to elucidate the key mechanisms and to quantify their relative importance.

Acknowledgements

We acknowledge financia support for the DUKE-FACE experiment by the Offic of Science (BER), US Department of Energy, Grant No. DE-FG02-95ER62083 and the Australian Greenhouse Office

References

Aber, J., McDowell, W., Nadelhoffer, K., Magill, A., Berntson, G., Kamakea, M., McNulty, S., Currie, W., Rustad, L. & Fernandez, I. (1998) Nitrogen saturation in temperate forest ecosystems: hypotheses revisited. *BioScience*, **48**, 921–934.

Aber, J.D. & Melillo, J.M. (2001) *Terrestrial Ecosystems*, 2nd edn. Academic Press, San Diego, USA.

Aber, J.D., Melillo, J.M. & McClaugherty, C.A. (1990) Predicting long-term patterns of mass loss, nitrogen dynamics and soil organic matter formation from initial fin litter chemistry in temperate forest ecosystems. *Can. J. Bot.*, **68**, 2201–2208.

Abramowitz, G., Gupta, H., Pitman, A., Wang, Y.P. & Leuning, R. (2005) Neural error regression diagnosis (NERD): a tool for model bias identificatio and prognostic data assimilation. *J. Hydrometeorol.*, **7**, 160–177.

Ainsworth, E.A. & Long, S.P. (2005) What have we learned from 15 years of free-air CO_2 enrichment (FACE)? A meta-analytic review of the responses of photosynthesis, canopy properties, and plant production to rising CO_2. *New Phytol.*, **165**, 351–372.

Baldock, J.A., Oades, J.M., Waters, A.G., Peng, X., Vassallo, A.M. & Wilson, M.A. (1992) Aspects of the chemical structure of soil organic materials as revealed by solid-state ^{13}C NMR spectroscopy. *Biogeochemistry*, **16**, 1–42.

Berntson, G.M., & Bazzaz, F.A. (1996) Belowground positive and negative feedbacks on CO_2 growth enhancement. *Plant Soil*, **187**, 119–131.

Braswell, B.H., Sacks, W.J., Linder, E. & Schimel, D.S. (2005) Estimating diurnal to annual ecosystem parameters by synthesis of a carbon flu model with eddy covariance net ecosystem exchange observations. *Global Change Biol.*, **11**, 335–355.

Comins, H.N. & McMurtrie, R.E. (1993) Long-term response of nutrient-limited forests to CO_2-enrichment; equilibrium behaviour of plant-soil models. *Ecol. Appl.*, **3**, 666–681.

Cramer, W., Bondeau, A., Woodward, F.I., Prentice, I.C., Betts, R.A., Borvkin, V., Cox, P.M., Fisher, V., Foley, J.A., Friend, A.D., Kucharik, C., Lomas, M.R., Ramankutty, N., Sitch, S., Smith, B., White, A. & Young-Molling, C. (2001) Global response of terrestrial ecosystem structure and function to CO_2 and climate change: results from six dynamic global vegetation models. *Global Change Biol.*, **7**, 357–373.

Curtis, P.S. & Wang, X.Z. (1998) A meta-analysis of elevated CO_2 effects on woody plant mass, form, and physiology. *Oecologia*, **113**, 299–313.

Diaz, S., Grime, J., Harris, J. & McPherson, E. (1993) Evidence of a feedback mechanism limiting plant response to elevated carbon dioxide. *Nature*, **364**, 616–617.

Dogherty, J. (2001) *Model-independent Parameter Estimation*, 4th edn. Watermark Numerical Computing, Brisbane, Australia.

Draper, N. & Smith, H. (1981) *Applied Regression Analysis*, 2nd edn. Wiley, New York.

Ellsworth, D.S. (1999) CO_2 enrichment in a maturing pine forest: are CO_2 exchange and water status in the canopy affected? *Plant Cell Environ.*, **22**, 461–472.

Farquhar, G.D., Caemmerer, von S. & Berry, J.A. (1980) A biochemical model of photosynthetic CO_2 assimilation in leaves of C_3 species. *Planta*, **149**, 78–90.

Finzi, A.C., De Lucia, E.H., Hamilton, J.G., Richter, D.D. & Schlesinger, W.H. (2002) The nitrogen budget of a pine forest under free air CO_2 enrichment. *Oecologia*, **132**, 567–578.

Gordon, W.S., Famiglietti, J.S., Fowler, N.L., Kittel, T.G.F. & Hibbard, K.A. (2004) Validation of simulated runoff from six terrestrial ecosystem models: results from VEMAP. *Ecol. Appl.*, **14**, 527–545.

Hanson, P.J., Amthor, J.S., Wullschleger, S.D., Wilson, K.B., Grant, R.E., Hartley, A., Hui, D., Hunt, E.R., Johnson, D.W., Kimball, J.S., King, A.W., Luo, Y., McNulty, S.G., Sun, G., Thornton, P.E., Wang, S., Williams, M., Baldocchi, D.D. & Cushman, R.M. (2004) Oak forest carbon and water simulations: model intercomparisons and evaluations against independent data. *Ecol. Monogr.*, **74**, 443–489.

Hendry, G.R., Ellsworth, D.S., Lewin, K.F. & Nagy, J. (1999) A free-air enrichment system for exposing tall forest vegetation to elevated CO_2. *Global Change Biol.*, **5**, 293–309.

Hudson, R.J.M., Gherini, S.A. & Goldstein, R.A. (1994) Modeling the global carbon cycle: nitrogen fertilization of the terrestrial biosphere and the 'missing' CO_2 sink. *Global Biogeochem. Cycles*, **8**, 307–333.

Hungate, B.A., Dukes, J.S., Shaw, M.R., Luo, Y. & Field, C.B. (2003) Nitrogen and climate change. *Science*, **302**, 1512–1513.

Kelly, R.H., Parton, W.J., Hartman, M.D., Stretch, L.K., Ojima, D.S. & Schimel, D.S. (2000) Intra-annual and interannual variability of ecosystem processes in shortgrass steppe. *J. Geophys. Res.*, **105**, 20093–20100.

Kirschbaum, M.U.F. (1995) The temperature dependence of soil organic matter decomposition, and the effect of global warming on soil organic C storage. *Soil Biol. Biochem.*, **27**, 753–760.

Kirschbaum, M.U.F., King, D.A., Comins, H.N., McMurtrie, R.E., Medlyn, B.E., Pongracic, S., Murty, D., Keith, H., Raison, R.J. & Khanna, P.K. (1994) Modelling forest response to increasing CO_2 concentration under nutrient-limited conditions. *Plant Cell Environ.*, **17**, 1081–1099.

Kirschbaum, M.U.F., Medlyn, B.E., King, D.A., Khanna, P.K., Raison, R.J., Pongracic, S., Snowdon, P. & Murty, D. (1998) Modelling forest response to increasing CO_2 concentration in relation to various factors affecting nutrient supply. *Global Change Biol.*, **4**, 23–41.

Knorr, W., Prentice, L.C., House, J.I. & Holland, E.A. (2005) Long-term sensitivity of soil carbon turnover to warming. *Nature*, **433**, 298–301.

Luo, Y., Bo, S., Currie, W.S., Dukes, J.S., Finzi, A., Hartwig, U., Hungate, B., McMurtrie, R., Oren, R., Parton, W.J., Pataki, D., Shaw, R., Zak, D.R. & Field, C.B. (2004) Progressive nitrogen limitation of ecosystem responses to rising atmospheric CO_2. *BioScience*, **54**, 731–739.

Luxmoore, R.J., Hargrove, W.W., Tharp, M.L., Post, W.M., Berry, M.W., Minser, K.S., Cropper, W.P., Jr, Johnson, D.W., Zeide, B., Amateis, R.L., Burkhart, H.E., Baldwin, V.C., Jr & Peterson, K.D. (2000) Signal-transfer modeling for regional assessment of forest responses to environmental changes in the southeastern United States. *Environ. Model. Assess.*, **5**, 125–137.

McMurtrie, R.E. & Comins, H.N. (1996) The temporal response of forest ecosystems to doubled atmospheric CO_2 concentration, *Global Change Biol.*, **2**, 49–57.

McMurtrie, R.E., Medlyn, B.E. & Dewar, R.C. (2001) Increased understanding of nutrient immobilization in soil organic matter is critical for predicting the carbon sink strength of forest ecosystems over the next 100 years. *Tree Physiol.*, **21**, 831–839.

McMurtrie, R.E., Medlyn, B.E., Dewar, R.C. & Jeffreys, M. (2000) Effects of rising CO_2 on growth and carbon sequestration in forests: a modelling analysis of the consequences of altered litter quantity and quality. *Plant Soil*, **224**, 135–152.

Medlyn, B.E. & McMurtrie, R.E. (2005) Effects of CO_2 on plants at different timescales. In: *A History of Atmospheric CO_2 and Its Impacts on Plants, Animals, and Ecosystems* (eds J. Elheringer, T. Cerling & D. Dearing), pp. 441–467. Springer-Verlag, San Diego, USA.

Medlyn, B.E., McMurtrie, R.E., Dewar, R.C. & Jeffreys, M.P. (2000) Soil processes dominate the long-term response of forest net primary productivity to increased temperature and atmospheric CO_2 concentration. *Can. J. For. Res.*, **30**, 873–888.

Melillo, J., Callaghan, T., Woodward, F., Salati, E. & Sinha, S. (1991) Effects on ecosystems. In: *Climate Change: The IPCC Scientific Assessment* (eds J. Houghton, G. Jenkins & J. Ephraums), pp. 282–310. Cambridge University Press, Cambridge, UK.

Nightingale, J.M., Phinn, S.R. & Held, A.A. (2004) Ecosystem process models at multiple scales for mapping tropical forest productivity. *Prog. Phys. Geogr.*, **28**, 241–281.

Norby, R.J., Cotrufo, M.F., Ineson, P., O'Neill, E.G. & Canadell, J. (2001) Elevated CO_2, litter chemistry, and decomposition: a synthesis. *Oecologia*, **127**, 153–165.

Norby, R.J., Ledford, J., Reilly, C.D., Miller, N.E. & O'Neill, E.G. (2004) Fine-root production dominates response of a deciduous forest to atmospheric CO_2 enrichment. *Proc. Natl. Acad. Sci. USA*, **101**, 9689–9693.

Oren, R., Ellsworth, D.S., Johnson, K.H., Phillips, N., Ewers, B.E., Maier, C., Schäfer, K.V.R., McCarthy, H., Hendrey, G.R., McNulty, S.G. & Katul, G.G. (2001) Soil fertility limits carbon sequestration by a forest ecosystem in a CO_2-enriched atmosphere. *Nature*, **411**, 469–472.

Parton, W.J., Schimel, D.S., Cole, C.V. & Ojima, D.S. (1987) Analysis of factors controlling soil organic matter levels in Great Plains grasslands. *Soil Sci. Soc. Am. J.*, **51**, 1173–1179.

Parton, W.J., Scurlock, J.M.O., Ojima, D.S., Gilmanov, T.G., Scoles, T.G., Schimel, D.S., Kirchner, T., Menaut, J.-C., Seastedt, T., Garcia, E.M.K.A. & Kinyamario, J.I. (1993) Observations and modeling of biomass and soil organic matter dynamics for the grassland biome worldwide. *Global Biogeochem. Cycles*, **7**, 785–809.

Pepper, D.A., Del Grosso, S.J., McMurtrie, R.E. & Parton, W.J. (2005) Simulated carbon sink response of shortgrass steppe, tallgrass prairie and forest ecosystem to rising [CO_2], temperature and nitrogen input. *Global Biogeochem. Cycles*, **19**, GB1004.

Prentice, I.C., Farquhar, G.D., Fasham, M.J.R., Goulden, M.L., Heimann, M., Jaramillo, V.J., Kheshgi, H.S., Le Quéré, C., Scholes, R.J. & Wallace, D.W.R. (2001) The carbon cycle and atmospheric carbon dioxide. In: *Climate Change 2001: The Scientific Basis* (eds J.T. Houghton, Y. Ding, D.J. Griggs, M. Noguer, P.J. van der Linden, X. Dai, K. Maskell & C.A. Johnson), pp. 183–237. Cambridge University Press, Cambridge, UK.

Raupach, M.R., Rayner, P.J., Barrett, D.J., DeFries, R., Heimann, M., Ojima, D.S. Quegan, S. & Schumllius, C. (2005) Model-data synthesis in terrestrial carbon observations: methods, data requirements and data uncertainty specifications *Global Change Biol.*, **11**, 378–397.

Reich, P.B., Hobbie, S., Lee, T., Ellsworth, D., West, J.B., Tilman, D., Craine, J., Knops, J., Naeem, S. & Trost, J. (2006) Nitrogen limitation constrains sustainability of ecosystem response to CO_2. *Nature*, **440**, 922–925.

Schäfer, K.V.R., Oren, R., Ellsworth, D.S., Chun-Ta, L., Herricks, J.D., Finzi, A.C., Richter, D.D. & Katul, G.G. (2003) Exposure to an enriched CO_2 atmosphere alters carbon assimilation and allocation in a pine forest ecosystem. *Global Change Biol.*, **9**, 1378–1400.

Schimel, D.S., Emanuel, W., Rizzo, B., Smith, T., Woodward, F.I., Fisher, H., Kittel, T.G.F., McKeown, R., Painter, T., Rosenbloom, N., Ojima, D.S., Parton, W.J., Kicklighter, D.W., McGuire, A.D., Melillo, J.M., Pan, Y., Haxeltine, A., Prentice, C., Sitch, S., Hibbard, K., Nemani, R., Pierce, L., Running, S., Borchers, J., Chaney, J., Neilson, R. & Braswell, B.H. (1997) Continental scale variability in ecosystem processes: models, data, and the role of disturbance. *Ecol. Monogr.*, **67**, 251–271.

Wang, Y.P., Leuning, R., Cleugh, H.A. & Coppin, P.A. (2001) Parameter estimation in surface exchange models using nonlinear inversion: how many parameters can we estimate and which measurements are most useful. *Global Change Biol.*, **7**, 495–510.

Wang, Y.P. & McGregor, J.L. (2003) Estimating regional terrestrial carbon flu es for the Australian continent using multiple-constraint approach. II. Atmospheric constraint. *Tellus B*, **55**, 290–304.

Williams, M., Schwarz, P.A., Law, B.E., Irvine, J. & Kurpius, M.R. (2005) An improved analysis of forest carbon dynamics using data assimilation. *Global Change Biol.*, **11**, 89–105.

Zak, D.R., Pregitzer, K.S., Curtis, P.S., Teeri, J.A., Fogel, R. & Randlett, D. (1993) Elevated atmospheric CO_2 and feedback between carbon and nitrogen cycles. *Plant Soil*, **151**, 105–117.

9 Predicting the effect of climate change on global plant productivity and the carbon cycle

John Grace and Rui Zhang

9.1 Introduction

We live in a period of global warming. Temperatures have increased by at least 0.6°C since 1900, whilst in the Arctic the increases are about twice as much, and the Greenland ice-sheet is melting at alarming rates. The underlying cause of this warming trend is the anthropogenic emission of greenhouse gases, of which the most important is carbon dioxide, which is also the raw material for plant growth. Approximately half of all biomass consists of carbon. Plant growth and the carbon cycle are therefore inextricably linked.

The anthropogenic emissions of carbon dioxide are currently rising rapidly, with 7.0 Pg C year^{-1} coming from fossil fuel burning and cement-making, and a further 1–2 Pg C year^{-1} from tropical deforestation (IPCC, 2000; Marland *et al.*, 2003). About 3 Pg of carbon stays in the atmosphere as CO_2, where its concentration is increasing at a rate of 1–2 ppm by volume per year. The remainder is dissolved in the ocean or taken up by terrestrial vegetation. Current estimates suggest a sink of 1.9 Pg year^{-1} in the ocean (Le Quéré *et al.*, 2003) and a similar one on the land where photosynthesis currently exceeds autotrophic and heterotrophic respiration (Figure 9.1).

We know that the carbon cycle and the sink strengths are not in steady state from (i) analysis of concentrations of carbon dioxide and its isotopes in the atmosphere (Keeling *et al.*, 1996; Gurney *et al.*, 2002; Rödenbeck *et al.*, 2003), (ii) analysis of CO_2 flu es over vegetation using a range of techniques (Oechel *et al.*, 2000; Valentini *et al.*, 2000), (iii) theoretical and physiological analysis (Lloyd & Farquhar, 1996; Lloyd, 1999) and (iv) remote sensing of the 'greening' of the land in the northern regions (Myneni *et al.*, 1997, 1998). These studies show that photosynthesis now exceeds respiration on a global basis. This response is generally attributed to three sorts of changes in the environment that exert a positive effect on photosynthesis (Lloyd & Farquhar, 1996; Nemani *et al.*, 2003; Ciais *et al.*, 2005). These are (i) the global increase in CO_2, which stimulates photosynthesis, (ii) warming in those areas of the world that are currently rather cold for photosynthesis and (iii) the deposition of anthropogenic nitrogen as ammonium and nitrate acting as a fertilizer (Galloway *et al.*, 1995, 2004). There may additionally be regional increases in rainfall, which could be having a significan impact on photosynthesis, and trends that arise entirely from patterns of forest management.

In this chapter we firs review the concepts that underlie the measurement and modelling of the carbon cycle. Then we show how the constituent flu es may change

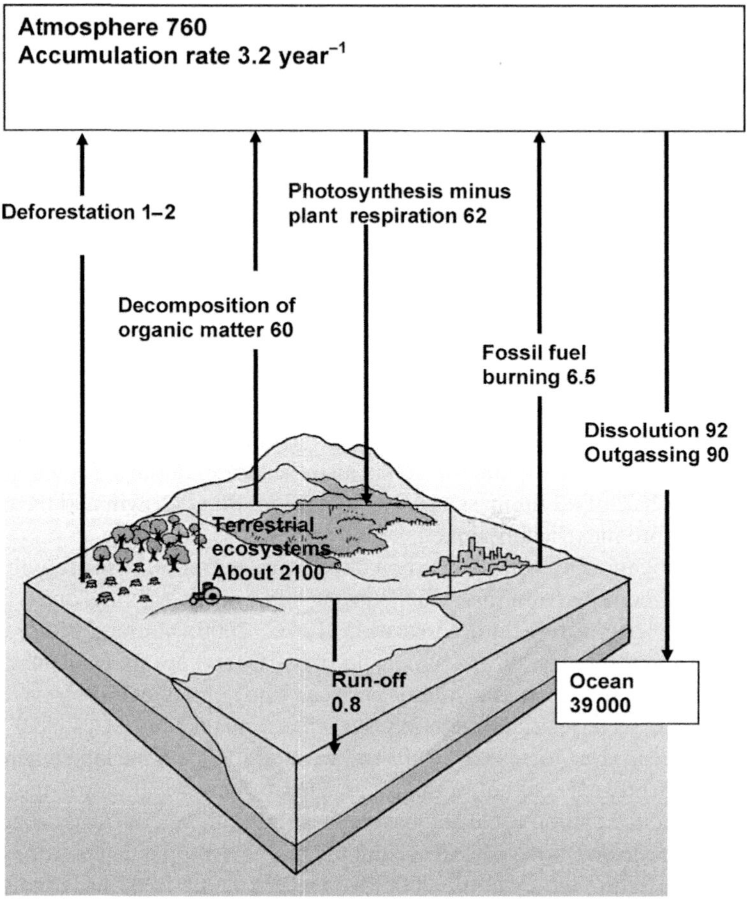

Figure 9.1 The global carbon cycle during the 1990s. Sources: IPCC (2000), Le Quéré *et al.* (2003). Boxes show carbon stocks in Petagrams (Pg $= 10^{15}$ g), arrows show annual flu es (Pg C year^{-1}). For a discussion of the uncertainty of each estimate, see Royal Society (2001).

in a warmer world in which both the CO_2 and the nitrogen supply are likely to be enhanced.

9.2 Definitions and conceptual framework

The transfers of carbon between the biosphere and atmosphere may be expressed by four related terms. These can be considered on a 'per area of land' basis or on a 'whole world' basis, as follows:

i. Gross primary productivity (GPP) is the rate of transfer of carbon from atmosphere to biosphere by photosynthesis. For the world as a whole it is about 120 Pg C year^{-1}, although this widely accepted value is sometimes questioned and may be higher (Saugier *et al.*, 2001). GPP includes the 'hidden' term of photorespiration, which always accompanies photosynthesis in C3 plants, but GPP excludes mitochondrial plant respiration that proceeds by day and night and is generally called 'autotrophic respiration'.

ii. Net primary productivity (NPP) is the rate of transfer of carbon from atmosphere to biosphere by net photosynthesis (GPP minus autotrophic respiration). In the 1960s, methods for estimating NPP by sequential harvesting were developed. These methods are usually rather poor as they cannot capture production processes belowground. They are, however, the basis of the commonly tabulated values of global NPP, from which it is estimated that the vegetation of the world produces somewhat more than 60 Pg C year^{-1} (Roy *et al.*, 2001). There are other gaseous flu es that are important in the carbon cycle, and that ought to be taken into account when measuring or modelling NPP; for example, plants emit a variety of volatile organic compounds, the most abundant of which, isoprene, is often about 1% of GPP (Guenther *et al.*, 1995).

iii. Net ecosystem productivity (NEP), otherwise known as net ecosystem exchange (NEE), is the rate of transfer of carbon from the atmosphere as photosynthesis minus autotrophic and heterotrophic respiration (i.e. the net of all CO_2 exchanges). The heterotrophs include bacteria, fungi and animals. NEP is thus the net carbon flu between the land surface and the atmosphere, if positive, carbon is accumulating. NEP is routinely measured from towers above the vegetation canopy, using the micrometeorological technique of eddy covariance at over 200 sites worldwide (Falge *et al.*, 2002). The measurement system has a 'flu footprint' of 0.1–1 km^2 (i.e. it measures the net flu of the ecosystem near the tower over an area of 0.1–1 km^2). The difference between NEP and NEE is one of sign convention. Ecologists generally use NEP (gains by the vegetation are shown as positive); micrometeorologists generally use NEE (gains by the vegetation are shown as negative). These flu es are the amount per area of land per unit of time, when 'amount' can be in molar or mass units.

iv. Net biome productivity (NBP) is the term introduced more recently to account for the flu es at a spatial scale from 1 km^2 upwards and a timescale from 1 year upwards where disturbance is significan (Steffen *et al.*, 1998).

These terms are interrelated as follows:

$$\mathrm{GPP} = P$$
$$\mathrm{NPP} = P - R_a$$
$$\mathrm{NEP} = P - R_a - R_h$$
$$\mathrm{NBP} = P - R_a - R_h - D$$

where P is the photosynthetic rate, R_a is the autotrophic respiration rate, R_h is the respiration rate by heterotrophs and D is the rate of change in C flu contributed by disturbance including cultivation and harvesting. It is sometimes useful to combine R_a and R_h as 'ecosystem respiration'.

There are some empirical relationships between these terms, which we should regard as 'rule of thumb' and provisional, but useful in modelling and upscaling. The relationship between GPP and NPP varies somewhat according to ecosystem, and there is still a degree of uncertainty because NPP is generally difficul to measure by either gravimetric or gas flu methods. However, in studies of forest ecosystems in the United States and Australasia, it was found that NPP/GPP was more or less constant at 0.47 (Waring *et al.*, 1998). Gifford (2003) found the value to be 0.47 ± 0.05 for a range of forests, and 0.58 ± 0.03 for whole plants in controlled environments (plus/minus symbol here denotes the standard deviation). Heterotrophic respiration is, however, more complex, as the process of decomposition is dependent on the action of a myriad of heterotrophic organisms, and the rate of CO_2 evolution is a function of temperature, moisture and the supply of organic matter. Not surprisingly, data from eddy covariance towers show GPP/NEE to be quite variable, usually in the range 0–0.4 (Falge *et al.*, 2002).

9.3 Empirical basis of our knowledge of carbon fluxes

9.3.1 NPP

First attempts to measure plant productivity on a large scale were made in the 1950s and 1960s, summarised in Rodin and Bazilevich's work *Production and Mineral Cycling in Terrestrial Vegetation* published in Moscow. In 1962 the planning began for the launch of the International Biological Programme by the International Council of Scientifi Unions, one of its aims being to understand the biological basis of plant productivity. A number of influentia publications ensued, including Eckardt (1986) and Whittaker and Likens (1975). The emphasis in most of these early studies was to measure NPP. It was difficul then, and still is. The methodology is gravimetric and consists of sampling at two points in time, usually with an interval of 1 year, to measure the increase in biomass ΔC and the litter collected in litter traps, L:

$$\text{NPP} = \Delta C + L$$

This approach does not usually include belowground litter, nor root exudates, although there are several additional techniques available to make such measurement. It is particularly unreliable for vegetation with fast turnover of components, such as tropical grassland and where there is substantial herbivory (see Long *et al.*, 1992). Generally the method is likely to underestimate the flu es. Despite this, the table of NPP values for the world, published by Whittaker and Likens (1975), is not very different from the one accepted now (Table 9.1) and leads us to the conclusion that global NPP is approximately 60 Pg C year^{-1}.

Table 9.1 The stocks of carbon and rates of formation of carbon as net primary productivity (Saugier *et al.*, 2001)

Biome	NPP ($t\ C\ ha^{-1}\ year^{-1}$)	Area (million km^2)	Total carbon pool (Pg C)	Total NPP ($Pg\ C\ year^{-1}$)
Tropical forests	12.5	17.5	553	21.9
Temperate forests	7.7	10.4	292	8.1
Boreal forests	1.9	13.7	395	2.6
Arctic tundra	0.9	5.6	117	0.5
Mediterranean Shrublands	5.0	2.8	88	1.4
Crops	3.1	13.5	15	4.1
Tropical savanna and grasslands	5.4	27.6	326	14.9
Temperate grasslands	3.7	15	182	5.6
Deserts	1.2	27.7	169	3.5
Ice		15.3		
Total		149.3		62.6

9.3.2 *NEP and NEE*

Very early on, several people saw the possibility of measuring ecosystem gas exchange using chambers or micrometeorological methods (Eckardt, 1986; Monteith, 1968). Chamber techniques involve covering the land surface cover with a transparent chamber and measuring the rate at which CO_2 is depleted (in daylight) or enriched (in dark). Chamber techniques are still widely used, especially for gas flu es from the soil. The disadvantages of chamber techniques are that (i) chambers develop an internal microclimate different from the climate outside, and so may need air-conditioning, and that (ii) they sample just a small patch of land, and therefore many of them are required to reduce the statistical sampling error to acceptable levels.

Micrometeorological techniques, on the other hand, have no effect on the vegetation being measured and they sample much larger areas of land, depending on the height of the sensors above the vegetation. The micrometeorological technique developed firs was the flu gradient method, which relies on the measurement of mean concentrations of CO_2 and wind speed at several heights above the canopy. This approach was found to give excellent estimates of the carbon flu es over short vegetation including crops (Biscoe *et al.*, 1975), but for many types of land cover the gradients found were very small and barely measurable. The preferred methodology is now eddy covariance, another micrometeorological technique, which exploits the capacity to collect automatically large quantities of data over long periods of time (effectively, continuously). The measurement approach was suggested over 50 years ago by Swinbank (1951) but rather surprisingly the practical development of it was delayed because gas analysers to measure rapid fluctuation in CO_2 concentration were not available, nor were data loggers to deal with the huge stream of data produced. The firs people to use eddy covariance to measure CO_2 flu es over landscape were probably Desjardins and Lemon (1974) and Leuning *et al.* (1982), but the 'modern era' of reliable, continuous measurements did not start until the 1990s (Moncrieff *et al.*, 1997; Aubinet *et al.*, 2000).

The eddy covariance technique is in principle deceptively simple. CO_2 and H_2O vapour flu es (mol m^{-2} s^{-1}) are calculated from knowledge of the instantaneous concentration χ' (mol m^{-3}) and the instantaneous vertical motion of the air, w' (m s^{-1}). These are measured by a sensor mounted in the turbulent airstream well above the canopy. The products of these terms are averaged to give the flu F:

$$F = \overline{w'\chi'}$$

Measurements are made many times per second, and the data are averaged over a period of 0.5 h to obtain a good statistical sample. In practice, there are various instrumental corrections that must be made as documented by Moncrieff *et al.* (1997) and Aubinet *et al.* (2000). Already there are over 200 stations in the world at which CO_2 and H_2O flu es are being measured by eddy covariance (Baldocchi *et al.*, 2001), and the sources of uncertainty have been widely discussed. These include (a) the problems of dealing with non-ideal sites that have significan slopes, (b) how to identify periods when the methodology may fail because of nocturnal stable conditions and (c) the methodology of 'gap filling for those periods when the records are considered to be unreliable (Saleska *et al.*, 2003; Kruijt *et al.*, 2004). Nevertheless, from NEE it is possible to estimate, with assumptions, the constituent terms GPP and total respiration of the ecosystem R_e. Examples for various forest biomes may be found in Falge *et al.* (2002) and Malhi *et al.* (1999).

To understand the functioning of photosynthesis and respiration in response to climatological variables, the estimation of ecosystem level CO_2 flux as outlined above, is often supplemented by leaf-scale measurements of net photosynthesis made inside the canopy (e.g. Carswell *et al.*, 2000) and soil respiration measurements at the soil surface.

It is worth noting that eddy covariance does not measure NPP, as this would require separation of autotrophic and heterotrophic respiration. Small chambers may be attached to plants to measure aboveground autotrophic respiration, but it is much more difficul to measure the belowground component of autotrophic respiration. In some cases, ingenious experimental manipulations have been made to partition 'soil respiration' into autotrophic and heterotrophic components, involving construction of trenches or killing trees to remove the heterotrophic component (Högberg *et al.*, 2001).

NEE data for a whole region may be conveniently summarised by a diagram in which the half-hour averages of NEE are plotted as an annual time course (Figure 9.2). When this is done for a region the size of Europe, different patterns emerge, related to the type of vegetation and the climate of the location (Valentini *et al.*, 2000). In northern Scandinavia we see that boreal (conifer) forest has substantially less activity than forest in warmer regions. Its dormant period is long, as a result of

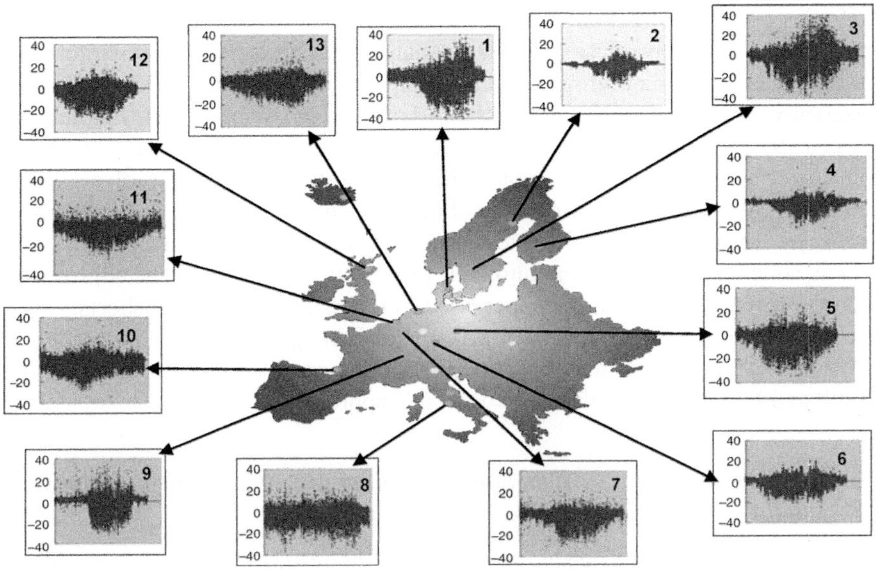

Figure 9.2 Net CO_2 flu measured at 13 European forests during the European project EUROFLUX. Each graph presents the annual CO_2 flu measured over the forest at half-hour intervals; the *y* axis is the CO_2 flu in units of μmol CO_2 m^{-2} s^{-1}, where negative means uptake from the atmosphere (occurring in the day, when photosynthesis exceeds respiration) and positive means net flu es from surface to the atmosphere (when respiration exceeds photosynthesis). The sites are: 1, Soroe, Denmark (deciduous broad-leaved forest); 2, Flakaliden, Sweden (coniferous forest); 3, Norunda, Sweden (mixed coniferous forest); 4, Hyytiälä, Finland (coniferous forest); 5, Tharandt, Germany (coniferous forest); 6, Bayreuth, Germany (coniferous forest); 7, Vielsalm, Belgium (mixed forest); 8, Castelporziano, Italy (evergreen forest); 9, Sarrebourg, France (deciduous broad-leaved forest); 10, Bordeaux, France (coniferous plantation); 11, Braschaat, Belgium (coniferous forest); 12, Aberfeldy, United Kingdom (coniferous forest); 13, Loobos, The Netherlands (coniferous forest).

long periods of sub-zero temperatures. Its maximum rates of NEE are never very large in this cold environment, because of the effect of low temperatures on photosynthesis (Lindroth *et al.*, 1998). On the other hand, if we consider coniferous forest in the more maritime areas such as Scotland and southwest France, there is activity in all months, and the rates achieved are very high (Figure 9.2). Turning to deciduous forests and mixed deciduous/coniferous forests, the abrupt onset of activity associated with bud-break is very clear (Figure 9.2).

9.3.3 GPP and NPP by remote sensing

There is, however, another approach to obtaining GPP and NPP, which involves remote sensing. Monteith (1972) pointed out that the relationship between biomass accumulated and solar radiation absorbed is generally linear. The slope of this linear relation is called the light-use efficien y, or the radiation-use efficien y, ε. As it is possible to remotely sense the solar radiation absorbed, it ought therefore be possible to estimate plant growth from space. First, it was thought that ε behaved conservatively (i.e. was nearly a constant), but in a review of 13 cases published between 1977 and 1985, Cannell *et al.* (1987) found that ε varied from 0.8 to 2.1 g biomass for every megajoule of solar radiation absorbed, and later Medlyn (1998) used a canopy model to explore the variability of ε in relation to the architecture of the canopy. Researchers in the remote sensing fiel have responded to the realisation that ε is not constant by building into their models an adjustable ε. The adjustment made to ε is typically in relation to increased CO_2 concentration, water shortage and nitrogen deposition (Prince & Goward, 1995; Landsberg & Waring, 1997; Potter & Klooster, 1997; Veroustraete *et al.*, 2002). Despite these uncertainties, global maps of GPP and NPP are generally based upon this approach.

9.3.4 Use of models to predict changes in plant growth and carbon fluxes at the large scale

Modelling approaches are needed to synthesise the results of empirical investigations. Such models need to enable 'scaling-up', as knowledge is inevitably sporadic in both space and time, so that we may see the bigger picture. Models generally integrate the concepts, knowledge, data and understanding that come from physiological work as well as from fiel measurements and satellite remote sensing. In the next section we look at modelling of flu es of carbon in relation to global change as it might occur over the next 100 years. In this time we expect the temperature to rise, the deposition of nitrogen to increase and atmospheric CO_2 concentration to increase. The magnitudes of these changes are not easily predicted, as they also depend on the behaviour of a complex system of socioeconomic factors. The Intergovernmental Panel on Climate Change (IPCC) accepts this difficult and adopts 'storylines and scenarios' to assist predictions (IPCC, 2001), based on running global circulation models (GCMs) in conjunction with specifie patterns of greenhouse gas emissions. These scenarios are (see also Chapter 1):

A1: A set of scenarios that describe a future world of very rapid economic growth, with global population peaking in mid-century and then declining. In this scenario, we expect the CO_2 to increase from its present value of 370–380 to 717 ppm by 2100 and the temperature to rise by 2.95°C.

A2: The underlying theme is self-reliance and preservation of local identities, with a continuously increasing population as fertility rates across regions converge. In this scenario we expect CO_2 to rise to 856 ppm by 2100 and temperature to rise by 3.8°C.

B1: Populations as in A1, but with the introduction of clean technologies, and the global development of sustainable economies. With this scenario, CO_2 will rise to 578 ppm and warming will be 1.98°C.

B2: Local development of sustainable economies, continuously increasing populations but at a lower rate than A2. In this case, we may expect CO_2 will rise to 715 ppm by 2100 and temperature will increase by 2.38°C.

These rises in temperature are global averages. Temperatures are expected to increase much more in the northern regions, perhaps twice as fast, as a result of changes in albedo and energy balance associated with melting of ice (IPCC, 2001).

We should keep these figure in mind when we look at approaches to modelling the impact of climate change on leaves and ecosystems.

9.4 Dependencies of fluxes on CO_2, light and nitrogen supply

9.4.1 Photosynthesis

Photosynthesis is a relatively well-understood process, which can be represented by biochemical equations. For C3 plants the equations were firs proposed by Farquhar *et al.* (1980) and were known as the 'Farquhar model'. The model describes the activity of rubisco in relation to its two competing substrates carbon dioxide and oxygen, in terms of classical Michaelis–Menten enzyme kinetics. For carboxylation activity we have

$$A_c = \frac{V_{cmax}\,(C - \Gamma)}{K_c(1 + O/K_o) + C} - R_d$$

where V_{cmax} is the maximum carboxylation rate, C and O are the carbon dioxide and oxygen concentrations in the chloroplasts, Γ is the CO_2 compensation point and K_c and K_o are the Michaelis constants for carboxylation and oxygenation respectively. R_d is the mitochondrial respiration.

Carboxylation is often limited by the rate that RuBP, the primary carbon dioxide acceptor, can be regenerated. This depends on supplies of ATP and NADP, which themselves require a f ow of electrons from the chloroplastic electron transport chain, and this is dependent on the absorption of light by photosystem II inside the chloroplasts. Under those conditions, the rate of carboxylation is expressed as

$$A_j = \frac{J\,(C - \Gamma)}{4(C + 2\Gamma)} - R_d$$

where J is the potential rate of electron transport and the factor 4 occurs because four electrons are required from the chloroplastic electron transport chain to generate one RuBP. The fl w of electrons depends on the absorption of useful light by photosystem II. This is represented by an empirical curvilinear equation in which θ expresses the curvature (it lies between 0 and 1 and is often about 0.7)

$$J = \frac{I + J_{max} - \sqrt{(I + J_{max})^2 - 4\theta I J_{max}}}{2\theta}$$

where I is the photon flu corrected for the absorptance of leaves and for the spectral quality (Lloyd *et al.*, 1995) and J_{max} is the maximum electron transport rate that can occur.

The actual rate of photosynthesis is then found as the minimum of A_c and A_j after selecting the appropriate values of the kinetic constants for K_c, K_o and Γ, usually assumed to be the same for all C3 species, and their temperature dependencies. On the other hand, V_{cmax}, J_{max} and R_d are related to the nitrogen content of the leaves, and are dependent not only on the species but also on the growing conditions and position in the canopy. For values of these constants, temperature dependences and a discussion of the physiological basis, the reader is referred to Wullschleger (1993) and von Caemmerer (2000). A complication in calculation of A_c and A_j using this approach is that the chloroplastic carbon dioxide concentration C is dependent on the degree of opening of the stomata. As they shut, the value of C declines as the inward diffusion of carbon dioxide is restricted. Thus, C is determined experimentally, and models have been developed to represent its behaviour.

Some calculations from the Farquhar model are shown in Figures 9.3 and 9.4. For the 'typical' parameter values shown in Figure 9.3a, we see that under normal CO_2 concentrations, the rate of photosynthesis rises rapidly from 0°C to a rather fla optimum between 20 and 30°C. This is very similar to what has been observed for many C3 species in laboratory and fiel experimentation. However, when the CO_2 is increased, the optimum shifts to the right, and the response between 0 and 20°C becomes much steeper. This is a well-known result, resulting from the effect of high

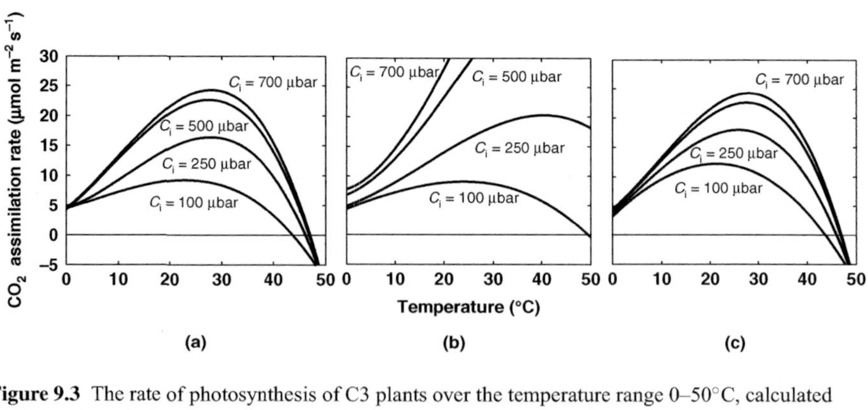

Figure 9.3 The rate of photosynthesis of C3 plants over the temperature range 0–50°C, calculated using the Farquhar model as a function of temperature and different chloroplast CO_2 partial pressures at PAR = 1000 μmol m^{-2} s^{-1}. (a) net CO_2 assimilation rate, (b) modelled RuBP saturated CO_2 assimilation rate, (c) modelled RuBP limited assimilation rate. According to the model, a is always the minimum of b and c.

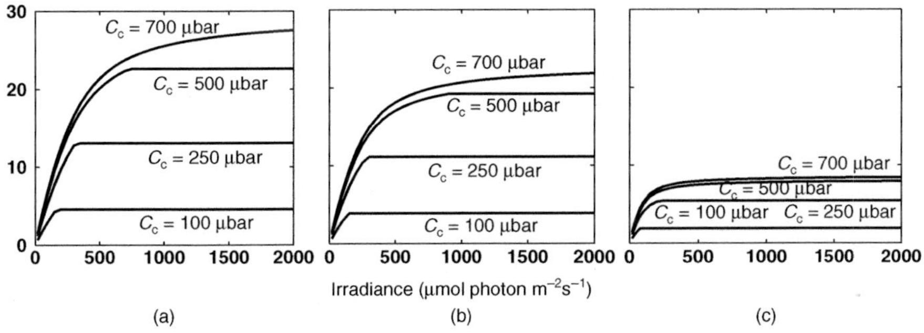

Figure 9.4 Modelled rate of CO_2 assimilation as a function of irradiance at four chloroplast CO_2 partial pressures C_c, 25°C and 200 mbar O_2, in crop (A), tropical forest (b) and boreal forest (c). Values of V_{cmax} and J_{max}, taken from Wullschleger (1993) to calculate the curve were as follows: a, V_{cmax} = 60 μmol m^{-2} s^{-1}, J_{max} = 137μmol m^{-2} s^{-1}; b, V_{cmax} = 51μmol m^{-2} s^{-1}, J_{max} = 107μmol m^{-2} s^{-1}; c, V_{cmax} = 25μmol m^{-2} s^{-1}, J_{max} = 40μmol m^{-2} s^{-1}.

temperatures on the balance between the carboxylation and oxygenase activities of rubisco (von Caemmerer, 2000). Another feature of Figure 9.3 is that the photosynthetic system tends to saturate at the higher CO_2 concentrations, as was also showed by Wullschleger (1993), and this is true irrespective of the light levels used in the simulation and irrespective of the values of the parameters V_{cmax} and J_{max} (Figure 9.4).

9.4.2 Autotrophic respiration

Respiration is not as well understood as photosynthesis. The classical view is that autotrophic respiration may be divided into 'maintenance respiration' and 'growth respiration' (McCree, 1970; Thornley, 1970). In this paradigm, it is assumed that maintenance respiration releases the energy required for protein turnover and maintenance of internal ion concentrations and gradients, and that it may also be involved in certain futile pathways and cycles (Cannell & Thornley, 2000). Its value has been found to lie between 0.012 and 0.1 g C (g C)$^{-1}$ day^{-1} and increases with an exponential relationship to temperature, often modelled to have a Q_{10} of 2.0. Growth respiration on the other hand depends on the growth rate and the chemical constitution of the tissue being constructed and is temperature insensitive. In models, this 'growth and maintenance' paradigm is almost always used, although it is probably an oversimplification and it certainly ignores species differences. Indeed, there is a current debate regarding the adequacy of this simple model of maintenance respiration. It is quite likely that over periods of days and weeks a process of acclimation occurs, and that maintenance respiration has some dependency on the supply of substrates (Atkin *et al.*, 2005; see also Chapter 3). Despite these uncertainties, theoretical and empirical studies suggest that respiration is likely to account for 0.32–0.5 of photosynthesis in whole plants, consistent with Gifford's review of empirical values of plant respiration (Cannell & Thornley, 2000a; Gifford, 2003).

9.4.3 Heterotrophic respiration

Heterotrophic respiration encompasses the flu es of CO_2 from bacteria, fungi and animals. Of these, the firs two are quantitatively the most important. When plant litter reaches the soil, it is broken down by soil organisms at a rate that is found experimentally to be strongly temperature dependent, approximating to an exponential relationship over the normal range of environmental temperatures (Lloyd & Taylor, 1994; Fang & Moncrieff, 2001). As in the case of autotrophic respiration, the Q_{10} is often assumed to be 2 (Lloyd & Taylor, 1994; Fang & Moncrieff, 2001) on the basis of short-term experiments, but may, however, be dependent on the long-term temperature. In long-term studies over a geographical range, different relationships may apply because the respiration depends on the supply of plant material as substrate for the microorganisms (Liski *et al.*, 1999; Giardina & Ryan, 2000; Grace & Rayment, 2000). Models of 'soil respiration' are being actively developed to take account of the wide divergence in rates as well as the variation in temperature dependency (Falloon & Smith, 2002; see also Chapter 3).

9.4.4 Ecosystem models

Whereas it is possible to investigate experimentally the CO_2 flu es from leaves and whole plants, it is much more difficul to carry out meaningful experiments upon ecosystems (see Osmond *et al.*, 2004), so the approach is often to model the responses, based on the mechanistic understanding outlined above.

In the past 20 years, the approaches and equations outlined above have been incorporated into most of the emerging models of ecosystem carbon flu es (Lloyd *et al.*, 1995; Williams *et al.*, 1996; Sellers *et al.*, 1997), and have been used in predictions of how carbon flu es will respond to warming and high CO_2. The Farquhar model is for C3 photosynthesis only; however, other analogous models have been developed for C4 plants (Collatz *et al.*, 1992) that may comprise about one-fift of global photosynthesis (Lloyd & Farquhar, 1994). These models enable us to make a forecast of how the net photosynthesis may change as temperature increases and CO_2 rises.

Although the Farquhar model works well as a description of photosynthesis of leaves, its use in ecosystem and global models is not straightforward for the following reasons:

i. The model contains parameters that must be fitte to observational data (du Pury & Farquhar, 1997) as there are differences between the photosynthetic attributes of species, for example, according to whether they are sun-loving or shade-tolerating, or trees as opposed to herbs. Moreover, plants acclimate on a daily basis, and ecosystems acclimate over longer periods (Oechel *et al.*, 2000), so model fittin can be a difficul process especially in stressed environments in which cold or dry seasons limit photosynthesis for reasons that are not dealt with by this model.

ii. To control the diffusional supply of CO_2 and to enable the internal CO_2 to be calculated, the Farquhar model must be coupled to a model of the behaviour of stomata, of which many ingenious representations have been proposed (Collatz *et al.*, 1991; Monteith, 1995; Mäkelä *et al.*, 1996; Williams *et al.*, 1996; Wang & Leuning, 1998). All of them may be described as 'empirical' or 'semi-empirical' because they do not have a mechanistic basis but they can nevertheless be fitte to observational data. In addition, there should be a submodel of aerodynamic conductance, as this term in many cases limits the rate of supply of CO_2, the loss of water by transpiration and the surface temperature of the canopy (Grace *et al.*, 1995).
iii. As conceived, the Farquhar model describes photosynthesis at the leaf level. Whilst it is possible to use a 'big-leaf model' at a large scale (Lloyd *et al.*, 1995), there is uncertainty about the errors that may result from this (de Pury & Farquhar, 1997), because photosynthesis exhibits a non-linear response to light, and the canopy contains leaves that are fully illuminated and fully or partially shaded. It is better to incorporate the model into a representation of canopy architecture where an understorey is represented (Williams *et al.*, 1996; Wang & Leuning, 1998).
iv. Photosynthesis is reduced by various forms of stress, most notably by low temperature and by water shortage. Ecosystem models should have some representation of the relevant processes. For example, many of them simulate the transport of water from the soil to the atmosphere and incorporate some mechanism whereby the plant responds to drought by closure of the stomata (Williams *et al.*, 1996; Wang & Leuning, 1998).
v. Leaf area index (LAI; total area of leaf per unit land surface area) is required when the model is applied at ecosystem scale. The index varies seasonally and between years, and is not very easy to measure. Recently, however, remote sensing products have become available to provide estimates of LAI at the large scale (e.g. the MODIS product at edcimswww.cr.usgs.gov/pub/imswelcome). However, efforts to predict phenology, canopy development and competitive interactions between plants from climatological data need to be tested against reliable and independent fiel data (Woodward & Lomas, 2004). It is another step to devise rules for the optimal allocation of carbon between transpiring foliage, conductive sapwood and absorbing roots. Magnani *et al.* (2002) show how this may be done using a heuristic approach whereby the algorithm tests the benefi to the plant of either making new roots, stem or leaves before allocating carbon to achieve the 'best' solution.

Despite these difficulties the Farquhar photosynthetic model is used now in most flu models that claim to be 'mechanistically based'. In this case we apply one such model, SPA (Williams *et al.*, 1996) to data from the boreal forest, one of the most important biomes as it constitutes a large fraction of the total global terrestrial carbon and is positioned in a part of the world which is expected to warm by up to 8°C over the next century. To do this, we have parameterised the model for Scots pine and adjusted parameter values by fittin it to flu data from a site at Hyytiälä, Finland. The data for this site are freely available on the Internet. The average temperature of the site is only 5.4°C. The model runs in hourly time steps, and for the 'standard run' we used the data from the site itself. To simulate warming we increased the temperature evenly in all seasons of the year, and made an appropriate adjustment to the water vapour pressure defici so that the air would not become too dry. Precipitation and solar radiation were left unchanged. We also increased the carbon dioxide concentration and the nitrogen concentration of the leaves to simulate the possible effect of these factors in the future. In the model, the temperature dependent processes are photosynthesis and respiration.

The effect of warming is seen here to be most profound on ecosystem respiration, although there is also an appreciable effect on GPP (Figure 9.5). However, GPP rises less sharply than R_e and so NEP progressively declines. Under both high and normal foliar nitrogen and carbon dioxide, we see that the NEP becomes negative when mean annual temperatures reach 10–14°C. (Figure 9.5). Although there are special considerations to be made in the very cold environments of the boreal forest (e.g. Mäkelä *et al.*, 2004), the result is rather general. If we were to repeat the exercise with other biomes, the result would inevitably be the same in general terms, as long as we were to assume that heterotrophic respiration increases with temperature.

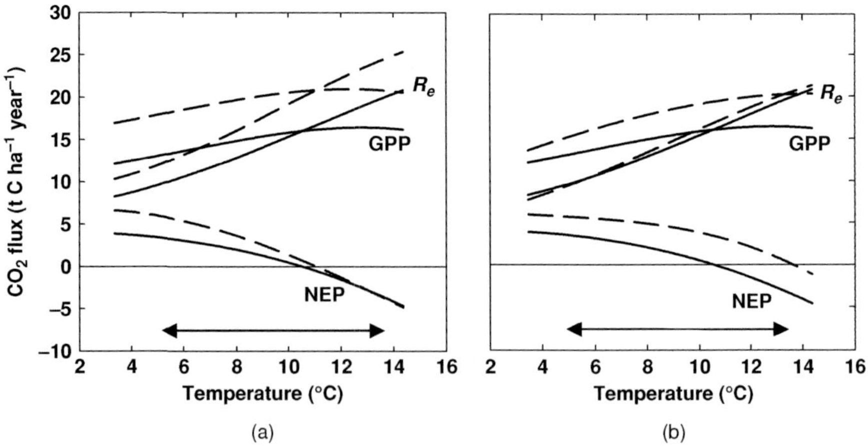

Figure 9.5 Possible effect of temperature change on gas exchange of a boreal forest, using parameter values taken from Hyytiälä, Finland. (a) Modelled annual GPP, NEP and *Re* as functions of temperature at normal foliage nitrogen concentration (solid line) and double foliage nitrogen (dashed line). The arrow indicates the temperature range from the current annual mean (5.4°C) to the possible mean of 13.4° given rapid warming for 100 years. (b) The same as in (a), but for doubled CO_2 concentration.

Such model calculations enable us to estimate the carbon balance of biomes in a future world of high CO_2, but much of the detail is still uncertain. The uncertainties relate especially to the effect of temperature on the breakdown of soil organic matter, as we have discussed above, and its consequences for N and other nutrient availability (as discussed in Chapter 8); and the interaction of elevated CO_2 with extremes of temperature where processes that have not been modelled may come into play. These processes include the down-regulation of photosynthesis at low temperatures when light-harvesting pigment systems are damaged by frost (Mäkelä *et al.*, 2004), and the effect of high CO_2 when temperatures are low enough to seriously restrict cell division and growth (Morison & Lawlor, 1999).

Very similar results to those in Figure 9.5 are seen in other models that incorporate the assumptions we have outlined above. We will mention two cases. Cramer *et al.* (2001) compared the results of six dynamic vegetation models run using climatic data, which represent what is expected to occur in the next one hundred years, according to predictions of the Hadley Centre climate model, assuming continuing emissions of fossil fuels according to the IS92a scenario. All of the models show that NPP at firs increases as a result of the fertilising effect of elevated CO_2 and a generally beneficia effect of global warming on NPP. The highest and lowest models are shown in Figure 9.6. Although they differ substantially in the total NPP, the rate of increase is the same, amounting to about 2% per decade at the present time (about 2.6 Pg per decade). Another study, based upon observational data, shows a similar increase in productivity: Nemani *et al.* (2003) show an increase in NPP of about 3.5% per decade from 1982 to 1999 (about 3.4 Pg per decade).

However, with all models reported by Cramer *et al.* (2001) the rate of increase in NPP fell as CO_2 saturation occurred and autotrophic respiration increased (Figure 9.6). NEP fell rapidly in one model but more gradually in others. In one of the models, the global NEP is negative at 2100, and thereafter the land becomes a source for carbon (Figure 9.6).

Such models (Cramer *et al.*, 2001) are run from climate data that have been generated using a GCM. A more ambitious idea is to couple ecosystem models to a GCM to achieve at least some of the feedbacks. For example, as the planetary surface warms and if it is true that ecosystem respiration increases sharply with temperature, this will enrich the atmosphere with further CO_2, and hence cause more warming. Such feedbacks, both positive and negative, are likely to be very important in the behaviour of the climate system. Cox *et al.* (2000) tried to do this, using a rather simple ecosystem model coupled to the Hadley Centre climate model HadCM3 (Figure 9.6). As a result of the positive feedback noted above, the warming was accelerated and the respiration was so great that the biosphere became a large source (NEP became negative as early as 2050, and the Amazon rainforest area declined as also suggested from rather different considerations by Nepstad *et al.*, 1999). In the only other attempt to couple a carbon cycle model to a climate model, the feedback was not so strong (Friedlingstein *et al.*, 2003).

It remains to be seen whether such predictions are well founded. One criticism of Cox *et al.* (2000) is that the heterotrophic respiration is modelled as a simple Q_{10}

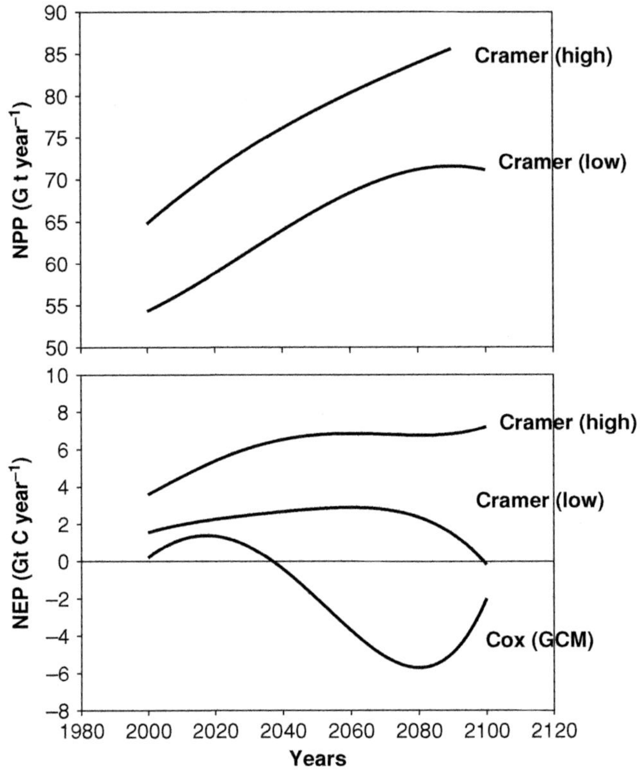

Figure 9.6 Projections of global NPP and NEP based on vegetation models presented by Cramer *et al.* (2001). In the upper panel is presented the result of the models which gave the highest and lowest estimates of NPP. Both increase over 100 years, but one of then reaches a plateau. In the lower panel, the global NEP is seen to decline to zero in the lower case. The completely different result from Cox *et al.* (2000) is shown as a contrast: here, NEP becomes negative within a few decades of the present day.

relationship to temperature. In reality, we do not know if this is what actually happens over long periods of time, as ultimately the easily used carbon substrates in the soil become depleted. In fact, when soil respiration across a broad latitudinal range is examined, it is found that respiration is not an exponential function of temperature, but closer to being a linear relation (Liski *et al.*, 1999; Giardina & Ryan, 2000; Grace & Rayment, 2000; see also Chapter 3). In addition, the interaction of soil carbon with nutrient cycles is critical to ecosystem productivity (see Chapter 8). The question of how soil organic matter behaves in the long-term, under warming, is still an open one.

9.5 Conclusions

1. It is widely accepted that CO_2 concentrations, temperatures and rates of nitrogen deposition will continue to increase over the century, despite the efforts to limit emissions (IPCC, 2001). The changes in temperature and

nitrogen deposition will increase photosynthesis, especially in northern forests that are warming faster, although they are currently limited by cold winter conditions and by nitrogen supply. A model analysis shows, however, that the response to elevated CO_2 is likely to reach a plateau when the partial pressure of chloroplastic CO_2 reaches 500 μbar. There are few experiments on mature trees, but a recent one showed that only one out of the three species tested showed an increase in growth in response to a doubling of CO_2 concentration (Körner *et al.*, 2005).

2. In other biomes the effect of climate change on photosynthesis is less positive. In the tropical biome, where temperatures are already high, the forests are likely to show a decline in photosynthesis, caused by water stress as warming and drying of the soil occurs during El Nino droughts, which are predicted to become more intense.
3. Heterotrophic respiration is likely to increase as a result in the increased supply of organic matter and also the rising temperature. The two climate models that have so far incorporated the coupling of carbon cycle models to a fully-developed GCM are in disagreement about how significan the temperature effect is likely to be (Cox *et al.*, 2000; Friedlingstein *et al.*, 2003). It may be that the higher rates of respiration will not endure once the readily available 'labile' fraction of organic matter is broken down. These uncertainties are being resolved by long-term experimentation with real soils (Eliasson *et al.*, 2005).
4. The disturbance flu es (D) will become increasingly important in the carbon budget of all biomes as human populations increase and the pressure on land rises. It will be necessary to curtail these losses as far as is possible, otherwise they will jeopardise carbon emission targets. They occur not only through tropical deforestation (currently between one-sixth and one-third of anthropogenic emissions), but also in modern agriculture where recent studies show losses to be as large as 0.6% per year (Bellamy *et al.*, 2005).

Acknowledgements

The authors are supported in part from NERC through the Centre for Terrestrial Carbon Dynamics and through CarboEurope-IP. In addition, Rui Zhang holds a scholarship from the Torrance Bequest.

References

Atkin, O.K., Bruhn, D., Hurry, V.M. & Tjoelker, M.G. (2005) The hot and the cold: unravelling the variable response of plant respiration to temperature. *Funct. Plant Biol.*, **32**, 87–105.

Aubinet, M., Grelle, A., Ibron, A. Rannik, U., Moncrieff, J., Foken, T., Kowalski, A.S., Martin, P.H., Berbigier, P.H. & Bernhofer, C. (2000) Estimates of the net carbon and water exchange of forests: the EUROFLUX methodology. *Adv. Ecol. Res.*, **30**, 113–175.

Baldocchi, D. Falge, E., Gu, L., Olson, R., Hollinger, D., Running, S., Anthoni, P., Bernhofer, Ch., Davis, K., Evans, R., Fuentes, J., Goldstein, A., Katul, G., Law, B., Lee, X., Malhi, Y., Meyers, T., Munger, W., Oechel, W., Paw, K.T., Pilegaard, K., Schmid, H.P., Valentini, R., Verma, S., Vesala, T., Wilson, K., & Wofsy, S. (2001) FLUXNET: a new tool to study the temporal and spatial variability of ecosystem-scale carbon dioxide, water vapour, and energy flu densities. *Bull. Am. Meteorol. Soc.*, **82**, 2415–2434.

Bellamy, P.H., Loveland, P.J., Bradley, R.I., Lark, R.M. & Kirk, G.J.D. (2005) Carbon losses from all soils across England and Wales 1978–2003. *Nature*, **437**, 245–248.

Biscoe, P.V., Scott, R.K. & Monteith, J.L. (1975) Barley and its environment. 3. Carbon budget of stand. *J. Appl. Ecol.*, **12**, 269–293.

Cannell, M.G.R., Milne, R., Sheppard, L.J. & Unsworth, M.H. (1987) Radiation interception and productivity of willow. *J. Appl. Ecol.*, **24**, 261–278.

Cannell, M.G.R. & Thornley, J.H.M. (2000) Modelling the components of plant respiration: some guiding principles. *Ann. Bot.*, **85**, 45–54.

Carswell, F.E., Meir, P., Wandelli, E.V., Bonates, L.C.M., Kruijt, B., Barbosa, E.M., Nobre A.D., Grace, J. & Jarvis, P.G. (2000) Photosynthetic capacity in a Central Amazonian rain forest. *Tree Physiol.*, **20**, 179–186.

Ciais, P., Janssens, I., Shvidenko, A., Wirth, C., Malhi, Y., Grace, J., Schulze, E.-D., Heimann, M., Phillips, O. & Dolman, A.J. (2005) The potential for rising CO_2 to account for the observed uptake of carbon by tropical, temperature and Boreal forest biomes. In: *The Carbon Balance of Forest Biomes* (eds H. Griffith & P.G. Jarvis), pp. 109–149. Taylor and Francis, Abingdon, UK.

Collatz, G.J., Ball, J.T., Grivet, C. & Berry, J.A. (1991) Physiological and environmental regulation of stomatal conductance, photosynthesis and transpiration – a model that includes a laminar boundary layer. *Agric. For. Meteorol.*, **54**, 107–136.

Collatz, G.J., Ribascarbo, M. & Berry, J.A. (1992) Coupled photosynthesis stomatal conductance model for leaves of C4 plants. *Aust. J. Plant Physiol.*, **19**, 519–538.

Cox, P.M., Betts, R.A., Jones, C.D., Spall, S.A., & Totterdell I.J. (2000) Acceleration of global warming due to carbon-cycle feedbacks in a coupled climate model. *Nature*, **408**, 184–187.

Cramer, W., Bondeau, A., Woodward, F.I., Prentice, I.C., Betts, R.A., Brovkin, V., Cox, P.M., Fisher, V., Foley, J.A., Friend, A.D., Kucharik, C., Lomas, M.R., Ramankutty, N., Sitch, S., Smith, B., White, A. & Young-Molling, C. (2001) Global response of terrestrial ecosystem structure and function to CO_2 and climate change: results from six dynamic global vegetation models. *Global Change Biol.*, **7**, 357–373.

De Pury, D.G.G. & Farquhar, G.D. (1997) Simple scaling of photosynthesis from leaves to canopies without the errors of big-leaf models. *Plant Cell Environ.*, **20**, 537–557.

Desjardins, R.L. & Lemon, E.R. (1974) Limitations of an eddy-correlation technique for the determination of the carbon dioxide and sensible heat flu es. *Boundary-Layer Meteorol.*, **5**, 475–488.

Eckardt, F.E. (1986) Techniques de measure de la photosynthèse sur le terrain basée sur l'emploi d'enceintes climatisées. In: *Functioning of Terrestrial Ecosystems at the Primary Production Level* (ed. F.E. Eckardt), pp. 289–320. UNESCO, Rome.

Eliasson, P.E., McMurtrie, R.E., Pepper, D.A., Stromgren, M., Linder, S. & Agren, G.I. (2005) The response of heterotrophic CO_2 flu to soil warming. *Global Change Biol.*, **11**, 167–181.

Falge, E., Baldocchi, D., Tehnunen, J., Aubinet, M., Bakwin, P., Berbigier, P., Bernhofer, C., Burba, G., Clement, R., Davis, K.J., Elbers, J.A., Goldstein A.H., Grelle, A., Granier, A., Guðmundsson, J., Hollinger, D., Kowalski, A.S., Katul, G., Law, B.E., Malhi, Y., Meyers, T., Monson, R.K., Munger, J.W., Oechel, W., Paw, U. K.T., Pilegaard, K., Rannik, U., Rebmann, C., Suyker, A., Valentini, R., Wilson, K. & Wofsy, S. (2002) Seasonality of ecosystem respiration and gross primary production as derived from FLUXNET measurements. *Agric. For. Meteorol.*, **113**, 53–74.

Falloon, P. & Smith, P. (2002) Simulating SOC changes in long-term experiments with RothC and CENTURY: model evaluation for a regional scale application. *Soil Use Manage.*, **18**, 101–111.

Fang, C. & Moncrieff, J.B. (2001) The dependence of soil CO_2 efflu on temperature. *Soil Biol. Biochem.*, **33**, 155–165.

Farquhar, G.D., Caemmerer, S.V. & Berry, J.A. (1980) A biochemical model of photosynthetic CO_2 assimilation in leaves of C-3 species. *Planta*, **149**, 78–90.

Friedlingstein, P., Dufresne, J.L., Cox, P.M. & Rayner, P. (2003) How positive is the feedback between climate change and the carbon cycle? *Tellus B*, **55**, 692–700.

Galloway, J.N., Schlesinger, W.H., Levym, H., II, Michaels, A. & Schnoor, L.J. (1995) Nitrogen fixation anthropogenic enhancement and environmental response. *Global Biogeochem. Cycles*, **9**, 235–252.

Galloway, J.N., Dentener, F.J., Capone, D.G., Boyer, E.W., Howarth, R.W., Seitzinger, S.P., Asner, G.P., Cleveland, C.C., Green, P.A., Holland, E.A., Karl, D.M., Michaels, A.F., Porter, J.H., Townsend, A.R. & Vorosmarty, C.J. (2004) Nitrogen cycles: past, present, and future. *Biogeochemistry*, **70**, 153–226.

Giardina, C.P. & Ryan, M.G. (2000) Evidence that decomposition rates of organic carbon in mineral soil do not vary with temperature. *Nature*, **404**, 858–861.

Gifford, R.M. (2003) Plant respiration in productivity models: conceptualisation, representation and issues for global terrestrial carbon-cycle research. *Funct. Plant Biol.*, **30**, 171–186.

Grace, J., Lloyd, J., McIntyre, J., Miranda, A.C., Meir, P., Miranda, H., Nobre, C., Moncrieff, J.M., Massheder, J., Malhi, Y., Wright, I.R. & Gash, J. (1995) Carbon dioxide uptake by an undisturbed tropical rain forest in South-West Amazonia 1992–1993. *Science*, **270**, 778–780.

Grace, J. & Rayment, M. (2000) Respiration in the balance. *Nature*, **404**, 819–820.

Guenther, A., Hewitt, C.N., Erickson, D., Fall, R., Geron, C., Graedel, T., Harley, P., Klinger, L., Lerdau, M., McKay, W.A., Pierce, T., Scholes, B., Steinbrecher, R., Tallamraju, R., Taylor, J. & Zimmerman, P. (1995) A global model of natural volatile organic compound emission. *J. Geophys. Res. (Atmospheres)*, **100** (D5), 8873–8892.

Gurney, K.R., Law, R.M., Denning, A.S., Rayner, P.J., Baker, D., Bousquet, P., Bruhwiler, L., Chen, Y.H., Ciais, P., Fan, S., Fung, I.Y., Gloor, M., Heimann, M., Higuchi, K., John, J., Maki, T., Maksyutov, S., Maserie, K., Peylin, P., Prather, M., Pak, B.C., Randerson, J., Sarmiento, J., Taguchi, S., Takahashi, T. & Yuen, C.W. (2002) Towards robust regional estimates of CO_2 sources and sinks using atmospheric transport models. *Nature*, **415**, 626–630.

Högberg, P., Ekblad, A., Buchmann, N., Taylor, A.F., Ekblad, A., Hogberg, M.N., Nyberg, G., Ottosson-Lofvenius, M. & Read, D.J. (2001) Large scale forest girdling shows that current photosynthesis drives soil respiration. *Nature*, **411**, 789–792.

IPCC (2000) *Land Use, Land-use Change and Forestry*. Cambridge University Press, Cambridge, UK.

IPCC (2001) *Climate Change 2001, the Scientific Basis*. Cambridge University Press, Cambridge, UK.

Keeling, R.F., Piper, S.C. & Heimenn, M. (1996) Global and hemispheric CO_2 sinks deduced from changes in atmospheric O_2 concentration. *Nature*, **381**, 218–221.

Körner, C., Asshoff, R., Bignucola, O., Hättenschwiler, S., Keel, S.G., Peláez-Riedl, Pepin, S., Siegwolf, R.T.W. & Zotz, G. (2005) Carbon flu and growth in mature deciduous forest trees exposed to elevated CO_2. *Science*, **309**, 1360–1362.

Kruijt, B., Elbers, J.A., von Randow, C., Araujo, A.C., Oliveira, P.J., Culf, A., Manzi, A.O., Nobre, A.D., Kabat, P. & Moors, E.J. (2004) The robustness of eddy correlation flu es for Amazon rain forest conditions. *Ecol. Appl.*, **14**, S101–S113.

Landsberg, J.J. & Waring, R.H. (1997) A generalised model of forest productivity using simplifie concepts of radiation-use efficien y, carbon balance and partitioning. *For. Ecol. Manag.*, **95**, 209–228.

Le Quéré, C., Aumont, O., Bopp, L., Bousquet, P., Ciais, P., Francey, R., Heimann, M., Keeling, C.D., Keeling, R.F., Kheshgi, H., Peylin, P., Piper, S.C., Prentice, I.C. & Rayner, P.J. (2003) Two decades of ocean CO_2 sink and variability. *Tellus B*, **55**, 649–656.

Leuning, R., Denmead, O.T., Lang, A.R.G. & Otaki, E. (1982) Effects of heat and water vapour transport on eddy covariance measurements of CO_2 flu es. *Boundary-Layer Meteorol.*, **23**, 209–222.

Lindroth, A., Grelle, A. & Moren, A.S. (1998) Long-term measurements of boreal forest carbon balance reveal large temperature sensitivity. *Global Change Biol.*, **4**, 443–450.

Liski, J., Ilvesniemi, H., Mäkelä, A. & Westman, C.J. (1999) CO_2 emissions from soil in response to climatic warming are overestimated. *Ambio*, **28**, 171–174.

Lloyd, J. (1999) The CO_2 dependence of photosynthesis, plant growth responses to elevated CO_2 concentrations and their interaction with soil nutrient status. II. Temperate and boreal forest productivity and the combined effects of increasing CO_2 concentrations and increased nitrogen deposition at a global scale. *Funct. Ecol.*, **13**, 439–459.

Lloyd, J. & Farquhar, G.D. (1994) ^{13}C discrimination during CO_2 assimilation by the terrestrial biosphere. *Oecologia*, **99**, 201–215.

Lloyd, J. & Farquhar, G.D. (1996) The CO_2 dependence of photosynthesis, plant growth responses to elevated CO_2 concentrations, and their interaction with soil nutrient status I. General principles and forest ecosystems. *Funct. Ecol.*, **10**, 4–23.

Lloyd, J., Grace, J., Miranda, A.C., Meir, P., Wong, S.C., Miranda, H., Wright, I., Gash, J.H.C. & McIntyre, J. (1995) A simple calibrated model of amazon rainforest productivity based on leaf biochemical properties. *Plant Cell Environ.*, **18**, 1129–1145.

Lloyd, J. & Taylor, J.A. (1994) On the temperature dependence of soil respiration. *Funct. Ecol.*, **8**, 315–323.

Long, S.P., Roberts, M.J. & Jones, M.B. (1992) *Primary Productivity of Tropical Ecosystems*. Chapman & Hall, London.

Magnani, F., Grace, J. & Borghetti, M. (2002) Adjustment of tree structure in response to the environment under hydraulic constraints. *Funct. Ecol.*, 16, 385–393.

Mäkelä, A., Berninger, F. & Hari, P. (1996) Optimal control of gas exchange during drought@ theoretical analysis. *Ann. Bot.*, **77**, 461–467.

Malhi, Y., Baldocchi, D.D. & Jarvis, P.G. (1999) The carbon balance of tropical, temperate and boreal forests. *Plant Cell Environ.*, **22**, 715–740.

Marland, G., Boden, T.A. & Andres, R.J. (2003) *Global, Regional, and National Fossil Fuel CO_2 Emissions*. Carbon Dioxide Information Analysis Center, Oak Ridge National Laboratory, U.S. Department of Energy, Oak Ridge, TN.

McCree, K.J. (1970) An equation for the respiration of white clover plants grown under controlled conditions. In: *Prediction and Measurement of Photosynthetic Productivity* (ed. I. Setlik), pp. 221–229. Pudoc, Wageningen, The Netherlands.

Medlyn, B.E. (1998) Physiological basis of the light use efficien y model. *Tree Physiol.*, **18**, 167–176.

Moncrieff, J.B., Massheder, J.M., Verhoef, A., Elbers, J, Heutsunkveld, B.H., Scott, S., de Bruin, H., Kabat, P. Soegaard, H. & Jarvis, P.G. (1997) A system to measure surface flu es of energy, momentum and carbon dioxide. *J. Hydrol.*, **188–189**, 589–611.

Monteith, J.L. (1968) Analysis of the photosynthesis and respiration of fiel crops from vertical flu es of carbon dioxide. In: *Functioning of Terrestrial Ecosystems at the Primary Production Level* (ed. F.E. Eckardt), pp. 349–354. UNESCO, Rome.

Monteith, J.L. (1972) Solar radiation and productivity in tropical ecosystems. *J. Appl. Ecol.*, **9**, 747–766.

Monteith, J.L. (1995) A reinterpretation of stomatal responses to humidity. *Plant Cell Environ.*, **18**, 357–364.

Morison, J.I.L. & Lawlor, D.W. (1999) Interaction between increasing CO_2 concentration and temperature on plant growth. *Plant Cell Environ.*, **22**, 659–682.

Myneni, R.B., Keeling, C.D., Tucker, C.J., Asrar, G. & Nemani, R.R. (1997) Increased plant growth in the northern high latitudes from 1981 to 1991. *Nature*, **386**, 698–702.

Myneni, R.B., Tucker, C.J., Asrar, G. & Keeling, C.D. (1998) Interannual variations in satellite-sensed vegetation index data from 1981 to 1991. *J. Geophys. Res.*, **103**, 6145–6160.

Nagy, L., Grabherr, G., Körner, Ch. & Thompson, D.B.A. (2003) *Alpine biodiversity in Europe. Ecological Studies 167*. Springer-Verlag, Berlin.

Nemani, R.R., Keeling, C.D., Hashimoto, H., Jolly, W.H., Piper, S.C., Tucker, C.J., Myneni, R.B. & Running, S.W. (2003) Climate-driven increases in global net primary productivity from 1982 to 1999. *Science*, **300**, 1560–1563.

Nepstad, D.C., Verissimo, A., Alencar, A., Nobre, C., Lima, E., Lefebvre, P., Schlesinger, P., Potter, C.,

Moutinho, P., Mendoza, E., Cochrane, M. & Brooks, V. (1999) Large scale impoverishment of Amazonian forests by logging and fire *Nature*, **398**, 505–508.

Oechel, W.C., Vourlitis, G.L., Hastings, S.J., Zulueta, R.C., Hinzman, L. & Kane, D. (2000) Acclimation of ecosystem CO_2 exchange in the Alaskan Arctic in response to decadal climate warming. *Nature*, **406**, 978–981.

Osmond, B., Ananyev, G., Berry, J., Langdon, C., Kolber, Z., Lin, G.H., Monson, R., Nichol, C., Rascher, U., Schurr, U., Smith, S. & Yakir, D. (2004) Changing the way we think about global change research: scaling up in experimental ecosystem science. *Global Change Biol.*, **10**, 393–407.

Potter, C.S. & Klooster, S.A. (1997) Global model estimates of carbon and nitrogen storage in litter and soil pools: response to change in vegetation quality and biomass allocation. *Tellus B*, **49**, 1–17.

Prince, S.D. & Goward, S.M. (1995) Global primary production: a remote sensing approach. *J. Biogeogr.*, **22**, 815–835.

Rödenbeck, C., Houweling, S., Gloor, M. & Heimann, M. (2003) CO_2 flu history 1982–2001 inferred from atmospheric data using a global inversion of atmospheric data. *Atmos. Chem. Phys. Discuss.*, **3**, 2575–2659.

Roy, J., Saugier, B. & Mooney, H.A. (2001) *Terrestrial Global Productivity*. Academic Press, San Diego, CA.

Royal Society (2001) *The Role of Land Carbon Sinks in Mitigating Global Climate Change.* Royal Society, London.

Saleska, S.R., Miller, S.D., Matross, D.M., Goulden, M.L., Wofsy, S.C., da Rocha, H.R., de Camargo, P.B., Crill, P., Daube, B.C., de Freitas, H.C., Hutyra, L., Keller, M., Kirchhoff, V., Menton, M., Munger, J.W., Pyle E.H., Rice, A.H. & Silva H. (2003) Carbon in Amazon forests: unexpected seasonal flu es and disturbance-induced losses. *Science*, **302**, 1554–1557.

Saugier, B., Roy, J. & Mooney, H.A. (2001) Estimations of global terrestrial productivity: converging toward a single number? In: *Terrestrial Global Productivity* (eds J. Roy, B. Saugier & H.A. Mooney), pp. 543–557. Academic Press, San Diego, CA.

Sellers, P.J., Dickinson, R.E., Randall, D.A., Betts, A.K., Hall, F.G., Berry, J.A., Collatz, G.J., Denning, A.S., Mooney, H.A., Nobre, C.A., Sato, N., Field, C.B. & HendersonSellers, A. (1997) Modeling the exchanges of energy, water, and carbon between continents and the atmosphere. *Science*, **275**, 502–509.

Steffen, W., Noble, I., Canadell, J., Apps, M., Schulze, E.-D., Jarvis, P.G., Baldocchi, D., Ciais, P., Cramer, W., Ehleringer, J., Farquhar, G., Field, C.P., Ghazi, A., Gifford, R., Heimann, M., Houghton, R., Kabat, P., Kšrner, C., Lambin, E., Linder, S., Mooney, H.A., Murdiyarso, D., Post, W.H., Prentice, I.C., Raupach, M.R., Schimel, D.S., Shvidenko A. & Valentini, R. (1998) The terrestrial carbon cycle: implications for the Kyoto Protocol. *Science*, **280**, 1393–1394.

Swinbank, W.C. (1951) The measurement of vertical transfer of heat and water vapor by eddies in the lower atmosphere. *J. Meteorol.*, **8**, 135–145.

Thornley, J.H.M. (1970) Respiration, growth and maintenance of plants. *Nature*, **227**, 304–305.

Valentini, R., Matteucci, G., Dolman, A.J. *et al.* (2000) Respiration as the main determinant of carbon balance in European forests. *Nature*, **404**, 861–865.

Veroustraete, F., Sabbe, H. & Eerens, H. (2002) Estimation of carbon mass flu es over Europe using the C-Fix model and Euroflu data. *Remote Sensing Environ.*, **83**, 376–399.

Von Caemmerer, S. (2000) *Biochemical Models of Leaf Photosynthesis.* CSIRO Publishing, Collingwood, Australia.

Wang, Y.-P. & Leuning, R. (1998) A two-leaf model for canopy conductance, photosynthesis and partitioning of available energy. *Agric. For. Meteorol.*, **91**, 89–111.

Waring, R.H., Landsberg, J.J. & Williams, M. (1998) Net primary production of forests: a constant fraction of gross primary production? *Tree Physiol.*, **18**, 129–134.

Whittaker, R.H. & Likens, G.E. (1975) The biosphere and man. In: *Primary Productivity of the Biosphere* (eds R.H. Whittaker & G.E. Likens), pp. 305–328. Springer-Verlag, Berlin.

Williams, M., Rastetter, E.B., Fernandes, D.N., Goulden, M.L., Wofsy, S.C., Shaver, G.R., Melillo, J.M., Munger, J.W., Fan, S.-M. & Nadelhoffer, K.J. (1996) Modelling the soil-plant-atmosphere continuum in a *Quercus-Acer* stand at Harvard Forest: the regulation of stomatal conductance by light, nitrogen and soil/plant hydraulic properties. *Plant Cell Environ.*, **19**, 911–927.

Woodward, F.I. & Lomas, M.R. (2004) Vegetation dynamics – simulating responses to climatic change. *Biol. Rev.*, **79**, 643–670.

Wullschleger, S.D. (1993) Biochemical limitations to carbon assimilation in C3 plants- a retrospective analysis of the A/C_i curves from 109 species. *J. Exp. Bot.*, **44**, 907–920.

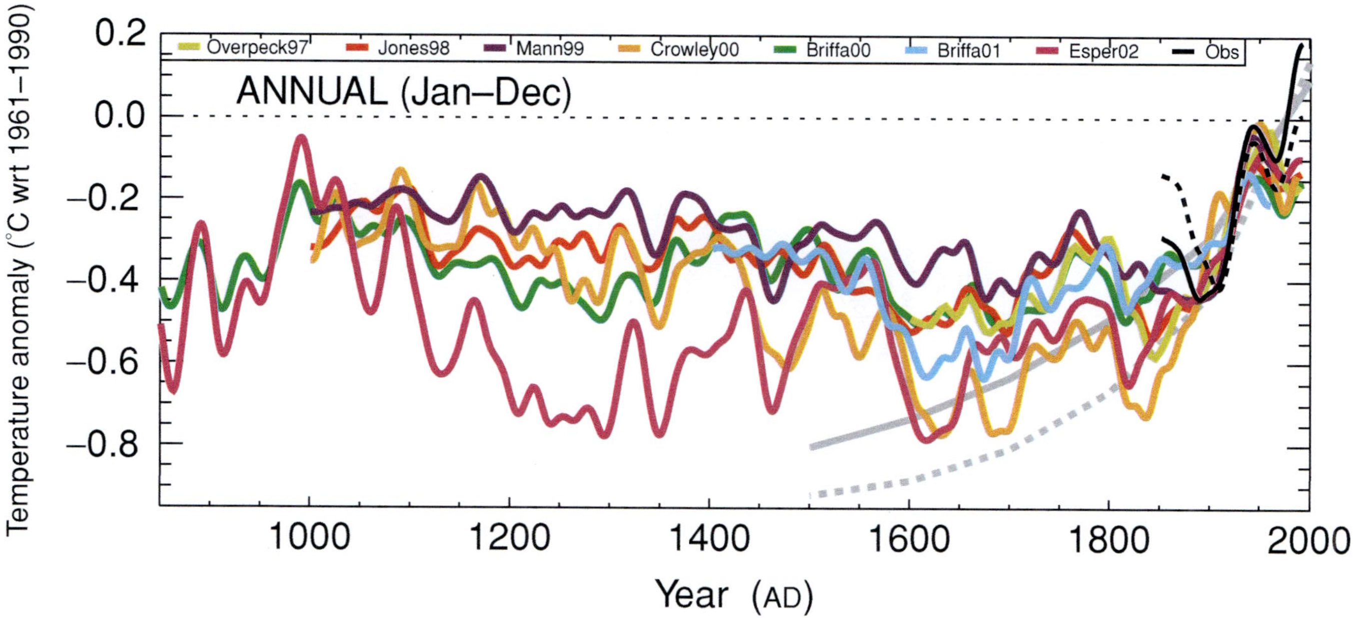

Colour Plate 1 Reconstructions of Northern Hemisphere mean annual temperatures over the last 1000 years, using various proxy data (tree rings, lake sediments, ice cores), together with the instrumental record. Data are smoothed with a 40-year low pass filte . (From IPCC, 2001a.)

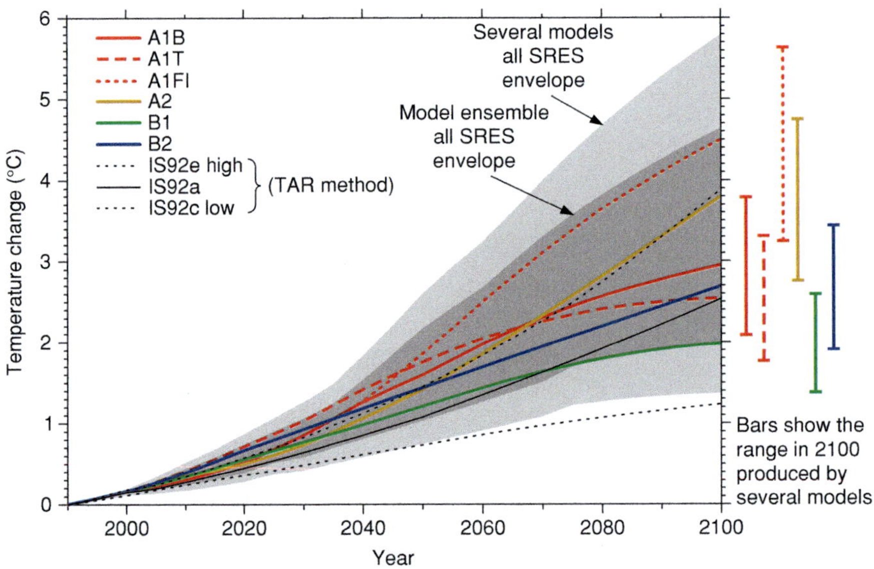

Colour Plate 2 Mean global temperature change in twenty-firs century, relative to 1990, for the different SRES and IS92 scenarios from IPCC (2001a). Lines are for individual scenarios identifie in the legend; the shading represents the range of outcome for different model sensitivities (see IPCC, 2001a, Chapter 9 for details).

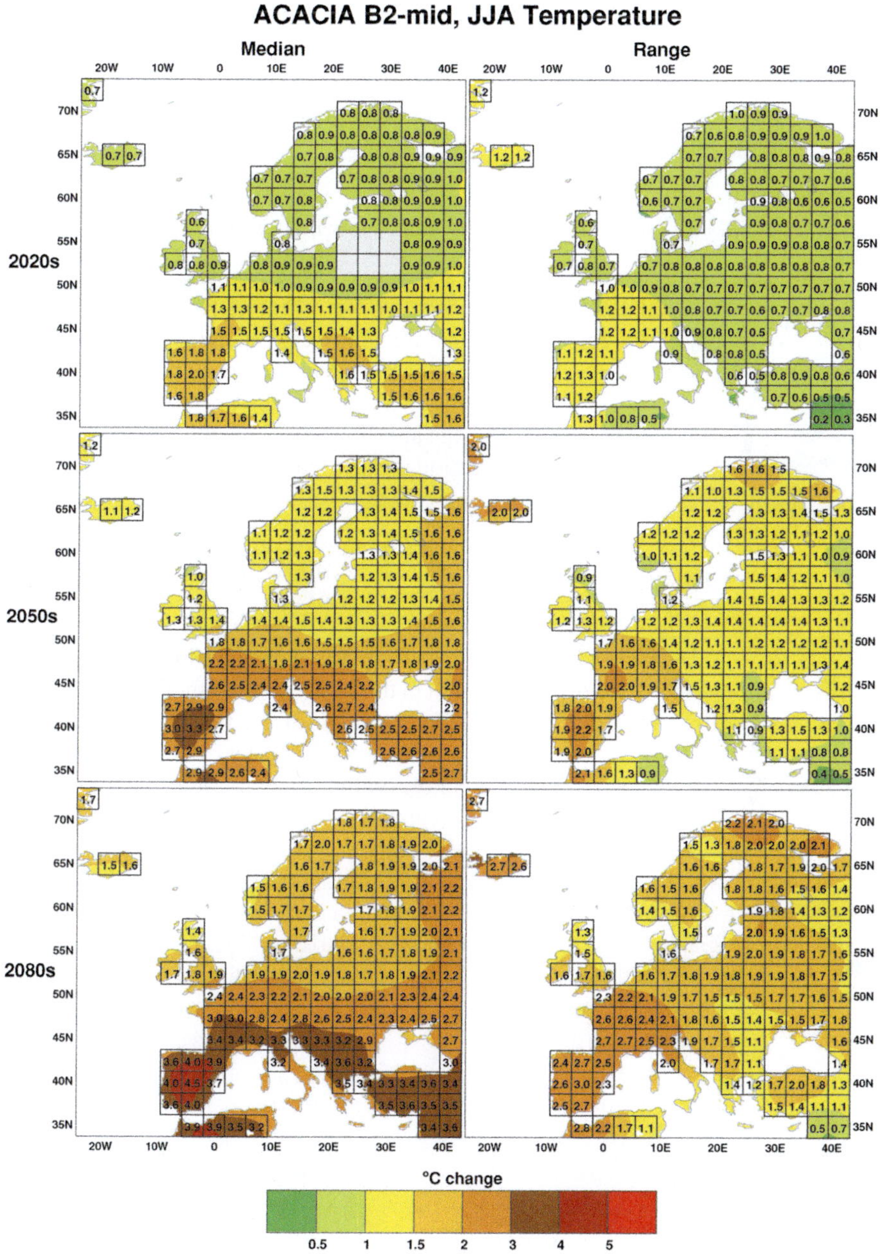

(a)

Colour Plate 3 Predictions of European summer (June, July and August) (a) temperature and (b) precipitation changes compared with the simulated 1961–1990 mean, from the Europe ACACIA methodology. The left panels show the median value predicted from eight AOGCMs using the SRES B2 scenario, and the right panels show the absolute range across the models. (From IPCC, 2001b, Chapter 13).

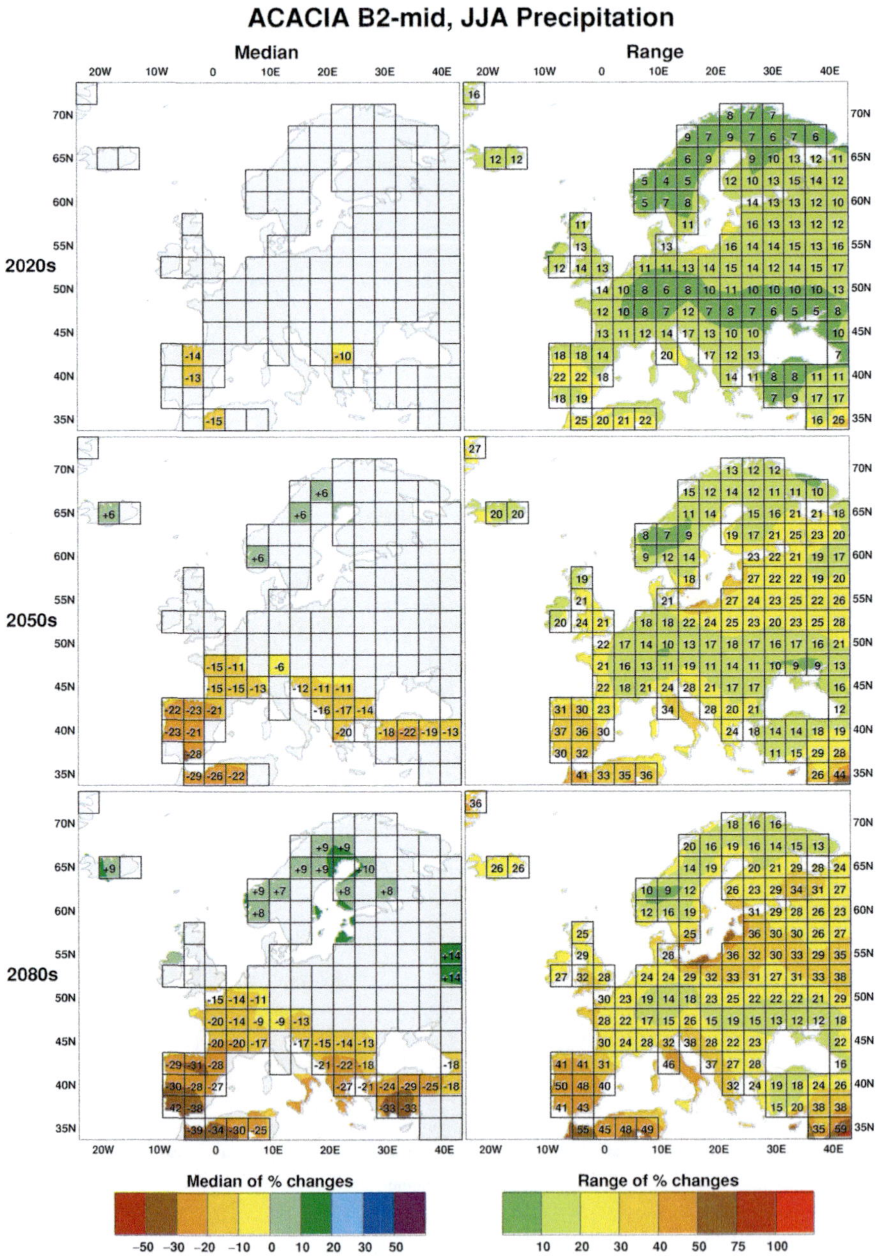

(b)

Colour Plate 3 (*Continued*)

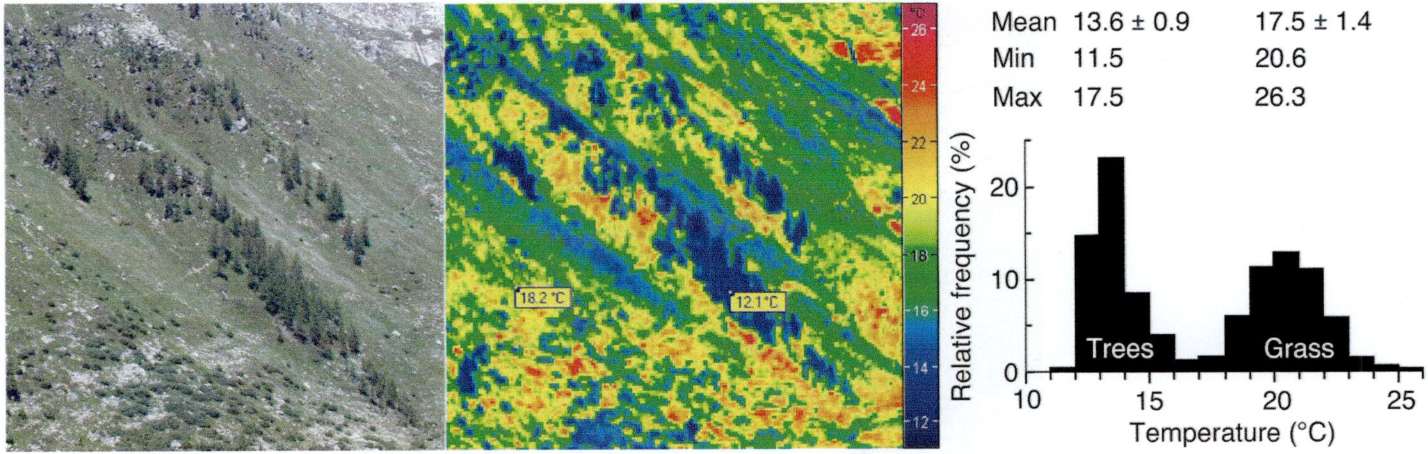

Colour Plate 4 A thermal image of a fragmented section of the treeline ecotone in the Alps (Furka Pass) at 2100-m elevation, which illustrates the significanc of plant morphology for the actual temperature experienced. The camera, with a resolution of 76 000 measurement points and 0.1-K resolution, images the warm grass and shrub vegetation, due to solar heating, and the cool tree canopies, which are aerodynamically coupled closely to air temperature, which was 10°C when the image was taken at noon in mid July (Körner, C. & Leuzinger, S., 2005, unpublished results).

Colour Plate 5 The effect of insufficien chilling temperature during a mild winter on the development of cereals. The picture was taken in late June. Left, a cultivar (Greina) which did not receive the needed chill dose to induce ears (vernalisation) and got trapped in a vegetative stage; right, a 'spring variety' (cv. Fiorina) which has little chilling requirement, and developed normally. The photograph was taken on June 13 in Basel.

Colour Plate 6 An example for photoperiod control of development in the genus *Taraxacum*. The picture was taken on April 1. The lowland species is fully developed and fl wering, the alpine species still 'waits' for the 'right' photoperiod, which is a signal of reduced risk of freezing damage (from Körner, 2003a, with permission by Springer-Verlag, Berlin). Such a blocking of development by photoperiod is seen in *Oxyria digyna* (Keller & Körner, 2003) for instance.

Index

Printed and bound by CPI Group (UK) Ltd, Croydon, CR0 4YY

16/09/2026

14963094-0001